Birkhäuser

Progress in Mathematics

Volume 361

Progress in Mathematics is a series of books intended for professional mathematicians and scientists, encompassing all areas of pure mathematics. This distinguished series, which began in 1979, includes research level monographs, polished notes arising from seminars or lecture series, graduate level textbooks, and proceedings of focused and refereed conferences. It is designed as a vehicle for reporting ongoing research as well as expositions of particular subject areas.

Ferran Cedó · Leandro Vendramin

Groups, Radical Rings, and the Yang-Baxter Equation

A Combinatorial Approach to Solutions

Birkhäuser

Ferran Cedó
Departament de Matemàtiques
Universitat Autònoma de Barcelona
Bellaterra (Barcelona), Spain

Leandro Vendramin
Mathematics and Data Science
Vrije Universiteit Brussel
Brussels, Belgium

ISSN 0743-1643 ISSN 2296-505X (electronic)
Progress in Mathematics
ISBN 978-3-032-01523-5 ISBN 978-3-032-01524-2 (eBook)
https://doi.org/10.1007/978-3-032-01524-2

Mathematics Subject Classification: 81R50, 16T25

This book is published under the imprint Birkhäuser, www.birkhauser-science.com by the registered company Springer Nature Switzerland AG
The registered company address is: Gewerbestrasse 11, 6330 Cham, Switzerland

Competing Interests The authors have no competing interests to declare that are relevant to the content of this manuscript.

FSB

Ferran Sunyer i Balaguer (1912–1967) was a self-taught Catalan mathematician who, in spite of a serious physical disability, was very active in research in classical mathematical analysis, an area in which he acquired international recognition. His heirs created the Fundació Ferran Sunyer i Balaguer inside the Institut d'Estudis Catalans to honor the memory of Ferran Sunyer i Balaguer and to promote mathematical research.

Each year, the Fundació Ferran Sunyer i Balaguer and the Institut d'Estudis Catalans award an international research prize for a mathematical monograph of expository nature. The prize-winning monographs are published in this series. Details about the prize and the Fundació Ferran Sunyer i Balaguer can be found at https://ffsb.espais.iec.cat/en/the-ferran-sunyer-i-balaguer-prize/.

This book has been awarded the Ferran Sunyer i Balaguer 2025 prize.

The members of the scientific committee of the 2025 prize were:

Ruth Kellerhals
Université de Fribourg
Jiang-Hua Lu
University of Hong Kong
Marta Sanz-Solé
Universitat de Barcelona
Kristian Seip
Norwegian University of Science and Technology (NTNU)
Yuri Tschinkel
Courant Institute for Mathematical Sciences
New York University

Ferran Sunyer i Balaguer Prize winners since 2015

2015 The scientific committee decided not to award the prize

2016	Vladimir Turaev and Alexis Virelizier *Monoidal Categories and Topological Field Theory*, PM 322
2017	Antoine Chambert-Loir, Johannses Nicaise and Julien Sebag *Motivic Integration*, PM 325
2018	Michael Ruzhansky and Durvudkhan Suragan *Hardy Inequalities on Homogeneous Groups*, PM 327
2019	The scientific committee decided not to award the prize
2020	Urtzi Buijs, Yves Félix, Aniceto Murillo and Daniel Tanré *Lie Models in Topology*, PM 335 Giovanni Catino and Paolo Mastrolia *A Perspective on Canonical Riemannian Metrics*, PM 336
2021	Tim Browning *Cubic Forms and the Circle Method*, PM 343
2022	Pascal Auscher and Moritz Egert *Boundary Value Problems and Hardy Spaces for Elliptic Systems with Block Structure*, PM 346
2023	Xavier Fernández-Real and Xavier Ros-Oton *Integro-Differential Equations*, PM 350
2024	Antonio Córdoba *Suprematism in Harmonic Analysis*, PM 356

Contents

Chapter 1
Introduction

The Yang–Baxter equation (YBE) first appeared in the works of Yang [115] and Baxter [18], and it is one of the fundamental equations in mathematical physics. Subsequently, it has become a key tool in pure mathematics, especially in the theory of quantum groups.

Attempts to construct solutions to the Yang–Baxter equation (YBE) led to the theory of quantum groups. These solutions, which are deformations of the flip map, have a significant impact on low-dimensional topology. There is a discrete analog to this situation. Since the class of linear YBE solutions, meaningful for physicists, is extremely broad and currently beyond the reach of mathematicians, Drinfeld proposed in [53] tackling the problem in two main steps:

First, classify set-theoretic solutions, as they are more accessible yet still highly non-trivial. Second, study linear deformations of set-theoretic solutions using an appropriate cohomology theory, and describe the variety of solutions attainable this way.

Considerable progress on both steps has been achieved in the last few years.

Etingof, Schedler, and Soloviev [56] and Gateva-Ivanova and Van den Bergh [62] initiated the study of set-theoretic non-degenerate and involutive solutions to the YBE by introducing various associated algebraic structures: the structure monoid, the structure group, and the permutation group associated with a solution. This study was expanded to the class of non-degenerate solutions to the YBE with the works of Lu, Yan, and Zhu [87] and Soloviev [107].

Rump [99] introduced new algebraic structures for studying set-theoretic non-degenerate and involutive solutions to the YBE: *braces* and *cycle sets*. Guarnieri and Vendramin [64] extended the concept of braces to *skew braces* to study the broader class of set-theoretic non-degenerate solutions to the YBE. The structure group and the permutation group of a non-degenerate solution to the YBE admit a natural structure of a skew brace.

In the last 20 years, the theory on non-degenerate set-theoretic solutions to the YBE and the theory of skew braces have developed sufficiently to motivate the writing

F. Cedó and L. Vendramin, *Groups, Radical Rings, and the Yang-Baxter Equation*, Progress in Mathematics 361, https://doi.org/10.1007/978-3-032-01524-2_1

of this book. The theory of skew braces is closely related to group theory and, in some sense, to the theory of non-commutative rings.

We aimed for this book to be as self-contained as possible and to serve as a useful tool for introducing new researchers to the basic topics of this theory.

In Chap. 2, we recall the definitions and the most basic results of group theory, rings, and modules (without proofs). Chapter 3 explains the basic results of non-commutative ring theory. Chapters 5 and 6 present classical results on nilpotent groups and solvable groups. Chapter 7 is also on group theory but more specialized. The remaining results of group theory, necessary for the exposition and demonstration of the main results of the theory of skew braces and also important results on set-theoretic non-degenerate solutions to the YBE, are introduced as needed in Chaps. 10, 11, and 15.

Chapter 4 introduces the YBE and the class of set-theoretic non-degenerate solutions. Chapter 8 is dedicated to the basic definitions of skew braces theory. Chapter 9 studies the structure of skew braces associated with a set-theoretic non-degenerate solution to the YBE. Chapter 10 shows that the structure group of a finite, set-theoretic, non-degenerate, and involutive solution is a Bieberbach group. Chapter 11 proves that the structure group of a finite, set-theoretic, non-degenerate, and involutive solution is a Garside group. Chapters 12 and 13 study the main results on left and right nilpotent skew braces, respectively. Chapter 14 is dedicated to the abundant class of multipermutation solutions. Chapter 15 compares results on factorizations of groups and on factorization of skew braces. Chapter 16 studies results on indecomposable and simple solutions of the YBE. Chapter 17 reduces the classification of set-theoretic, non-degenerate, and involutive solutions to the YBE to the classification of abelian-type skew braces. Finally, Chap. 18 is an introduction to the study of simple abelian-type skew braces.

Each chapter ends with a list of exercises to work on the presented topics, some open problems, and comments on the authorship and development of the results presented, except when they are classical results, in which case, a book where they can be found is cited.

This work was partially supported by project OZR3762 of Vrije Universiteit Brussel and FWO Senior Research Project G004124N.

We thank our colleagues Emiliano Acri, Özgür Akgün, David Bachiller, Ilaria Colazzo, Davide Ferri, Tatiana Gateva-Ivanova, Leandro Guarnieri, Eric Jespers, Olexandr Konovalov, Łukasz Kubat, Victoria Lebed, Rafał Łutowksi, Jan Okniński, Silvia Properzi, Agata Smoktunowicz, Arne Van Antwerpen and Charlotte Verwimp, who have assisted us in obtaining many of the results on braces and the YBE presented in this book. We also appreciate the support and patience of our families, Josep Maria and Roser, Dino Vendramin, without whom it would have been much more difficult to write this book.

October 2024

Els Guiamets	Brussels
F. Cedó	L. Vendramin

Chapter 2
Preliminaries

In this chapter, we briefly discuss the necessary material needed in the subsequent chapters of the book. We provide a concise overview of fundamental concepts concerning groups, rings, and modules. The chapter primarily focuses on introducing these concepts and their basic properties, with minimal emphasis on providing proofs.

2.1 Groups

A *semigroup* is a set S with an associative operation

$$*\colon S \times S \longrightarrow S, \quad (x, y) \mapsto x * y,$$

that is $(x * y) * z = x * (y * z)$ for all $x, y, z \in S$.

If there exists an element $e \in S$ such that $e * x = x * e = x$ for all $x \in S$, then the semigroup $(S, *)$ is called a *monoid* and e is the *neutral element* for $*$. The neutral element in a monoid $(S, *)$ is unique.

A *group* is a monoid $(G, *)$ such that for every $x \in G$ there exists $x' \in G$ satisfying

$$x * x' = x' * x = e,$$

where $e \in G$ is the neutral element for $*$. The element x' is called the *symmetric element* of x for $*$ and it is unique.

The standard notation for the operation of a general semigroup is multiplicative terminology. Thus if $(S, \cdot)$ is a semigroup, then we write the multiplication $x \cdot y = xy$ and we say that xy is the product of x and y. If $(S, \cdot)$ is a monoid, then 1 denotes its neutral element and it is called the *unit-element* of S. If $(S, \cdot)$ is a group, then the symmetric element x' of $x \in S$ is called the *inverse* of x and it is denoted by x^{-1}.

F. Cedó and L. Vendramin, *Groups, Radical Rings, and the Yang-Baxter Equation*, Progress in Mathematics 361, https://doi.org/10.1007/978-3-032-01524-2_2

In some chapters, we also will use additive terminology for general semigroups, monoids, and groups. Thus if $(S, +)$ is a semigroup, the addition of elements $x + y$ is called the sum of x and y. If $(S, +)$ is a monoid, then 0 denotes its neutral element and is called the *zero* of S. If $(S, +)$ is a group, then the symmetric element x' of $x \in S$ is called the *opposite* of x and it is denoted by $-x$.

In the remainder of this chapter, we shall use the multiplicative terminology for general semigroups, monoids, and groups.

A semigroup S is said to be *commutative* if $xy = yx$ for all $x, y \in S$. A commutative group is also called an *abelian group*.

Example 2.1.1 We denote by $\mathbb{N}$ the set of non-negative integers. Then $(\mathbb{N}, +)$ is a commutative monoid and $0 \in \mathbb{N}$ is its neutral element.

Example 2.1.2 $(\mathbb{Z}, +), (\mathbb{Q}, +), (\mathbb{R}, +), (\mathbb{C}, +), (\mathbb{Q} \setminus \{0\}, \cdot), (\mathbb{R} \setminus \{0\}, \cdot), (\mathbb{C} \setminus \{0\}, \cdot)$ are abelian groups.

Example 2.1.3 The *symmetric group* on a set X is the set

$$\mathbb{S}_X = \{f : X \to X : f \text{ is bijective}\}$$

with the composition of maps. For every positive integer n, the symmetric group of degree n is $\mathbb{S}_n = \mathbb{S}_{\{1,2,\ldots,n\}}$. Note that $\mathbb{S}_n$ is not abelian for $n > 2$.

A *subgroup* of a group G is a non-empty subset H of G such that $xy^{-1} \in H$ for all $x, y \in H$. Note that every subgroup H of G also is a group with the operation of G restricted to H. The notation $H \leq G$ will mean that H is a subgroup of G.

Proposition 2.1.4 *Let $(H_i)_{i \in I}$ be a non-empty family of subgroups of a group G. Then*

$$\bigcap_{i \in I} H_i$$

is a subgroup of G. □

Let S be a subset of a group G. The *subgroup of G generated by S* is

$$\langle S \rangle = \bigcap_{S \subseteq H \leq G} H.$$

Note that if S is a non-empty subset of G, then

$$\langle S \rangle = \{x_1^{\varepsilon_1} \cdots x_n^{\varepsilon_n} : n \geq 0,\ \varepsilon_i = \pm 1,\ x_i \in S,\ 1 \leq i \leq n\}.$$

We say that G is *finitely generated* if there exists a finite subset $F = \{x_1, \ldots, x_n\}$ of G such that $G = \langle F \rangle$. In this case we also write $G = \langle x_1, \ldots, x_n \rangle$. We say that G is *cyclic* if there exists $x \in G$ such that $G = \langle x \rangle$.

Let H be a subgroup of a group G. The *left cosets* of H in G are the subsets of G of the form

$$xH = \{xh : h \in H\}$$

for $x \in G$. A subset of G containing just one element from each left coset of H in G is called a *left transversal* of H in G. Note that $xH = yH$ if and only if $x^{-1}y \in H$. Right cosets and right transversals are defined similarly. We say that H is a *normal subgroup* of G if $xH = Hx$ for all $x \in G$. In this case, the set $G/H = \{xH : x \in G\}$, with the operation defined by the rule $(xH) \cdot (yH) = (xy)H$ for all $x, y \in G$, is a group called the *quotient group* of G by the normal subgroup H. The notation $H \trianglelefteq G$ will mean that H is a normal subgroup of G. Note that every subgroup of an abelian group is normal.

Example 2.1.5 The subgroups of the group $(\mathbb{Z}, +)$ are of the form

$$n\mathbb{Z} = \{nz : z \in \mathbb{Z}\}$$

for a non-negative integer n. Note that $\mathbb{Z}$ is a cyclic group.

Example 2.1.6 The group $\mathbb{Z}/n\mathbb{Z} = \{0, 1, \ldots, n-1\}$ of integers modulo n is a cyclic group.

Example 2.1.7 Let G be a group and let H be a subgroup of G. The *normalizer* of H in G is the set

$$N_G(H) = \{x \in G : xH = Hx\}.$$

Note that $H \trianglelefteq N_G(H) \leq G$.

Let G be a group. If G is finite, then the number of elements of G is called the *order* of G and is denoted by $|G|$. If G is not finite, we say that G has infinite order and write $|G| = \infty$. Let $x \in G$. The order of x is $o(x) = |\langle x \rangle|$. If $o(x)$ is finite, then it coincides with the smallest positive integer n such that $x^n = 1$. Let H be a subgroup of G. If the number of left cosets of H in G is finite, then it coincides with the number of right cosets of H in G, and this number is called the *index* of H in G, denoted by $(G : H)$.

Theorem 2.1.8 (Lagrange's theorem) *Let G be a finite group and H be a subgroup of G. Then $|G| = (G : H)|H|$.* □

A *homomorphism of groups* is a map $f\colon G_1 \to G_2$, where G_1 and G_2 are groups, and such that $f(xy) = f(x)f(y)$ for all $x, y \in G_1$. A bijective homomorphism is called an *isomorphism*. Two groups G_1 and G_2 are isomorphic if there exists an isomorphism $f\colon G_1 \to G_2$. The notation $G_1 \cong G_2$ will mean that G_1 and G_2 are isomorphic. An *automorphism* of a group G is an isomorphism $G \to G$. The set $\mathrm{Aut}(G)$ of all automorphisms of a group G is a group with the usual composition of maps; it is called the *automorphism group* of G.

Example 2.1.9 Let G be a group and H be a subgroup of G. The inclusion mapping $\iota\colon H \to G, h \mapsto h$, is an injective homomorphism of groups.

Example 2.1.10 Let G be a group. For every $x \in G$, the map $\varphi_x : G \to G$ defined by $\varphi_x(y) = xyx^{-1}$, for all $y \in G$, is an automorphism of G. The map $\varphi : G \to \mathrm{Aut}(G)$ defined by $\varphi(x) = \varphi_x$, for all $x \in G$, is a homomorphism of groups. The group $\mathrm{Inn}(G)$ of *inner automorphisms* of G is defined as the image of φ.

The *kernel* of a homomorphism of groups $f : G_1 \to G_2$ is the set

$$\ker(f) = \{x \in G_1 : f(x) = 1\}.$$

The kernel $\ker(f)$ of f is a normal subgroup of G_1. The image

$$\mathrm{im}(f) = \{f(g) : g \in G_1\}$$

of f is a subgroup of G_2.

Example 2.1.11 Let G be a group and N be a normal subgroup of G. The canonical or natural map $\pi : G \to G/N$, $x \mapsto xN$, is a surjective homomorphism of groups with kernel equal to N.

Theorem 2.1.12 (First isomorphism theorem) *For any homomorphism of groups $f : G_1 \to G_2$ there exists a unique isomorphism $\tilde{f} : G_1/\ker(f) \to \mathrm{im}(f)$ such that the diagram*

$$\begin{array}{ccc} G_1 & \xrightarrow{f} & G_2 \\ {\scriptstyle\pi}\downarrow & & \uparrow{\scriptstyle\iota} \\ G_1/\ker(f) & \xrightarrow{\tilde{f}} & \mathrm{im}(f) \end{array}$$

is commutative, that is $f = \iota \tilde{f} \pi$, where ι is the inclusion mapping and π is the natural homomorphism. □

Let H and K be subgroups of a group G. If $H \leq N_G(K)$, then

$$KH = \{xy : x \in K,\ y \in H\}$$

is a subgroup of G. Furthermore, $K \trianglelefteq KH$.

Theorem 2.1.13 (Second isomorphism theorem) *Let G be a group and H and K be subgroups of G such that $H \leq N_G(K)$. Then $K \cap H \trianglelefteq H$ and*

$$H/(K \cap H) \cong (KH)/K.$$ □

Theorem 2.1.14 (Third isomorphism theorem) *Given a group G and $N \trianglelefteq G$, the map $H \mapsto H/N$ yields a bijective correspondence between (normal) subgroups of G containing N and (normal) subgroups of G/N. Furthermore, if $N \leq H \trianglelefteq G$, then*

$$(G/N)/(H/N) \cong G/H.$$ □

Let n be a positive integer. The elements of $\mathbb{S}_n$ are called *permutations*. We denote each element $\sigma \in \mathbb{S}_n$ by

$$\sigma = \begin{pmatrix} 1 & 2 & \dots & n \\ \sigma(1) & \sigma(2) & \dots & \sigma(n) \end{pmatrix}.$$

A permutation $\sigma \in \mathbb{S}_n$ is a *cycle* of length r if there exist r distinct elements $a_1, \dots, a_r \in \{1, \dots, n\}$ such that $\sigma(a_i) = a_{i+1}$ for $1 \le i \le r-1$, $\sigma(a_r) = a_1$ and $\sigma(b) = b$ for all $b \in \{1, \dots, n\} \setminus \{a_1, \dots, a_r\}$. In this case, we will denote σ by

$$\sigma = (a_1 \ \dots \ a_r).$$

A *transposition* is a cycle of length two. We say that the cycles $(a_1 \ \dots \ a_r)$ and $(b_1 \ \dots \ b_s)$ are disjoint if

$$\{a_1, \dots, a_r\} \cap \{b_1, \dots, b_s\} = \emptyset.$$

Note that if σ and τ are disjoint cycles, then $\sigma\tau = \tau\sigma$.

Theorem 2.1.15 *Every permutation $\sigma \in \mathbb{S}_n$ can be written as a product of pair-wise disjoint cycles. This factorization is unique except for the order in which the factors occur.* □

Theorem 2.1.16 *Let $n > 1$ be an integer. There exists a unique surjective group homomorphism* $\operatorname{sign}\colon \mathbb{S}_n \to (\{1, -1\}, \cdot)$ *such that* $\operatorname{sign}(\tau) = -1$ *for every transposition $\tau \in \mathbb{S}_n$.* □

The group homomorphism sign is the *sign homomorphism* of $\mathbb{S}_n$. The kernel of sign is the *alternating group* $\mathbb{A}_n$. We say that the elements of $\mathbb{A}_n$ are the *even* permutations of $\mathbb{S}_n$ and the elements of $\mathbb{S}_n \setminus \mathbb{A}_n$ are the *odd* permutations of $\mathbb{S}_n$.

Theorem 2.1.17 *Let $n > 1$ be an integer. Then the following statements hold:*

(1) $\mathbb{S}_n = \langle (ij) : 1 \le i < j \le n \rangle$.
(2) $\mathbb{S}_n = \langle (12), (13), \dots, (1n) \rangle$.
(3) $\mathbb{S}_n = \langle (12), (23), \dots, (n-1\, n) \rangle$.
(4) $\mathbb{S}_n = \langle (12), (12 \cdots n) \rangle$.
(5) $\mathbb{A}_n = \langle (ijk) : 1 \le i < j < k \le n \rangle$.
(6) $\mathbb{A}_n = \langle (12k) : 3 \le k \le n \rangle$. □

Theorem 2.1.18 (Cayley's theorem) *Let G be a finite group of order n. Then G is isomorphic to a subgroup of $\mathbb{S}_n$.* □

Let p be a prime. A *p-group* is a group of order a power of p. Let G be a finite group of order $n = p^\alpha m$, where $\gcd(p, m) = 1$. Any subgroup of G of order p^α is called a *Sylow p-subgroup* of G.

Let $\operatorname{Syl}_p(G)$ be the set of Sylow p-subgroups of G.

Theorem 2.1.19 (Sylow) *Let G be a finite group and p a prime. The following statements hold:*

(1) $\mathrm{Syl}_p(G) \neq \emptyset$.
(2) Every p-subgroup of G is contained in a Sylow p-subgroup of G.
(3) If $P_1, P_2 \in \mathrm{Syl}_p(G)$, then there exists $x \in G$ such that $xP_1x^{-1} = P_2$.
(4) $|\mathrm{Syl}_p(G)| \equiv 1 \bmod p$. □

Let G be a group. The *center* of G is

$$Z(G) = \{x \in G : xy = yx, \quad \text{for all } y \in G\}.$$

Theorem 2.1.20 *Let p be a prime. Any finite non-trivial p-group has a non-trivial center.* □

2.2 Rings

A *ring* is a set R with two operations

$$R \times R \to R, \quad (x, y) \mapsto x + y, \qquad R \times R \to R, \quad (x, y) \mapsto xy,$$

satisfying the following properties:

(1) $(R, +)$ is an abelian group.
(2) $(ab)c = a(bc)$ for all $a, b, c \in R$.
(3) $a(b + c) = ab + ac$ and $(a + b)c = ac + bc$ for all $a, b, c \in R$.

If $(R, \cdot)$ has a unit-element, then this is denoted by 1 and it is said that R is a unitary ring.

Convention 2.2.1 From now on, a ring will mean a unitary ring unless otherwise specified.

A *commutative ring* is a ring R such that $ab = ba$ for all $a, b \in R$.

Let R be a ring. We say that $a \in R$ is a *zero-divisor* if there exists $b \in R \setminus \{0\}$ such that $ab = 0$ or $ba = 0$. We say that $a \in R$ is an *idempotent* if $a^2 = a$. We say that $a \in R$ is *nilpotent* if there exists a positive integer n such that $a^n = 0$. An idempotent $a \in R$ is *non-trivial* if $a \notin \{0, 1\}$.

An *integral domain* is a ring R with $0 \neq 1$ and without nonzero zero-divisors. A *division ring* is a ring R such that $(R \setminus \{0\}, \cdot)$ is a group. A *field* is a commutative division ring. Any division ring is an integral domain.

Example 2.2.2 The zero ring is $\{0\}$ with $0 + 0 = 0 = 0 \cdot 0$. This is the only ring such that $0 = 1$.

Example 2.2.3 On the one hand, $\mathbb{Z}$ is a commutative integral domain that is not a field. On the other hand, $\mathbb{Q}$, $\mathbb{R}$, and $\mathbb{C}$ are fields.

Example 2.2.4 If $\{R_i\}_{i\in I}$ is a non-empty family of rings, then $\prod_{i\in I} R_i$, with the addition and the multiplication defined component-wise, is a ring, called the *Cartesian product* of the rings R_i. If $|I| \geq 2$ and the rings R_i are nonzero, then $\prod_{i\in I} R_i$ has non-trivial idempotents. For example,

$$e_i = (\delta_{ij})_{j\in I} \quad \text{where} \quad \delta_{ij} = \begin{cases} 0 & \text{if } j \neq i, \\ 1 & \text{if } j = i. \end{cases}$$

Example 2.2.5 Let R be a ring. The *power series ring* over R is the set

$$R[\![X]\!] = \left\{ \sum_{i=0}^{\infty} a_i X^i : a_i \in R \right\}$$

with the addition and the multiplication defined by the rules

$$\sum_{i=0}^{\infty} a_i X^i + \sum_{i=0}^{\infty} b_i X^i = \sum_{i=0}^{\infty} (a_i + b_i) X^i,$$

and

$$\left(\sum_{i=0}^{\infty} a_i X^i \right) \left(\sum_{i=0}^{\infty} b_i X^i \right) = \sum_{i=0}^{\infty} c_i X^i,$$

where $c_i = \sum_{j=0}^{i} a_j b_{i-j}$.

Example 2.2.6 Let R be a ring and n be a positive integer. The full $n \times n$ *matrix ring* is the set

$$M_n(R) = \left\{ \begin{pmatrix} a_{1,1} & a_{1,2} & \dots & a_{1,n} \\ a_{2,1} & a_{2,2} & \dots & a_{2,n} \\ \vdots & \vdots & \ddots & \vdots \\ a_{n,1} & a_{n,2} & \dots & a_{n,n} \end{pmatrix} : a_{i,j} \in R \right\}$$

with the addition defined component-wise and the multiplication defined by the rule

$$\begin{pmatrix} a_{1,1} & a_{1,2} & \dots & a_{1,n} \\ a_{2,1} & a_{2,2} & \dots & a_{2,n} \\ \vdots & \vdots & \ddots & \vdots \\ a_{n,1} & a_{n,2} & \dots & a_{n,n} \end{pmatrix} \begin{pmatrix} b_{1,1} & b_{1,2} & \dots & b_{1,n} \\ b_{2,1} & b_{2,2} & \dots & b_{2,n} \\ \vdots & \vdots & \ddots & \vdots \\ b_{n,1} & b_{n,2} & \dots & b_{n,n} \end{pmatrix} = \begin{pmatrix} c_{1,1} & c_{1,2} & \dots & c_{1,n} \\ c_{2,1} & c_{2,2} & \dots & c_{2,n} \\ \vdots & \vdots & \ddots & \vdots \\ c_{n,1} & c_{n,2} & \dots & c_{n,n} \end{pmatrix},$$

where $c_{i,j} = \sum_{k=1}^{n} a_{i,k} b_{k,j}$.

A *subring* of a ring R is a subset S of R such that $a - b \in S$ and $ab \in S$ for all $a, b \in S$ and $1 \in S$.

Example 2.2.7 $\mathbb{Z} \subseteq \mathbb{Q} \subseteq \mathbb{R} \subseteq \mathbb{C}$ is a chain of subrings.

Example 2.2.8 Let R be a ring. The *polynomial ring* over R is the following subring of the power series ring over R:

$$R[X] = \left\{ \sum_{i=0}^{\infty} a_i X^i \in R[\![X]\!] : a_i \neq 0 \text{ for finitely many non-negative integers } i \right\}.$$

The elements of $R[X]$ are usually written as finite sums of the form $\sum_{i=0}^{n} a_i X^i$ and are called polynomials. Note that R is a subring of $R[X]$.

Let R be a ring. A *right ideal* of R is an additive subgroup I of R such that $xa \in I$ for all $x \in I$ and $a \in R$. Similarly, one defines a *left ideal* of R. An *ideal* of R is, by definition, an additive subgroup of R that is both a left and right ideal of R.

If I and J are left (right) ideals of a ring R, then

$$IJ = \left\{ \sum_{i=1}^{n} a_i b_i : n \in \mathbb{N},\ a_i \in I,\ b_i \in J \right\}$$

also is a left (right) ideal of R.

Example 2.2.9 Let R be a ring. Then $\{0\}$ and R are ideals of R. The ideal $\{0\}$ is called the *zero ideal* and is sometimes written as 0.

Any ideal of R different from R is said to be *proper*.

Example 2.2.10 The ideals of the ring $\mathbb{Z}$ are of the form $n\mathbb{Z}$ for some integer $n \geq 0$.

Example 2.2.11 Let $\{I_j\}_{j \in J}$ be a non-empty family of ideals of a ring R. Then $\bigcap_{j \in J} I_j$ is an ideal of R.

If S is a subset of a ring R, then the ideal of R generated by S is the intersection of all ideals I of R containing S. If $\{I_j\}_{j \in J}$ is a non-empty family of ideals of a ring R, then its sum $\sum_{j \in J} I_j$ is the ideal of R generated by $\cup_{j \in J} I_j$.

Let I be an ideal of a ring R. The *quotient ring* of R by the ideal I is the additive quotient group R/I with the multiplication of left cosets defined by

$$(a + I)(b + I) = (ab) + I$$

for $a, b \in R$. Since I is an ideal of R, the multiplication is well-defined. It follows that R/I with the addition and the multiplication of left cosets is a ring.

Let R and S be rings. A map $f \colon R \to S$ is said to be a *homomorphism of rings* if $f(a + b) = f(a) + f(b)$ and $f(ab) = f(a)f(b)$ for all $a, b \in R$ and $f(1) = 1$. The image $\text{im}(f)$ of f is a subring of S and the kernel $\ker(f) = \{r \in R : f(r) = 0\}$ of f is an ideal of R.

Example 2.2.12 Let R be a ring. The map $f\colon \mathbb{Z} \to R$ defined by

$$f(z) = \begin{cases} 1 + \cdots + 1 \ (z \text{ times}) & \text{if } z > 0, \\ 0 & \text{if } z = 0, \\ (-1) + \cdots + (-1) \ (-z \text{ times}) & \text{if } z < 0, \end{cases}$$

is a homomorphism of rings. There exists a unique non-negative integer n such that $\ker(f) = n\mathbb{Z}$. This integer n is said to be the *characteristic* of R.

Example 2.2.13 Let R be a ring, and S be a subring of R. The inclusion $\iota\colon S \to R$, $a \mapsto a$, is an injective homomorphism of rings. Let I be an ideal of R. The natural map $\pi : R \to R/I$, $a \mapsto a + I$, is a surjective homomorphism of rings with kernel equal to I.

An *isomorphism* of rings is a bijective homomorphism. Two rings R and S are isomorphic if an isomorphism $f\colon R \to S$ exists. The notation $R \cong S$ will mean that R and S are isomorphic. An *automorphism* of a ring R is an isomorphism from R to itself.

Theorem 2.2.14 (First isomorphism theorem) *For any homomorphism of rings $f\colon R \to S$ there exists a unique isomorphism $\tilde{f}\colon R/\ker(f) \to \operatorname{im}(f)$ such that the diagram*

$$\begin{array}{ccc} R & \xrightarrow{\ f\ } & S \\ {\scriptstyle \pi}\big\downarrow & & \big\uparrow{\scriptstyle \iota} \\ R/\ker(f) & \xrightarrow{\ \tilde{f}\ } & \operatorname{im}(f) \end{array}$$

is commutative, that is $f = \iota\tilde{f}\pi$, where ι is the inclusion mapping and π is the natural homomorphism. □

Theorem 2.2.15 (Second isomorphism theorem) *Let R be a ring, S a subring, and I an ideal of R. Then $S + I$ is a subring of R, $I \cap S$ is an ideal of S and*

$$S/(I \cap S) \cong (S + I)/I. \qquad \square$$

Theorem 2.2.16 (Third isomorphism theorem) *Let R be a ring and I an ideal of R. Then there is a natural bijection between the subrings (respectively ideals) of R containing I and the subrings (respectively ideals) of R/I. Furthermore, if J is an ideal of R containing I, then*

$$(R/I)/(J/I) \cong R/J. \qquad \square$$

Let $(A, \leq)$ be a *partially ordered set*, this means that A is a set together with a reflexive, transitive, and anti-symmetric binary relation R on $A \times A$, where $a \leq b$ if and only if $(a, b) \in R$. Recall that the relation is reflexive if $a \leq a$ for all $a \in A$, and the relation is transitive if $a \leq b$ and $b \leq c$ imply that $a \leq c$ and the relation is anti-symmetric if $a \leq b$ and $b \leq a$ imply $a = b$.

The elements $a, b \in A$ are said to be *comparable* if $a \leq b$ or $b \leq a$. An element $a \in A$ is said to be *maximal* if $c \leq a$ for all $c \in A$ that is comparable with a. An *upper bound* for a non-empty subset $B \subseteq A$ is an element $d \in A$ such that $b \leq d$ for all $b \in B$. A *chain* in A is a subset B such that every pair of elements of B is comparable. *Zorn's lemma* states the following property:

If A is a non-empty partially ordered set such that every chain in A has an upper bound in A, then A contains a maximal element.

2.3 Modules

Let R be a ring. A *(left) R-module* is an abelian group $(M, +)$ with a multiplication by elements of R

$$R \times M \to M, \quad (a, m) \mapsto am,$$

satisfying the following properties:

(1) $(a + b)m = am + bm$ for all $m \in M$ and $a, b \in R$.
(2) $a(m_1 + m_2) = am_1 + am_2$ for all $m_1, m_2 \in M$ and $a \in R$.
(3) $(ab)m = a(bm)$ for all $m \in M$ and $a, b \in R$.
(4) $1m = m$ for all $m \in M$.

Right R-modules are defined similarly.

Convention 2.3.1 From now on, a module will mean a left module unless otherwise specified.

Example 2.3.2 Let K be a field. The K-modules are the K-vector spaces.

Example 2.3.3 The $\mathbb{Z}$-modules are the abelian groups.

Let M be an R-module. A *submodule* of M is a subgroup N of M such that $am \in N$ for all $m \in N$ and $a \in R$.

Example 2.3.4 Let $\{N_j\}_{j \in J}$ be a non-empty family of submodules of an R-module M. Then $\bigcap_{j \in J} N_j$ is a submodule of M.

If S is a subset of an R-module M, then the submodule of M generated by S is the intersection of all submodules N of M containing S. If $\{N_j\}_{j \in J}$ is a non-empty family of submodules of an R-module M, then its sum $\sum_{j \in J} N_j$ is the submodule of M generated by $\cup_{j \in J} N_j$. If $(m_i)_{i \in I}$ is a family of elements of M, the submodule generated by this family is $\sum_{i \in I} Rm_i$.

Let N be a submodule of M. The *quotient module* of M by the submodule N is the quotient additive group M/N with the multiplication by elements of R defined by

$$a(m + N) = am + N$$

for all $m \in M$ and $a \in R$. This multiplication is well-defined and, moreover, M/N is an R-module.

Let M and N be R-modules. A map $f : M \to N$ is said to be a *homomorphism* of R-modules if

$$f(a_1m_1 + a_2m_2) = a_1 f(m_1) + a_2 f(m_2)$$

for all $m_1, m_2 \in M$ and $a_1, a_2 \in R$. If f is injective, surjective, or bijective, we say that f is a monomorphism, an epimorphism, or an isomorphism, respectively. If $M = N$, we say that f is an endomorphism of M. A bijective endomorphism is an automorphism. We say that M and N are isomorphic if an isomorphism from M to N exists. The notation $M \cong N$ will mean that M and N are isomorphic.

Let $f : M \to N$ be a homomorphism of R-modules. Then $\operatorname{im}(f)$ is a submodule of N and $\ker(f) = \{m \in M : f(m) = 0\}$ is a submodule of M.

Theorem 2.3.5 (First isomorphism theorem) *Let R be a ring. For any homomorphism of R-modules $f : M \to N$ there exists a unique isomorphism of R-modules $\tilde{f} : M/\ker(f) \to \operatorname{im}(f)$ such that the diagram*

$$\begin{array}{ccc} M & \xrightarrow{\ f\ } & N \\ {\scriptstyle\pi}\downarrow & & \uparrow{\scriptstyle\iota} \\ M/\ker(f) & \xrightarrow{\ \tilde{f}\ } & \operatorname{im}(f) \end{array}$$

is commutative, that is, $f = \iota\tilde{f}\pi$, where ι is the inclusion mapping and π is the natural homomorphism. □

Theorem 2.3.6 (Second isomorphism theorem) *Let R be a ring. Let M be an R-module. If N and L are submodules of M, then*

$$N/(N \cap L) \simeq (N + L)/L. \qquad \square$$

Theorem 2.3.7 (Third isomorphism theorem) *Let R be a ring. Given an R-module M and a submodule N of M, there is a natural bijection between the submodules of M containing N and the submodules of M/N. Furthermore, if H is a submodule of M containing N, then*

$$(M/N)/(H/N) \cong M/H. \qquad \square$$

2.4 Free Groups and Free Monoids

Let F be a group and X a set. It is said that F is a *free group* on X if there exists a function $\sigma : X \to F$ such that for every function $\alpha : X \to G$, where G is a group, there exists a unique group homomorphism $f : F \to G$ such that $f\sigma = \alpha$. Note that, in this case, σ is injective and F also is a free group on $\sigma(X)$ with respect to the inclusion map $\sigma(X) \to F$. A group is *free* if it is free on some set.

Theorem 2.4.1 *Let X be a set. Then there exists a free group F on X.* □

Proposition 2.4.2 *Let G be a group and X a subset of G. If every element $g \in G$ can be written uniquely in the form*

$$g = x_1^{n_1} \cdots x_s^{n_s}, \tag{2.1}$$

for some $s \geq 0$, where $x_1, \dots, x_s \in X$, $n_1, \dots, n_s \in \mathbb{Z} \setminus \{0\}$ and $x_i \neq x_{i+1}$ for all $1 \leq i < s$, then G is free on X. □

If G is a free group on a subset $X \subseteq G$, then we say that X is a *basis* of G, and the right-hand side of (2.1) is called the *normal form* of $g \in G$ for the basis X. All the bases of G have the same size, which is then called the *rank* of the free group G.

A *free presentation* of a group G is a surjective homomorphism $\pi : F \to G$, where F is a free group. The kernel of π is the subgroup of the *relators* of the presentation. Let Y be a basis of F and let S be a subset of $\ker(\pi)$ such that $\ker(\pi)$ is the smallest normal subgroup of F containing S, i.e., $\ker(\pi)$ is the intersection of all normal subgroups containing S. Every $s \in S$ has a normal form $w(s)$ with respect to Y. Then we write

$$G = \mathrm{gr}(Y : w(s) = 1 \text{ for all } s \in S)$$

and say that the right-hand side of this equality is a presentation of G with generators Y and defining relations $w(s) = 1$ for all $s \in S$. If $Y = \{y_1, \dots, y_n\}$, then we also write

$$G = \mathrm{gr}(y_1, \dots, y_n : w(s) = 1 \text{ for all } s \in S).$$

If, for example, $y_1 y_2 y_1^{-1} y_2^{-1} \in S$, then the defining relation $y_1 y_2 y_1^{-1} y_2^{-1} = 1$ can also be written in the form $y_1 y_2 = y_2 y_1$.

Example 2.4.3 Let C_n be a cyclic group of order n. Then $C_n = \mathrm{gr}(x : x^n = 1)$ is a presentation of C_n.

Proposition 2.4.4 *Let $G = \mathrm{gr}(Y : w(s) = 1$ for all $s \in S)$. Let H be a group and let $\alpha : Y \to H$ be a function such that for every $s \in S$, if $w(s) = y_1^{n_1} \cdots y_k^{n_k}$, then $\alpha(y_1)^{n_1} \cdots \alpha(y_k)^{n_k} = 1$. Then there exists a unique group homomorphism $f : G \to H$ such that $f(y) = \alpha(y)$ for all $y \in Y$.* □

A *homomorphism of monoids* is a map $f : M_1 \to M_2$, where M_1 and M_2 are monoids, and such that $f(xy) = f(x)f(y)$ for all $x, y \in M_1$ and $f(1) = 1$. A bijective homomorphism is called an *isomorphism*. Two monoids M_1 and M_2 are isomorphic if there exists an isomorphism $f : M_1 \to M_2$. The notation $M_1 \cong M_2$ will mean that M_1 and M_2 are isomorphic.

Let X be a set. We denote by $FM(X)$ the set of all the words $x_1 \cdots x_n$ with letters $x_1, \dots, x_n \in X$. The empty word is denoted by 1. We define an operation on $FM(X)$ by the rule

$$(x_1 \cdots x_n) \cdot (y_1 \cdots y_m) = x_1 \cdots x_n y_1 \cdots y_m$$

for all $x_1, \dots, x_n, y_1, \dots, y_m \in X$. It follows that $FM(X)$ with this operation is a monoid. This is the *free monoid* on X.

Proposition 2.4.5 *Let X be a set. Let M be a monoid and let $\alpha\colon X \to M$ be a function. Then there exists a unique homomorphism of monoids $f\colon FM(X) \to M$ such that $f(x) = \alpha(x)$ for all $x \in X$.* □

Let M be a monoid. A *congruence* on M is an equivalence relation $\sim$ on M such that, for $a, b \in M$,

$$a \sim b \Rightarrow \forall c, d \in M,\ cad \sim cbd.$$

Let $\bar{a} = \{b \in M : b \sim a\}$, where $\sim$ is a congruence on the monoid M. We define an operation on $M/\!\sim\, = \{\bar{a} : a \in M\}$ by the rule $\bar{a} \cdot \bar{b} = \overline{ab}$ for all $a, b \in M$. It is easy to check that this operation is well-defined and that $M/\!\sim$ with this operation is a monoid. The natural map $\pi\colon M \to M/\!\sim, a \mapsto \bar{a}$, is a surjective homomorphism of monoids.

Example 2.4.6 Let $f\colon M_1 \to M_2$ be a homomorphism of monoids. Let ρ be the binary relation on M_1 defined by $a\rho b$ if and only if $f(a) = f(b)$. Then ρ is a congruence on M_1 and the map $\tilde{f}\colon M_1/\rho \to M_2$, defined by $\tilde{f}(\bar{a}) = f(a)$ for all $a \in M_1$, is an injective homomorphism of monoids.

Example 2.4.7 Let X be a set. Let $X' = \{x' : x \in X\}$ be a copy of X and let Y be the disjoint union of X and X'. Let $G = \mathrm{gr}(X)$ be the free group on X. Let $\alpha\colon Y \to G$ be the function defined by $\alpha(x) = x$ and $\alpha(x') = x^{-1}$ for all $x \in X$. Let $f\colon FM(Y) \to G$ be the unique homomorphism of monoids such that $f(y) = \alpha(y)$ for all $y \in Y$. Let $\sim$ be the congruence on $FM(Y)$ such that $a \sim b$ if and only if $f(a) = f(b)$. Then the homomorphism $\tilde{f}\colon FM(Y)/\!\sim\, \to G$, defined by $\tilde{f}(\bar{a}) = f(a)$ for all $a \in FM(Y)$, is an isomorphism of groups. Note that $\sim$ is the smallest congruence on $FM(Y)$ such that $xx' \sim 1$ and $x'x \sim 1$ for all $x \in X$.

Let X be a set. Let $W_1 = (w_i)_{i \in I}$ and $W_2 = (u_i)_{i \in I}$ be two families of elements in $FM(X)$. The monoid M presented with set of generators X and set of relations $\{(w_i, u_i) : i \in I\}$ is $FM(X)/\!\sim$, where $\sim$ is the smallest congruence on $FM(X)$ such that $w_i \sim u_i$ for all $i \in I$. We will denote M by

$$M = \langle X : w_i = u_i \text{ for all } i \in I \rangle.$$

2.5 Free Modules and Abelian Groups

In this section, we use the additive notation for abelian groups.

Let R be a ring. An R-module F is *free* if there exists a family $(x_i)_{i \in I}$ of elements of F such that for every element $x \in F$ there exist unique elements $r_i \in R$, only finitely many of the r_i are nonzero, satisfying

$$x = \sum_{i \in I} r_i x_i.$$

In this case, we say that $(x_i)_{i \in I}$ is a *basis* of the free module F.

Example 2.5.1 Let R be a ring and I be a set. For every $i \in I$, let $R_i = R$. Let $M = \prod_{i \in I} R_i$ be the Cartesian product of the R_i. Then M with the sum and the multiplication by elements of R defined component-wise is an R-module. Let

$$e_i = (\delta_{ij})_{j \in I} \in M, \quad \delta_{ij} = \begin{cases} 1 & \text{if } j = i, \\ 0 & \text{otherwise.} \end{cases}$$

Then the submodule

$$F = \bigoplus_{i \in I} R_i = \sum_{i \in I} Re_i$$

of M is a free R-module and $(e_i)_{i \in I}$ is a basis of F. We say that F is the *direct sum* of the R_i.

Proposition 2.5.2 *Every R-module is a homomorphic image of a free R-module.* □

A *free abelian group* is a free $\mathbb{Z}$-module. An abelian group G is *torsion-free* if G has no nonzero elements of finite order.

Proposition 2.5.3 *A finitely generated abelian group G is torsion-free if and only if it is a free abelian group.* □

Theorem 2.5.4 *Let G be a finitely generated abelian group. Then there exist non-negative integers n, m and integers $d_i > 1$ for $1 \leq i \leq m$ such that*

$$G \cong \mathbb{Z}^n \oplus \bigoplus_{i=1}^{m} \mathbb{Z}/d_i\mathbb{Z}$$

and $d_{j+1} \mid d_j$ for all $1 \leq j < m$. Furthermore, n, m, and d_i are the only integers satisfying these conditions. The integer n is called the rank *of G, and $d_1, \ldots, d_m$ are the* invariant factors *of G.* □

2.6 Exercises

Problem 2.6.1 Prove all the results stated in this chapter.

2.7 Notes

The material in this chapter is basic and standard; see, for example, [49, 50].

Chapter 3
The Jacobson Radical

In this chapter, we develop the basic theory of non-commutative rings, putting special attention on the Jacobson radical.

3.1 Primitive Rings

Let M be an R-module and S a subset of M. The *annihilator* of S is

$$\mathrm{Ann}_R(S) = \{a \in R : as = 0 \text{ for all } s \in S\}.$$

If $S = \{m\}$, then we will write $\mathrm{Ann}_R(m) = \mathrm{Ann}_R(\{m\})$.

It is an exercise to check that $\mathrm{Ann}_R(S)$ is a left ideal of R. Furthermore, if S is a submodule of M, then $\mathrm{Ann}_R(S)$ is an ideal of R.

Let T be a subset of a right R-module, then the annihilator of T is

$$\mathrm{Ann}^r_R(T) = \{a \in R : ta = 0 \text{ for all } t \in T\}.$$

Definition 3.1.1 A *faithful* R-module is an R-module M such that $\mathrm{Ann}_R(M) = \{0\}$.

Similarly, one defines faithful right R-modules.

Example 3.1.2 Let K be a field and $R = M_2(K)$. Then

$$I = \begin{pmatrix} K & K \\ 0 & 0 \end{pmatrix} = \left\{ \begin{pmatrix} a & b \\ 0 & 0 \end{pmatrix} : a, b \in K \right\}$$

is a right ideal of R, which is not a left ideal. It is an exercise to show that

$$\mathrm{Ann}_R(I) = \begin{pmatrix} 0 & K \\ 0 & K \end{pmatrix} \quad \text{and} \quad \mathrm{Ann}^r_R(I) = \{0\}.$$

F. Cedó and L. Vendramin, *Groups, Radical Rings, and the Yang-Baxter Equation*, Progress in Mathematics 361, https://doi.org/10.1007/978-3-032-01524-2_3

Hence I is a faithful right R-module.

Lemma 3.1.3 *Let M be an R-module. Then M is a faithful $(R/\operatorname{Ann}_R(M))$-module.*

Proof We know that $\operatorname{Ann}_R(M)$ is an ideal of R, and thus $R/\operatorname{Ann}_R(M)$ is a ring. We define on M the multiplication by elements of $R/\operatorname{Ann}_R(M)$ by

$$(a + \operatorname{Ann}_R(M))m = am$$

for all $m \in M$ and $a \in R$. This multiplication is well-defined. Let $m \in M$ and $a, a' \in R$ be such that $a + \operatorname{Ann}_R(M) = a' + \operatorname{Ann}_R(M)$. Since $a - a' \in \operatorname{Ann}_R(M)$, $(a - a')m = 0$, that is $am = a'm$. A straightforward calculation shows that M with the addition and this multiplication is a faithful $R/\operatorname{Ann}_R(M)$-module. □

Definition 3.1.4 A module M is *simple* if $M \neq \{0\}$ and $\{0\}$ and M are the only submodules of M.

Lemma 3.1.5 *An R-module M is simple if and only if there exists a maximal left ideal I of R such that $M \cong R/I$.*

Proof Suppose first that M is simple. Let $m \in M \setminus \{0\}$. Then $M = Rm$. Let $f: R \to M$ be the map defined by $f(a) = am$. It is clear that f is an epimorphism of R-modules. By the first isomorphism theorem, $M \cong R/\ker(f)$. Furthermore, if J is a left ideal of R such that $\ker(f) \subseteq J \subseteq R$, then $J/\ker(f)$ is isomorphic to $\{0\}$ or M. Hence J is equal to $\ker(f)$ or R, and thus $\ker(f)$ is a maximal left ideal of R.

The converse is clear. □

Note that Zorn's lemma implies that every unitary ring has simple modules, see Exercise 3.6.1.

Definition 3.1.6 Let $(M_i)_{i\in I}$ be a family of submodules of an R-module M. We say that the sum $\sum_{i\in I} M_i$ is a *direct sum* if for every $j \in I$, $M_j \cap \sum_{i\in I\setminus\{j\}} M_i = \{0\}$. We will write $\bigoplus_{i\in I} M_i$ to denote the direct sum of the submodules M_i of M.

Definition 3.1.7 An R-module is *semisimple* if it is the sum of its simple submodules.

Proposition 3.1.8 *Let M be a semisimple module. Let N be a submodule of M. Then*

(1) There exists a submodule L of M such that $M = N \oplus L$,
(2) N and M/N are semisimple, and
(3) M is a direct sum of simple submodules.

Proof Let $I = \{S : S \text{ is a simple submodule of } M\}$. Let

$$C = \{J : J \subseteq I, \text{ such that } N + \sum_{S \in J} S \text{ is a direct sum }\}.$$

Note that $\emptyset \in C$. Then, applying Zorn's lemma to C ordered by the inclusion, we get that there exists a maximal element $J \in C$. Let

$$M_1 = N \oplus \bigoplus_{S \in J} S.$$

We shall see that $M_1 = M$. Suppose that $M_1 \neq M$. Then there exists a simple submodule $S' \in I$ such that $S' \nsubseteq M_1$. Since S' is simple, we have that $S' \cap M_1 = \{0\}$ and thus $J \cup \{S'\} \in C$, in contradiction with the maximality of J. Hence $M_1 = M$ and (1) follows. Note that for $N = \{0\}$, we have that $M = \bigoplus_{S \in J} S$ is a direct sum of simple submodules, and this proves (3).

Let $I' = \{S : S \text{ is a simple submodule of } N\}$. $N_1 = \sum_{S \in I'} S$. By (1), there exists a submodule L of M such that $M = N_1 \oplus L$. Hence $N = N_1 \oplus (N \cap L)$. Suppose that $N_1 \neq N$. Let $a \in N \cap L$ be a nonzero element. By Zorn's lemma there exists a submodule N_2 of $N \cap L$ maximal with the property that $a \notin N_2$. By (1), there exists a submodule L_1 of M such that $M = N_2 \oplus L_1$. Hence $N \cap L = N_2 \oplus (N \cap L \cap L_1)$. Since $N \cap L \cap L_1$ is not simple, there exists a nonzero proper submodule N_3 of $N \cap L \cap L_1$. Then, by the maximality of N_2, $a \in N_2 \oplus N_3$. By (1) there exists a submodule L_2 of M such that $M = N_3 \oplus L_2$. Hence $N \cap L \cap L_1 = N_3 \oplus (N \cap L \cap L_1 \cap L_2)$. By the maximality of N_2, $a \in N_2 \oplus (N \cap L \cap L_1 \cap L_2)$. Hence there exist $a_1, a_1' \in N_2$, $a_2 \in N_3$ and $a_3 \in N \cap L \cap L_1 \cap L_2$ such that

$$a = a_1 + a_2 = a_1' + a_3.$$

Since $N \cap L = N_2 \oplus N_3 \oplus (N \cap L \cap L_1 \cap L_2)$, $a_2 = a_3 = 0$ and $a = a_1 = a_1' \in N_2$, a contradiction. Therefore $N_1 = N$ is semisimple. Since $M = N \oplus L$, we have that $M/N \cong L$ also is semisimple, and this proves (2). □

Convention 3.1.9 From now on, a ring will mean a nonzero ring unless otherwise specified.

Definition 3.1.10 A ring R is *left primitive* if there exists a faithful simple R-module.

Similarly one defines *right primitive* rings.

Definition 3.1.11 A ring R is *simple* if $\{0\}$ and R are the only ideals of R.

Note that every simple ring is (left and right) primitive.

Example 3.1.12 Let K be a field. Then $M_n(K)$ is a simple ring for all positive integers n. Furthermore,

$$\begin{pmatrix} K & K & \dots & K \\ 0 & 0 & \dots & 0 \\ \vdots & \vdots & & \vdots \\ 0 & 0 & \dots & 0 \end{pmatrix}$$

is a minimal right ideal of $M_n(K)$ and thus it is a faithful simple right $M_n(K)$-module. Similarly,

$$\begin{pmatrix} K & 0 & \dots & 0 \\ K & 0 & \dots & 0 \\ \vdots & \vdots & & \vdots \\ K & 0 & \dots & 0 \end{pmatrix}$$

is a minimal left ideal of $M_n(K)$ and thus, it is a faithful simple $M_n(K)$-module.

Proposition 3.1.13 *Let R be a commutative ring. Then the following statements are equivalent:*

(1) R is primitive.
(2) R is simple.
(3) R is a field.

Proof The implications (3) $\Longrightarrow$ (2) $\Longrightarrow$ (1) are trivial. Let us prove that (1) $\Longrightarrow$ (3). Suppose that R is primitive. Let V be a faithful simple R-module. By Lemma 3.1.5, $V \cong R/I$ for some maximal ideal I of R. Now $\{0\} = \operatorname{Ann}_R(V) = \operatorname{Ann}_R(R/I) = I$, and thus $\{0\}$ is a maximal ideal of R. If $r \in R \setminus \{0\}$, then $rR = Rr = R$. Hence R is a field. □

The next result shows an example of a primitive ring that is not simple.

Proposition 3.1.14 *Let K be a field and V an infinite-dimensional K-vector space. Then* $\operatorname{End}_K(V)$ *is a primitive ring that is not simple.*

Proof Let $R = \operatorname{End}_K(V)$. Define on V a multiplication by elements of R by the rule $f \cdot v = f(v)$ for $f \in R$ and $v \in V$. Then V is an R-module. It is clear that for every $v \in V \setminus \{0\}$ and $w \in V$ there exists $f \in R$ such that $f(v) = w$. Hence V is a simple R-module. Clearly, $\operatorname{Ann}_R(V) = \{0\}$ and thus V is a faithful R-module. Hence R is left primitive.

Let $V^* = \operatorname{Hom}_K(V, K)$ the dual space of V. Define on V^* a multiplication by elements of R by the rule $\omega \cdot f = \omega f$ for $f \in R$ and $\omega \in V^*$. Then V^* is a faithful simple right R-module. Hence R is right primitive.

Note that $I = \{f \in R : \dim_K(\operatorname{im}(f)) < \infty\}$ is a nonzero proper ideal of R. Hence R is not simple. □

Bergman constructed in [20, 21] the first example of a right primitive ring that is not left primitive.

Proposition 3.1.15 *Let R be a left primitive ring. Then $M_n(R)$ is left primitive for all positive integer n.*

Proof Let V be a faithful simple R-module. Then $M_{n\times 1}(V)$ is a faithful simple $M_n(R)$-module. □

Note that every division ring is a simple ring. The following result gives us a method to construct division rings.

Lemma 3.1.16 (Schur) *Let V be a simple R-module. Then $\mathrm{End}_R(V)$ is a division ring.*

Proof Let $f \in \mathrm{End}_R(V) \setminus \{0\}$. Since V is simple and $\mathrm{im}(f) \neq \{0\}$, it follows that $\mathrm{im}(f) = V$. Since $\ker(f) \neq V$, $\ker(f) = \{0\}$. Hence f is an automorphism of V, and therefore $\mathrm{End}_R(V)$ is a division ring. □

The following theorem goes back to Chevalley and Jacobson.

Theorem 3.1.17 (Density theorem) *Let V be a simple R-module and $D = \mathrm{End}_R(V)$. Then V is a D-vector space. Furthermore, if $x_1, \ldots, x_n \in V$ are D-linearly independent and $y_1, \ldots, y_n \in V$, then there exists $r \in R$ such that $rx_i = y_i$ for all $i = 1, \ldots, n$.*

Proof By Schur's lemma, D is a division ring. We define on V a multiplication by elements of D by

$$f \cdot v = f(v)$$

for $f \in D$ and $v \in V$. Then V with the addition, and this multiplication is a D-vector space.

We claim that there exists $r \in R$ such that $rx_i = y_i$ for all $i = 1, \ldots, n$. We proceed by induction on n. For $n = 1$, since V is a simple R-module, $Rx_1 = V$ and thus there exists $r \in R$ such that $rx_1 - y_1$.

Suppose that $n > 1$ and that the result is true for $n - 1$. By the induction hypothesis, $V^{n-1} = R(x_1, \ldots, x_{n-1})$. We shall show that there exists $r_n \in R$ such that $r_n x_n \neq 0$ and $r_n x_i = 0$ for all $i = 1, \ldots, n-1$. Suppose that there is no such element. In this case, the map

$$\varphi\colon R(x_1, \ldots, x_{n-1}) \to V, \quad r(x_1, \ldots, x_{n-1}) \mapsto rx_n,$$

is well-defined. Since $\mathrm{Hom}_R(V^{n-1}, V) \cong (\mathrm{End}_R(V))^{n-1} = D^{n-1}$, we may identify $\mathrm{Hom}_R(V^{n-1}, V)$ with D^{n-1}, and then there exist $d_1, \ldots, d_{n-1} \in D$ such that $\varphi = (d_1, \ldots, d_{n-1})$. Thus $x_n = \varphi(x_1, \ldots, x_{n-1}) = \sum_{i=1}^{n-1} d_i x_i$, a contradiction because $x_1, \ldots, x_n$ are D-linearly independent. Hence there exists $r_n \in R$ such that $r_n x_n \neq 0$ and $r_n x_i = 0$ for all $i = 1, \ldots, n-1$. Similarly one can see that for every $j \in \{1, \ldots, n\}$, there exists $r_j \in R$ such that $r_j x_j \neq 0$ and $r_j x_i = 0$ for all $i \neq j$. Since V is a simple R-module, there exists $s_j \in R$ such that $s_j r_j x_j = y_j$. Let $r = \sum_{j=1}^{n} s_j r_j$. One checks that $rx_i = y_i$ for all $i = 1, \ldots, n$. Therefore the result follows by induction. □

Theorem 3.1.18 *Let R be a left primitive ring. Then one of the following conditions holds:*

(1) $R \cong M_n(D)$ for some division ring D and some positive integer n.
(2) There exists a division ring D such that for every positive integer m there exist a subring S_m of R and an surjective homomorphism $\varphi_m \colon S_m \to M_m(D)$.

Proof Let V be a faithful simple R-module and $D = \mathrm{End}_R(V)$. By Schur's lemma, D is a division ring. Moreover, V is a D-vector space. Let $\psi\colon R \to \mathrm{End}_D(V)$, $\psi(r)(v) = rv$ for $r \in R$ and $v \in V$. It is clear that ψ is a homomorphism of rings.

Suppose that $\dim_D(V) < \infty$ and let $v_1, \dots, v_n$ be a D-basis of V. Let $f \in \mathrm{End}_D(V)$. By the density theorem, there exists $r \in R$ such that $f(v_i) = rv_i$ for all $i = 1, \dots, n$. Hence $\psi(r) = f$ and thus ψ is surjective. Since V is a faithful R-module, ψ is injective. Therefore $R \cong \mathrm{End}_D(V) \cong M_n(D)$.

Suppose now that V is an infinite-dimensional D-vector space. Let $(v_i)_{i \geq 1}$ be a family of D-linearly independent vectors of V. Let

$$V_m = Dv_1 + \cdots + Dv_m \quad \text{and} \quad S_m = \{r \in R : rV_m \subseteq V_m\}.$$

It is clear that S_m is a subring of R. By the density theorem, the restriction map $\psi|_{S_m} \colon S_m \to \mathrm{End}_D(V_m)$ is surjective. Hence there exists a surjective homomorphism from S_m to $M_m(D)$, and the result follows. □

3.2 Radicals

Definition 3.2.1 Let R be a ring. The *Jacobson radical* of R is

$$J(R) = \bigcap_{V \text{ simple } R\text{-module}} \mathrm{Ann}_R(V).$$

Since each $\mathrm{Ann}_R(V)$ is an ideal of R, $J(R)$ also is an ideal of R.

Theorem 3.2.2 (Jacobson) *Let R be a ring and $a \in R$. Then the following conditions are equivalent:*

(1) $a \in J(R)$.
(2) $a \in M$, for every maximal left ideal M of R.
(3) $R(1 - xa) = R$ for all $x \in R$.
(4) $1 - xay$ is invertible for all $x, y \in R$.
(5) The dual left-right conditions of (1) – (4).

Proof (1) $\Rightarrow$ (2). Let M be a maximal left ideal of R and suppose that $a \in J(R)$. By Lemma 3.1.5, $V = R/M$ is a simple R-module. Thus $a \in \mathrm{Ann}_R(V) \subseteq M$.

(2) $\Rightarrow$ (3). Let $x \in R$ and suppose that $a \in M$, for every maximal left ideal M of R. Suppose that $R(1 - xa) \neq R$. By Zorn's lemma, there exists a maximal left

ideal M of R such that $R(1 - xa) \subseteq M$. But $a \in M$, hence $1 = 1 - xa + xa \in M$, a contradiction. Therefore $R(1 - xa) = R$.

(3) $\Rightarrow$ (4). Let $x, y \in R$ and suppose that $R(1 - xa) = R$ for all $x \in R$. Suppose that $R(1 - xay) \neq R$. By the implication (1) $\Rightarrow$ (3), there exists a simple R-module V such that $ayV \neq 0$. Hence $aV \neq 0$. Let $v \in V$ such that $av \neq 0$. Since V is simple, there exists $z \in R$ such that $zav = v$. Thus $(1 - za)v = 0$ and, since $v \neq 0$, we have that $R(1 - za) \neq R$, a contradiction. Therefore $R(1 - xay) = R$. Hence there exists $b \in R$ such that $(1 - b)(1 - xay) = 1$. Thus $b = bxay - xay = (bx - x)ay$ and, then we get that $R(1 - b) = R$. Hence $1 - b$ is invertible and $(1 - b)^{-1} = 1 - xay$.

(4) $\Rightarrow$ (1). Let V be a simple R-module and suppose that $1 - xay$ is invertible for all $x, y \in R$. Suppose that $aV \neq 0$. Hence as above, there exist $v \in V \setminus \{0\}$ and $z \in R$ such that $(1 - za)v = 0$. But $1 - za$ is invertible, a contradiction, therefore $a \in \mathrm{Ann}_R(V)$ for all simple R-module V, that is $a \in J(R)$.

Since condition (4) is left-right symmetric, the result follows. □

Definition 3.2.3 Let $(R_i)_{i\in I}$ be a non-empty family of rings. Let

$$\pi_j \colon \prod_{i\in I} R_i \to R_j \quad (j \in I),$$

be the natural maps. A subring S of $\prod_{i\in I} R_i$ is said to be a *subdirect product* of the rings R_i if for every $j \in I$ the restriction $\pi_j|_S$ of π_j to S is surjective.

Let R be a ring and $I = \{\mathrm{Ann}_R(V) : V \text{ is a simple } R\text{-module}\}$. Define

$$\begin{aligned} \varphi \colon R &\to \textstyle\prod_{A\in I} R/A \\ r &\mapsto (r + A)_{A\in I}. \end{aligned}$$

It is clear that φ is a homomorphism of rings and $\ker(\varphi) = J(R)$. Hence, by the first isomorphism theorem, there is an injective homomorphism

$$\begin{aligned} \overline{\varphi} \colon R/J(R) &\to \textstyle\prod_{A\in I} R/A \\ r + J(R) &\mapsto (r + A)_{A\in I}. \end{aligned}$$

Thus $R/J(R) \cong \varphi(R)$, which is a subdirect product of the rings R/A, and these rings are left primitive rings by Lemma 3.1.3.

Definition 3.2.4 A ring is said to be *semiprimitive* if it is isomorphic to a subdirect product of left primitive rings.

Proposition 3.2.5 *Let R be a ring. Then the following conditions are equivalent.*

(1) R is semiprimitive.
(2) $J(R) = \{0\}$.
(3) R is isomorphic to a subdirect product of right primitive rings.

Proof Since (2) is left-right symmetric, it is enough to prove that (1) and (2) are equivalent. Above we have seen that (2) implies (1).

We may assume that R is a subdirect product of the rings $\{R_i\}_{i\in I}$ and that R_j is left primitive for all $j \in I$. Let V_j be a faithful simple R_j-module. Let $\pi_j : \prod_{i\in I} R_i \to R_j$ be the natural map. Then V_j is an R-module via π_j, that is

$$r \cdot v := \pi_j(r)v$$

for all $v \in V_j$ and all $r \in R$. It is clear that V_j is a simple R-module. Since V_j is a faithful R_j-module, we have that $\mathrm{Ann}_R(V_j) = \ker(\pi_j|_R)$. Since R is a subring of $\prod_{i\in I} R_i$, we have that $\bigcap_{j\in I} \mathrm{Ann}_R(V_j) = \{0\}$. Since $J(R) \subseteq \bigcap_{j\in I} \mathrm{Ann}_R(V_j)$, we get that $J(R) = \{0\}$, and the result follows. □

Note that every left (or right) primitive ring is semiprimitive. A consequence of Proposition 3.2.5 is the following result.

Corollary 3.2.6 *If R is a ring, then $J(R/J(R)) = \{0\}$.* □

Proposition 3.2.7 (Nakayama's lemma) *Let R be a ring and $S \subseteq J(R)$. Let M be a finitely generated R-module. If $SM = M$, then $M = \{0\}$.*

Proof Let $m_1 = 0, m_2, \dots, m_n \in M$ such that $M = Rm_1 + Rm_2 + \cdots + Rm_n$. We shall prove that $M = \{0\}$ by induction on n. For $n = 1$, it is clear. Suppose that $n > 1$ and the result holds for $n - 1$. We know that there exist $s_1, s_2, \dots, s_n \in S$ such that

$$m_n = s_1 m_1 + s_2 m_2 + \cdots + s_n m_n.$$

Hence $(1 - s_n)m_n = s_1 m_1 + \cdots + s_{n-1} m_{n-1}$ and, since $1 - s_n$ is invertible, we get that $m_n = (1 - s_n)^{-1} s_1 m_1 + \cdots + (1 - s_n)^{-1} s_{n-1} m_{n-1}$. Hence $M = Rm_1 + \cdots + Rm_{n-1}$ and by the induction hypothesis, $M = \{0\}$. Therefore the result follows by induction. □

Note that if a is a nilpotent element of a ring R, then there exists a positive integer n such that $a^n = 0$, and then

$$(1 - a)(1 + a + \cdots + a^{n-1}) = 1.$$

Definition 3.2.8 A *nil left (right) ideal* of a ring R is a left (right) ideal I of R such that every $a \in I$ is nilpotent.

By Theorem 3.2.2 and the above remark, it is clear that every nil left (right) ideal of R is contained in $J(R)$.

Definition 3.2.9 A left (right) ideal I of R is *nilpotent* if there exists a positive integer n such that $I^n = \{0\}$.

Definition 3.2.10 An ideal J of R is said to be a *nilradical* if it is a nil ideal and R/J has no nonzero nilpotent ideal.

Proposition 3.2.11 *Let R be a ring. Then there exists a unique maximal nil ideal $\mathcal{U}(R)$ of R. Furthermore, $\mathcal{U}(R)$ is a nilradical and it is called the upper nilradical of R.*

Proof First, we shall see that the sum of two nil ideals is a nil ideal. Let I_1, I_2 be nil ideals of R. Let $x \in I_1$ and $y \in I_2$. There exists a positive integer n such that $x^n = 0$. Hence $(x+y)^n \in I_2$. Thus $x + y$ is nilpotent and $I_1 + I_2$ is a nil ideal. Let $C = \{I : I \text{ is a nil ideal of } R\}$. Let $\mathcal{U}(R) = \sum_{I \in C} I$. It is clear that $\mathcal{U}(R)$ is the unique maximal nil ideal of R. Let I be a nilpotent ideal of $R/\mathcal{U}(R)$ and let $\pi : R \to R/\mathcal{U}(R)$ be the natural map. Then there exists a positive integer m such that $I^m = \{0\}$. Hence, $\pi^{-1}(I)^m = \mathcal{U}(R)$. Thus $\pi^{-1}(I)$ is a nil ideal and, therefore $\pi^{-1}(I) = \mathcal{U}(R)$. Hence $I = \{0\}$, and the result follows. □

Definition 3.2.12 A ring R is said to be *prime* if for ideals I and J of R, $IJ = \{0\}$ implies that $I = \{0\}$ or $J = \{0\}$.

Note that the previous definition is equivalent to $aRb \neq \{0\}$ for all $a, b \in R \setminus \{0\}$.

Example 3.2.13 The commutative prime rings are precisely the commutative integral domains.

Proposition 3.2.14 *Every left primitive ring is prime.*

Proof Let V be a faithful simple R-module. Let I, J be nonzero ideals of R. We shall see that $IJ \neq \{0\}$. Since V is faithful and simple, $IV = V$ and $JV = V$. Hence $IJV = IV = V$ and thus $IJ \neq \{0\}$. □

Definition 3.2.15 An ideal P of a ring R is said to be *prime* if R/P is a prime ring.

Let $C = \{P : P \text{ prime ideal of } R\}$. Define

$$\varphi : R \to \prod_{P \in C} R/P$$

by $\varphi(r) = (r + P)_{P \in C}$ for all $r \in R$. It is clear that φ is a homomorphism of rings and $\ker(\varphi) = \bigcap_{P \in C} P$. The *Baer–McCoy radical* or *(Baer's) lower nilradical* of R is

$$\mathcal{B}(R) = \bigcap_{P \in C} P.$$

$\mathcal{B}(R)$ is also called the *prime radical* of R. A ring R is said to be *semiprime* if it is isomorphic to a subdirect product of prime rings. Note that $\mathcal{B}(R) = \ker(\varphi)$. Hence, if $\mathcal{B}(R) = \{0\}$, then R is semiprime. In fact, we have the following result.

Theorem 3.2.16 *Let R be a ring. Then the following conditions are equivalent.*

(1) R is semiprime.
(2) $\mathcal{B}(R) = \{0\}$.
(3) For an ideal I of R, $I^2 = \{0\}$ implies $I = \{0\}$.

Proof (1) $\Rightarrow$ (2) We may assume that R is a subdirect product of prime rings $(R_i)_{i\in I}$. Let $\pi_j\colon \prod_{i\in I} R_i \to R_j$ be the natural map for all $j \in I$. Then $P_j = \ker(\pi_j|_R)$ is a prime ideal of R because $R/P_j \cong R_j$. Hence $\bigcap_{i\in I} P_i = \{0\}$. Since $\mathcal{B}(R) \subseteq \bigcap_{i\in I} P_i$, we have that $\mathcal{B}(R) = \{0\}$.

(2) $\Rightarrow$ (1) We have seen this above.

(2) $\Rightarrow$ (3) Suppose that $\mathcal{B}(R) = \{0\}$. Let I be an ideal of R such that $I^2 = \{0\}$. Let P be a prime ideal of R. Since $I^2 \subseteq P$, we have that $I \subseteq P$. Hence $I \subseteq \mathcal{B}(R) = \{0\}$.

(3) $\Rightarrow$ (2) Let $a \in R \setminus \{0\}$. We shall prove that there exists a prime ideal P of R such that $a \notin P$ and thus $\mathcal{B}(R) = \{0\}$.

By (3), we have that $aRa \neq \{0\}$. We shall construct inductively a sequence $a_0 = a, a_1, \ldots, a_n, \ldots$ of nonzero elements of R such that $a_n \in a_{n-1}Ra_{n-1}$ for all positive integer n. Suppose we have constructed $a_0, a_1, \ldots, a_m$ for some $m \geq 0$. Then by (3), $a_mRa_m \neq \{0\}$. Let $a_{m+1} \in a_mRa_m \setminus \{0\}$. Hence we get $a_0, a_1, \ldots, a_n, \ldots$ nonzero elements in R such that $a_n \in a_{n-1}Ra_{n-1}$. Let $S = \{a_n : n \geq 0\}$. Let

$$C = \{I : I \text{ ideal of } R \text{ such that } I \cap S = \emptyset\}.$$

It is clear that $\{0\} \in C$ and that any chain in C (ordered by inclusion) has an upper bound in C. By Zorn's lemma, there exists a maximal element P in C. We shall see that P is prime. Let I, J be ideals of R such that $I \nsubseteq P$ and $J \nsubseteq P$. Since P is maximal in C, there exist non-negative integers m, n such that $a_m \in I + P$ and $a_n \in J + P$. We may assume that $m \leq n$. Hence $a_n \in (I + P) \cap (J + P)$. Since $a_{n+1} \in a_nRa_n$, we have that $a_{n+1} \in (I + P)(J + P) \subseteq IJ + P$. Since $P \cap S = \emptyset$, $IJ \nsubseteq P$. Therefore the result follows. □

Proposition 3.2.17 *Let R be a ring. Then $\mathcal{B}(R)$ is a nilradical of R, and if I is a nilradical of R, then $\mathcal{B}(R) \subseteq I$.*

Proof Let I be a nilradical of R. By definition R/I has no nonzero nilpotent ideals. By Theorem 3.2.16, R/I is semiprime. Hence I is the intersection of the prime ideals of R containing I, and thus $\mathcal{B}(R) \subseteq I$. By Proposition 3.2.11, R has nilradicals, and thus $\mathcal{B}(R)$ is a nil ideal of R. By Theorem 3.2.16, $R/\mathcal{B}(R)$ has no nonzero nilpotent ideals. Hence $\mathcal{B}(R)$ is a nilradical of R, and the result follows. □

Let R be a ring. Let $\mathcal{N} = \{I : I \text{ nilpotent ideal of } R\}$. The *nilpotent radical* of R is $N(R) = \sum_{I\in\mathcal{N}} I$. It is clear that $N(R) \subseteq \mathcal{B}(R)$, but in general, this inclusion is not an equality, that is, in general, $N(R)$ is not a nilradical of R. Furthermore, $N(R)$, in general, is not a nilpotent ideal.

Note that if R is a commutative ring, then

$$N(R) = \mathcal{B}(R) = \mathcal{U}(R) = \{x \in R : x \text{ is nilpotent}\}$$

is the unique nilradical of R, and thus, in this case, $N(R)$ is said to be the nilradical of R.

3.3 Artinian Rings

An R-module M is said to be *Artinian* if every descending chain of submodules of M is stationary, that is, if

$$M_1 \supseteq M_2 \supseteq \cdots \supseteq M_n \supseteq \ldots$$

is a descending chain of submodules of M, there exists a positive integer n_0 such that $M_n = M_{n_0}$ for all $n \geq n_0$.

The proof of the next result is easy and we left it to the reader.

Proposition 3.3.1 *An R-module M is Artinian if and only if every non-empty set of submodules of M has a minimal element.* □

A ring R is said to be *right Artinian* if it is Artinian as right R-module. Similarly one defines left Artinian rings. We say that the ring R is Artinian if it is right and left Artinian.

Example 3.3.2 Let R be a ring. Suppose that R has a subring which is a division ring D. Then every R-module also is a D-vector space. Hence every R-module finitely dimensional as D-vector space is an Artinian R-module. For example, if D is a division ring, then $M_n(D)$ is an Artinian ring for all positive integer n.

Note that if e is an idempotent element in a ring R, then eRe is a ring (possibly zero) and e is its unit-element.

Lemma 3.3.3 *Let R be a ring and $e \in R$ be an idempotent. Then $eRe \cong \mathrm{End}_R(eR)$.*

Proof Consider the map $\varphi\colon eRe \to \mathrm{End}_R(eR)$ defined by $\varphi(ere)(es) = eres$ for all $r, s \in R$. It is easy to check that φ is an injective homomorphism of rings. Let $f \in \mathrm{End}_R(eR)$. Note that $f(e) = f(e^2) = f(e)e \in eRe$. One can verify that $\varphi(f(e)) = f$, and thus φ is an isomorphism and the result follows. □

Theorem 3.3.4 (Wedderburn–Artin) *The following conditions are equivalent.*

(1) R is an Artinian simple ring.
(2) R is a left Artinian simple ring.
(3) R is a left Artinian and a left primitive ring.
(4) $R \cong M_n(D)$, for some division ring D and some positive integer n.
(5) The left-right dual condition of (1)–(4).

Furthermore, in (4), n is uniquely determined by R, and D is determined up to isomorphism.

Proof It is clear that (1) $\Rightarrow$ (2) and (2) $\Rightarrow$ (3).

(3) $\Rightarrow$ (4). Suppose that R is a left Artinian and left primitive ring. Let V be a faithful simple R-module and $D = \mathrm{End}_R(V)$. By Schur's lemma, D is a division ring. We know that V is a D-vector space. Suppose that V is an infinite-dimensional

D-vector space. Let $(v_n)_{n\geq 1}$ be a family of D-linearly independent vectors of V. Let $A_m = \mathrm{Ann}_R(\{v_1, \ldots, v_m\})$. It is clear that

$$A_1 \supseteq A_2 \supseteq \cdots \supseteq A_m \supseteq \ldots$$

By the Density theorem, these inclusions are strict, but this is a contradiction because R is left Artinian. Hence V is finite-dimensional D-vector space. Let $\dim_D(V) = n$. By the proof of Theorem 3.1.18, $R \cong M_n(D)$.

(4) $\Rightarrow$ (1) It is easy to check that $M_n(D)$ is an Artinian simple ring.

Let D, D' be division rings and n, m positive integers such that $M_n(D) \cong M_m(D')$. We shall prove that $n = m$ and $D \cong D'$. Let $\varphi\colon M_n(D) \to M_m(D')$ be an isomorphism. Let $E_{1,1} \in M_n(D)$ be the matrix with 1 in the $(1, 1)$-entry and zero in the other entries. Note that $E_{1,1}$ is an idempotent of $M_n(D)$, and it is easy to see that $E_{1,1}M_n(D)$ is a simple right $M_n(D)$-module and $\dim_D(E_{1,1}M_n(D)) = n$. By Lemma 3.3.3, we have:

$$\begin{aligned} D &\cong E_{1,1}M_n(D)E_{1,1} \cong \mathrm{End}_{M_n(D)}(E_{11}M_n(D)) \\ &\cong \mathrm{End}_{M_m(D')}(\varphi(E_{1,1})M_m(D')) \cong \varphi(E_{1,1})M_m(D')\varphi(E_{1,1}). \end{aligned}$$

Since $\varphi(E_{1,1})$ is an idempotent, using an automorphism of $M_m(D')$, we may assume that

$$\varphi(E_{1,1}) = \begin{pmatrix} I_r & 0 \\ 0 & 0 \end{pmatrix} \in M_m(D'),$$

where I_r is the $r \times r$ identity matrix. Since $\varphi(E_{1,1})M_m(D')$ is a simple $M_m(D')$-module, it is clear that $r = 1$. Hence $D \cong \varphi(E_{1,1})M_m(D')\varphi(E_{1,1}) \cong D'$. Furthermore,

$$n = \dim_D(E_{1,1}M_n(D)) = \dim_{D'}(\varphi(E_{1,1})M_m(D')) = m,$$

and the result follows. □

Theorem 3.3.5 *Let R be a right Artinian ring. Then $J(R)$ is nilpotent.*

Proof Let $J = J(R)$ and consider the chain

$$J \supseteq J^2 \supseteq \cdots \supseteq J^n \supseteq \ldots$$

Since R is right Artinian, there exists a positive integer n such that $J^n = J^m$ for all $m \geq n$. Let $I = J^n$. Suppose that $I \neq \{0\}$. Since R is right Artinian and $I = I^2$, there exists a right ideal M which is minimal with the property $MI \neq \{0\}$. Since $MI^2 = MI \neq \{0\}$, we have that $MI = M$. It is clear that M is generated by one element, hence, by Nakayama's lemma, $M = \{0\}$, in contradiction with $MI \neq \{0\}$. Therefore $I = J^n = \{0\}$ and the result follows. □

Theorem 3.3.6 (Wedderburn–Artin) *Let R be a ring. Then R is right (left) Artinian and semiprimitive if and only if*

$$R \cong M_{n_1}(D_1) \times \cdots \times M_{n_r}(D_r),$$

where the D_i are division rings. Furthermore, r is determined by R, and the pairs (D_i, n_i) are determined up to a permutation.

Proof Let $C = \{A : A = Ann^r_R(S)$ where S is a simple right R-module $\}$. Suppose that R is right Artinian and semiprimitive. By Theorem 3.2.2, $J(R) = \bigcap_{A \in C} A = \{0\}$. Let $\mathcal{A} = \{\bigcap_{A \in \mathcal{F}} A : \mathcal{F}$ is a finite subset of $C\}$. Since R is right Artinian, there exist $A_1, \ldots, A_r \in C$ such that $\bigcap_{i=1}^r A_i$ is a minimal element of $\mathcal{A}$. Suppose that $\bigcap_{i=1}^r A_i \neq \{0\}$ and let $a \in \bigcap_{i=1}^r A_i \setminus \{0\}$. Since $J(R) = \{0\}$, there exists $A \in C$ such that $a \notin A$. Hence $A \cap (\bigcap_{i=1}^r A_i)$ is strictly contained in $\bigcap_{i=1}^r A_i$, in contradiction with the minimality of $\bigcap_{i=1}^r A_i$. Hence $\bigcap_{i=1}^r A_i = \{0\}$.

We may assume that for every $j \in \{1, \ldots, r\}$

$$\bigcap_{1 \leq i \leq r,\, i \neq j} A_i \neq \{0\}.$$

Let

$$\varphi \colon R \to \prod_{i=1}^r R/A_i, \quad \varphi(x) = (x + A_1, \ldots, x + A_r).$$

It is clear that φ is an injective homomorphism of rings. We shall see that it is surjective. By Theorem 3.3.4, R/A_i is simple. Hence A_i is a maximal ideal of R. Thus $A_1 + \bigcap_{i=2}^r A_i = R$. Hence there exist $a_1 \in A_1$ and $b_1 \in \bigcap_{i=2}^r A_i$ such that $1 = a_1 + b_1$, and thus $\varphi(1 - a_1) = (1 + A_1, A_2, \ldots, A_r)$. Similarly, one can see that there exists $a_j \in A_j$ such that $\varphi(1 - a_j) = (A_1, \ldots, A_{j-1}, 1 + A_j, A_{j+1}, \ldots, A_r)$. Now it is easy to see that φ is surjective. Therefore φ is an isomorphism. By Theorem 3.3.4,

$$R \cong M_{n_1}(D_1) \times \cdots \times M_{n_r}(D_r),$$

where the D_i are division rings.

The converse is a consequence of the fact that every finite product of right Artinian rings is right Artinian. Note that r is the number of minimal ideals of R and that these minimal ideals are rings isomorphic to $M_{n_i}(D_i)$. By Theorem 3.3.4, the result follows. □

3.4 Noetherian Rings

An R-module M is said to be *Noetherian* if every ascending chain of submodules of M is stationary, that is, if

$$M_1 \subseteq M_2 \subseteq \cdots \subseteq M_n \subseteq \ldots$$

is an ascending chain of submodules of M, there exists a positive integer n_0 such that $M_n = M_{n_0}$ for all $n \geq n_0$.

The proof of the next result is easy, and we left it to the reader.

Proposition 3.4.1 *An R-module M is Noetherian if and only if every non-empty set of submodules of M has a maximal element.* □

Proposition 3.4.2 *An R-module M is Noetherian if and only if every submodule of M is finitely generated.*

Proof Suppose that M is a Noetherian R-module. Suppose that there exists a submodule N of M which is not finitely generated. By induction we shall construct a sequence $x_1, x_2, \dots, x_n, \dots \in N$ such that $x_{n+1} \notin Rx_1 + \dots + Rx_n$ for all positive integer n. Let $x_1 \in N$. Since $N \neq Rx_1$, there exists $x_2 \in N \setminus Rx_1$. Suppose that $n \geq 2$ and $x_1, \dots, x_n \in N$ satisfy that

$$x_{i+1} \notin Rx_1 + \dots + Rx_i$$

for all $1 \leq i < n$. Since $N \neq Rx_1 + \dots + Rx_n$, there exists $x_{n+1} \in N \setminus (Rx_1 + \dots + Rx_n)$. Hence by induction there exist $x_1, x_2, \dots, x_n, \dots \in N$ such that $x_{n+1} \notin Rx_1 + \dots + Rx_n$ for all positive integer n. Let $N_k = \sum_{i=1}^{k} Rx_i$. We have that the ascending chain

$$N_1 \subseteq N_2 \subseteq \dots \subseteq N_n \subseteq \dots$$

is not stationary, a contradiction. Hence every submodule of M is finitely generated.

Conversely, suppose that every submodule of M is finitely generated. Let

$$M_1 \subseteq M_2 \subseteq \dots \subseteq M_n \subseteq \dots$$

be an ascending chain of submodules of M. Let $N = \bigcup_{n=1}^{\infty} M_n$. Since N is a submodule of M, there exist $x_1, \dots, x_k \in N$ such that $N = \sum_{i=1}^{k} Rx_i$. Now there exists a positive integer n_0 such that $x_1, \dots, x_k \in M_{n_0}$. Hence $M_n = M_{n_0} = N$ for all $n \geq n_0$. Therefore M is Noetherian. □

Corollary 3.4.3 *Let M be an R-module and let N be a submodule of M. Then M is Noetherian if and only if N and M/N are Noetherian.*

Proof Suppose that M is Noetherian. By Proposition 3.4.2, every submodule of M is finitely generated. In particular, every submodule of N is finitely generated, and thus N is Noetherian. Since the submodules of M/N are of the form L/N, where L is a submodule of M containing N, it is clear that the submodules of M/N are finitely generated, and thus M/N is Noetherian.

Conversely, suppose that N and M/N are Noetherian. Let K be a submodule of M. By Proposition 3.4.2, $K \cap N$ is finitely generated and $K/(K \cap N) \cong (K + N)/N$ also is finitely generated. Let $x_1, \dots, x_s, y_1, \dots, y_t \in K$ such that

$$K \cap N = Rx_1 + \dots + Rx_s$$

$$\text{and} \quad K/(K \cap N) = R(y_1 + (K \cap N)) + \cdots + R(y_t + (K \cap N)).$$

It is easy to see that

$$K = Rx_1 + \cdots + Rx_s + Ry_1 + \cdots + Ry_t.$$

By Proposition 3.4.2, it follows that M is Noetherian. □

A ring R is said to be *right Noetherian* if it is Noetherian as a right R-module. Similarly, one defines left Noetherian rings. We say that the ring R is Noetherian if it is right and left Noetherian.

Example 3.4.4 $\mathbb{Z}$ is a Noetherian ring.

Example 3.4.5 Every division ring is Noetherian.

Theorem 3.4.6 *Let R be a right Noetherian ring. Then the polynomial ring $R[X]$ is right Noetherian.*

Proof Let I be a right ideal of $R[X]$. Let

$$J = \{a \in R : \text{ there exists } a_0 + a_1X + \cdots + a_nX^n \in I \text{ such that } a_n = a\}.$$

It is an easy exercise to check that J is a right ideal of R. Since R is right Noetherian, there exists $c_1, \ldots, c_m \in J$ such that

$$J = c_1R + \cdots + c_mR.$$

Let $p_i(X) = a_{i,0} + a_{i,1}X + \cdots + a_{i,n_i}X^{n_i} \in I$ such that $a_{i,n_i} = c_i$ for all $1 \le i \le m$. Let n be a positive integer such that $n \ge n_i$ for all $1 \le i \le m$. Since R is right Noetherian, by Corollary 3.4.3, it is easy to see by induction on n that R^{n+1} is a Noetherian right R-module. Consider the epimorphism of right R-modules

$$\pi : R^{n+1} \to R + RX + \cdots + RX^n, \quad \pi(a_0, a_1, \ldots, a_n) = a_0 + a_1X + \cdots + a_nX^n.$$

By Corollary 3.4.3, $R + RX + \cdots + RX^n$ is a Noetherian right R-module and $I \cap (R + RX + \cdots + RX^n)$ also is a Noetherian right R-module. Hence there exist $g_1(X), \ldots, g_s(X) \in I \cap (R + RX + \cdots + RX^n)$ such that

$$I \cap (R + RX + \cdots + RX^n) = \sum_{j=1}^{s} g_j(X)R.$$

We shall prove by induction on thedegree of $p(X) \in I$ that

$$p(X) \in \sum_{j=1}^{s} g_j(X)R[X] + \sum_{i=1}^{m} p_i(X)R[X].$$

We may assume that $\deg(p(X)) = t > n$ and that for every $q(X) \in I$ of degree $< t$,

$$q(X) \in \sum_{j=1}^{s} g_j(X)R[X] + \sum_{i=1}^{m} p_i(X)R[X].$$

Let $a_0, a_1, \ldots, a_t \in R$ such that $p(X) = a_0 + a_1 X + \cdots + a_t X^t$. Since $a_t \in J$, there exist $r_1, \ldots, r_m \in R$ such that $a_t = \sum_{i=1}^{m} c_i r_i$. Hence

$$q(X) = p(X) - \sum_{i=1}^{m} p_i(X)X^{t-n_i} r_i \in I$$

and $\deg q(X) < t$. By the inductive hypothesis,

$$q(X) \in \sum_{j=1}^{s} g_j(X)R[X] + \sum_{i=1}^{m} p_i(X)R[X].$$

Hence

$$p(X) \in \sum_{j=1}^{s} g_j(X)R[X] + \sum_{i=1}^{m} p_i(X)R[X].$$

By induction, I is finitely generated and therefore $R[X]$ is right Noetherian. □

Theorem 3.4.7 (Hopkins' theorem) *Let R be a ring. Then R is right Artinian if and only if R is right Noetherian, $J(R)$ is nilpotent, and $R/J(R)$ is a semisimple right R-module.*

Proof Suppose that R is right Artinian. By Theorem 3.3.5, $J(R)$ is nilpotent. Since $R/J(R)$ is right Artinian and semiprimitive, by Wedderburn–Artin theorem, it is easy to see that $R/J(R)$ is a semisimple R-module. Let n be a positive integer such that $J(R)^n = \{0\}$. Consider the descending chain

$$R \supseteq J(R) \supseteq \cdots \supseteq J(R)^n = \{0\}.$$

Note that $J(R)^i/J(R)^{i+1}$ is a right $R/J(R)$-module for all $1 \leq i < n$. It is an exercise to prove that every direct sum of semisimple right modules is semisimple. Since $J(R)^i/J(R)^{i+1}$ is an epimorphic image of a direct sum of copies of $R/J(R)$, we have that $J(R)^i/J(R)^{i+1}$ is a semisimple right R-module. Since R is right Artinian, $J(R)^i/J(R)^{i+1}$ is an Artinian right module. By Proposition 3.1.8, $J(R)^i/J(R)^{i+1}$ and its right submodules are finite direct sums of simple right R-modules. In particular, every right submodule of $J(R)^i/J(R)^{i+1}$ is finitely generated. Hence $J(R)^i/J(R)^{i+1}$ is a right Noetherian R-module. Hence R also is right Noetherian.

Conversely, assume that R is right Noetherian, $J(R)$ is nilpotent and $R/J(R)$ is a semisimple right R-module. There exists a positive integer n such that

$$R \supseteq J(R) \supseteq \cdots \supseteq J(R)^n = \{0\}.$$

As above $J(R)^i/J(R)^{i+1}$ is a semisimple right R-module. Since R is right Noetherian, by Proposition 3.1.8, $J(R)^i/J(R)^{i+1}$ and its right submodules are finite direct sums of simple right R-modules. Now it is clear that $J(R)^i/J(R)^{i+1}$ is a right Artinian R-module and therefore R is right Artinian. □

3.5 Rings Without Unity

Let S be a ring without unity. Consider $S^1 = \mathbb{Z} \times S$ with the addition defined component-wise and the multiplication defined by the rule

$$(z_1, a)(z_2, b) = (z_1z_2, z_1b + z_2a + ab),$$

for all $z_1, z_2 \in \mathbb{Z}$ and all $a, b \in S$. It is easy to check that S^1 is a ring and that $(1, 0)$ is its unit element. Furthermore, $\{0\} \times S$ is an ideal of S^1, which is naturally isomorphic to S (as rings without unity).

We say that a ring S without unity is a *(Jacobson) radical ring* if it is isomorphic to the Jacobson radical of a ring (with unity). Note that if S is a radical ring and R is a ring (with unity) such that $S \cong J(R)$, we have that for every $a \in S$ there exists a unique $b \in S$ such that $a + b + ab = a + b + ba = 0$. This can be proved as follows. Let $\psi\colon S \to R$ be an injective homomorphism of rings without unit such that $\mathrm{im}(\psi) = J(R)$, this exists because of $S \cong J(R)$. Then, for $a \in S$, $1 + \psi(a) \in R$ is invertible. Thus there exists $c \in R$ such that $(1 + \psi(a))(1 + c) = (1 + c)(1 + \psi(a)) = 1$. Note that this implies that $c = -\psi(a)c - \psi(a) \in J(R)$. Hence there exists $b \in S$ such that $\psi(b) = c$. Therefore

$$a + b + ab = a + b + ba = 0.$$

The uniqueness of b is clear. We shall study the maximal left ideals of $S^1 = \mathbb{Z} \times S$. Let M be a maximal left ideal of S^1. Since for all $a \in S$ there exists $b \in S$ such that $a + b + ab = a + b + ba = 0$, we have that $(1, a)(1, b) = (1, b)(1, a) = 1$. Note that if $(0, a) \notin M$, then there exist $(z_1, c) \in S^1$ and $(z_2, d) \in M$ such that

$$(z_1, c)(0, a) + (z_2, d) = (1, 0),$$

that is $z_2 = 1$ and $z_1a + ca + d = 0$, thus $(1, d) \in M$, a contradiction because $(1, d)$ is invertible in S^1. Hence $\{0\} \times S \subseteq M$. Now it is easy to see that $M = I \times S$, where I is a maximal ideal of $\mathbb{Z}$. Therefore $J(S^1) = \{0\} \times S \cong S$.

Thus we have proved the following result.

Proposition 3.5.1 *Let S be a ring without unity. Then the following conditions are equivalent.*

(1) S is a radical ring.

(2) *For all* $a \in S$ *there exists a unique* $b \in S$ *such that* $a + b + ab = a + b + ba = 0$.
(3) $S \cong J(S^1)$. □

Let S be a ring without unity (or with unity). Define on S the operation $\circ$ by

$$a \circ b = a + b + ab,$$

for all $a, b \in S$. It is easy to see that $(S, \circ)$ is a monoid with neutral element 0. Note that S is a radical ring if and only if $(S, \circ)$ is a group. If $a \in S$ is invertible in the monoid $(S, \circ)$, we shall denote by a' its inverse.

A *nil ring* is a ring S without unity such that every element of S is nilpotent. Note that every nil ring is a radical ring.

Example 3.5.2 Let K be a field. Then $XK[[X]]$ is a radical ring and it is not a nil ring.

3.6 Exercises

3.6.1 Prove that unitary rings have simple modules.
3.6.2 Let M be an R-module. Prove that if for every submodule N of M there exists a submodule L of M such that $M = N \oplus L$, then M is semisimple.
3.6.3 Prove Proposition 3.3.1.
3.6.4 Prove Proposition 3.4.1.
3.6.5 Let D be the ring of rationals with odd denominators. Let $R = \begin{pmatrix} D & \mathbb{Q} \\ 0 & \mathbb{Q} \end{pmatrix}$. Prove that R is right Noetherian and $J(R) = \begin{pmatrix} J(D) & \mathbb{Q} \\ 0 & 0 \end{pmatrix}$. Prove that R is not left Noetherian. Prove that $J(R)^n \supseteq \begin{pmatrix} 0 & \mathbb{Q} \\ 0 & 0 \end{pmatrix}$ and hence $\bigcap_n J(R)^n$ is nonzero.

3.7 Open Problems

Open problem 3.7.1 (*Kaplansky*) If a ring R is right primitive and Noetherian, is R left primitive?
Open problem 3.7.2 (*Jacobson*) Let R be a Noetherian ring. Is then

$$\bigcap_{n \geq 1} J(R)^n = \{0\}?$$

Open problem 3.7.3 (Köthe) Let R be a ring. Is the sum of two arbitrary nil left ideals of R nil?

3.8 Notes

The material on non-commutative ring theory is standard, see for example [25] or [49]. Radical rings were introduced by Jacobson in [70]. Nil rings were used by Zelmanov in his solution to Burnside's problem, see for example [116].

In 1957 [77] Kaplansky formulated the question of whether right primitive rings are left primitive. After the example of Bergman of a right primitive ring which is not left primitive [20, 21] and the examples of Jategaonkar [72] with the additional property of being principal left ideal domains, Kaplansky formulated Open problem 3.7.1 in [78].

Open problem 3.7.2 was originally formulated by Jacobson in 1956 [71] for one-sided Noetherian rings. In 1965 Herstein [65] found a counterexample in the case of one-sided Noetherian rings (see Exercise 3.6.5) and reformulated the conjecture as it appears here. For commutative Noetherian rings, Krull's intersection theorem gives a positive answer to this problem (see page 402 in [49]).

Open problem 3.7.3 is the well-known Köthe's conjecture. The conjecture was first formulated in 1930, see [81]. It is known to be true in several cases. In full generality, the problem is still open. In [82] Krempa proved that the following statements are equivalent:

(1) Köthe's conjecture is true.
(2) If R is a nil ring, then $R[X]$ is a radical ring.
(3) If R is a nil ring, then $M_2(R)$ is a nil ring.
(4) Let $n \geq 2$. If R is a nil ring, then $M_n(R)$ is a nil ring.

In 1956 Amitsur formulated the following conjecture, see for example [4]: If R is a nil ring, then $R[X]$ is a nil ring. In [101] Smoktunowicz found a counterexample to Amitsur's conjecture. This counterexample suggests that Köthe's conjecture might be false. A simplification of Smoktunowicz's example appears in [90]. See [102, 103] for more information on Köthe's conjecture and related topics.

Chapter 4
The Yang–Baxter Equation

The quantum Yang–Baxter equation is one of the important equations of mathematical physics first appearing in the papers of Yang [115] and Baxter [18]. A *solution to the quantum Yang–Baxter equation* is a bijective linear map $R\colon V \otimes V \to V \otimes V$, where V is a vector space, such that

$$R_{12}R_{13}R_{23} = R_{23}R_{13}R_{12},$$

where R_{ij} denotes the map $V \otimes V \otimes V \to V \otimes V \otimes V$ acting as R on the (i, j) tensor factor and as the identity on the remaining factor. Let $\varphi\colon V \otimes V \to V \otimes V$ be the linear map such that $\varphi(u \otimes v) = v \otimes u$ for all $u, v \in V$. Then one can check that $R\colon V \otimes V \to V \otimes V$ is a solution to the quantum Yang–Baxter equation if and only if $\overline{R} = \varphi R$ satisfies

$$\overline{R}_{12}\overline{R}_{23}\overline{R}_{12} = \overline{R}_{23}\overline{R}_{12}\overline{R}_{23}.$$

In this case, we say that $\overline{R}$ is a *solution to the Yang–Baxter equation*.

Note that if X is a basis of V and $r\colon X \times X \to X \times X$ is a bijective map such that

$$(r \times \mathrm{id}_X)(\mathrm{id}_X \times r)(r \times \mathrm{id}_X) = (\mathrm{id}_X \times r)(r \times \mathrm{id}_X)(\mathrm{id}_X \times r),$$

then r induces a solution to the Yang–Baxter equation. In this case, we say that r is a *set-theoretic* solution to the Yang–Baxter equation.

In this chapter we study a particular important class of set-theoretic solutions to the Yang–Baxter equation.

F. Cedó and L. Vendramin, *Groups, Radical Rings, and the Yang-Baxter Equation*, Progress in Mathematics 361, https://doi.org/10.1007/978-3-032-01524-2_4

4.1 Set-Theoretic Solutions

In [53], Drinfeld briefly discusses set-theoretic solutions to the Yang–Baxter equation. He not only observed that it makes sense to consider the Yang–Baxter equation in the category of sets but also that "maybe it would be interesting to study set-theoretical solutions."

Definition 4.1.1 A *set-theoretic solution* to the Yang–Baxter equation (YBE) is a pair (X, r), where X is a non-empty set and $r : X \times X \to X \times X$ is a bijective map that satisfies

$$(r \times \mathrm{id}_X)(\mathrm{id}_X \times r)(r \times \mathrm{id}_X) = (\mathrm{id}_X \times r)(r \times \mathrm{id}_X)(\mathrm{id}_X \times r),$$

where, if $r(x, y) = (\sigma_x(y), \tau_y(x))$, then

$$\begin{aligned} r \times \mathrm{id}_X &: X \times X \times X \to X \times X \times X, & (r \times \mathrm{id}_X)(x, y, z) &= (\sigma_x(y), \tau_y(x), z), \\ \mathrm{id}_X \times r &: X \times X \times X \to X \times X \times X, & (\mathrm{id}_X \times r)(x, y, z) &= (x, \sigma_y(z), \tau_z(y)). \end{aligned}$$

The solution (X, r) is said to be *finite* if X is a finite set (Fig. 4.1).

Example 4.1.2 Let X be a non-empty set. Then (X, id_{X^2}) is a set-theoretic solution to the YBE.

Example 4.1.3 Let X be a non-empty set. Then (X, r), where $r(x, y) = (y, x)$, is a set-theoretic solution to the YBE. This solution is known as the *trivial solution* over the set X.

Convention 4.1.4 *If (X, r) is a set-theoretic solution to the YBE, we always write*

$$r(x, y) = (\sigma_x(y), \tau_y(x)).$$

By definition, the map $r : X \times X \to X \times X$ is invertible. By convention, we always write

$$r^{-1}(x, y) = (\widehat{\sigma}_x(y), \widehat{\tau}_y(x)).$$

It follows that

$$x = \widehat{\sigma}_{\sigma_x(y)}\tau_y(x), \quad y = \widehat{\tau}_{\tau_y(x)}\sigma_x(y).$$

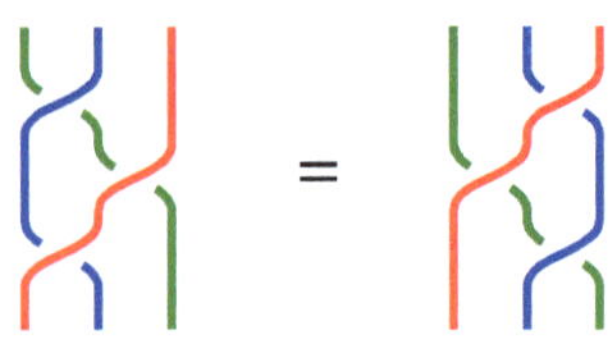

Fig. 4.1 The Yang–Baxter equation

Lemma 4.1.5 *Let X be a non-empty set and $r\colon X \times X \to X \times X$ be a bijective map, where $r(x, y) = (\sigma_x(y), \tau_y(x))$. Then (X, r) is a set-theoretic solution to the YBE if and only if*

$$\sigma_x\sigma_y = \sigma_{\sigma_x(y)}\sigma_{\tau_y(x)}, \qquad \sigma_{\tau_{\sigma_y(z)}(x)}\tau_z(y) = \tau_{\sigma_{\tau_y(x)}(z)}\sigma_x(y), \qquad \tau_z\tau_y = \tau_{\tau_z(y)}\tau_{\sigma_y(z)}$$

for all $x, y, z \in X$.

Proof We write $r_1 = r \times \mathrm{id}_X$ and $r_2 = \mathrm{id}_X \times r$. We first compute

$$\begin{aligned} r_1r_2r_1(x, y, z) &= r_1r_2(\sigma_x(y), \tau_y(x), z) = r_1(\sigma_x(y), \sigma_{\tau_y(x)}(z), \tau_z\tau_y(x)) \\ &= \left(\sigma_{\sigma_x(y)}\sigma_{\tau_y(x)}(z), \tau_{\sigma_{\tau_y(x)}(z)}\sigma_x(y), \tau_z\tau_y(x)\right). \end{aligned}$$

Then we compute

$$\begin{aligned} r_2r_1r_2(x, y, z) &= r_2r_1(x, \sigma_y(z), \tau_z(y)) = r_2(\sigma_x\sigma_y(z), \tau_{\sigma_y(z)}(x), \tau_z(y)) \\ &= \left(\sigma_x\sigma_y(z), \sigma_{\tau_{\sigma_y(z)}(x)}\tau_z(y), \tau_{\tau_z(y)}\tau_{\sigma_y(z)}(x)\right) \end{aligned}$$

and the result follows. □

Example 4.1.6 If (X, r) is a set-theoretic solution to the YBE, then so is (X, r^{-1}). By Lemma 4.1.5, the following formulas hold:

$$\widehat{\tau}_y\widehat{\tau}_x = \widehat{\tau}_{\tau_y(x)}\widehat{\tau}_{\sigma_x(y)}, \quad \widehat{\sigma}_x\widehat{\sigma}_y = \widehat{\sigma}_{\sigma_x(y)}\widehat{\sigma}_{\tau_y(x)}.$$

Example 4.1.7 Let $X = \{1, 2, 3, 4\}$ and $r(x, y) = (\sigma_x(y), \tau_y(x))$, where

$$\begin{aligned} &\sigma_1 = (132), &&\sigma_2 = (124), &&\sigma_3 = (143), &&\sigma_4 = (234), \\ &\tau_1 = (12)(34), &&\tau_2 = (12)(34), &&\tau_3 = (12)(34), &&\tau_4 = (12)(34). \end{aligned}$$

Then (X, r) is a solution to the YBE and $r^{-1}(x, y) = (\widehat{\sigma}_x(y), \widehat{\tau}_y(x))$ is given by

$$\begin{aligned} &\widehat{\sigma}_1 = (12)(34), &&\widehat{\sigma}_2 = (12)(34), &&\widehat{\sigma}_3 = (12)(34), &&\widehat{\sigma}_4 = (12)(34), \\ &\widehat{\tau}_1 = (142), &&\widehat{\tau}_2 = (123), &&\widehat{\tau}_3 = (243), &&\widehat{\tau}_4 = (134). \end{aligned}$$

Definition 4.1.8 A *homomorphism* between the set-theoretic solutions (X, r) and (Y, s) is a map $f\colon X \to Y$ such that the diagram

$$\begin{array}{ccc} X \times X & \xrightarrow{r} & X \times X \\ {\scriptstyle f\times f}\downarrow & & \downarrow{\scriptstyle f\times f} \\ Y \times Y & \xrightarrow[s]{} & Y \times Y \end{array}$$

is commutative, that is $s(f \times f) = (f \times f)r$. An *isomorphism* of solutions is a bijective homomorphism of solutions.

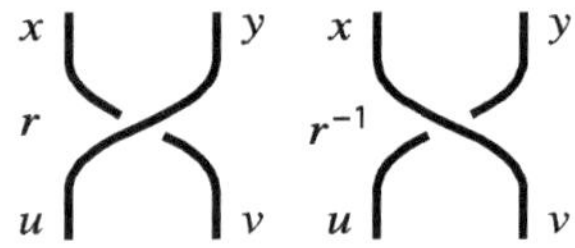

Fig. 4.2 Any row or column determines the whole square

Since we are interested in studying the combinatorics behind set-theoretic solutions to the YBE, it makes sense to study the following family of solutions.

Definition 4.1.9 We say that a set-theoretic solution (X, r) to the YBE is *non-degenerate* if all the maps σ_x and τ_x are permutations of X.

Convention 4.1.10 *A* solution *to the YBE will always mean a non-degenerate set-theoretic solution to the YBE.*

Lemma 4.1.11 *Let (X, r) be a solution to the YBE.*

(1) Given $x, u \in X$, there exist unique $y, v \in X$ such that $r(x, y) = (u, v)$.
(2) Given $y, v \in X$, there exist unique $x, u \in X$ such that $r(x, y) = (u, v)$.

Proof For the first claim take $y = \sigma_x^{-1}(u)$ and $v = \tau_y(x)$. For the second, $x = \tau_y^{-1}(v)$ and $u = \sigma_x(y)$. □

The bijectivity of r means that any row in Fig. 4.2 determines the whole square. Lemma 4.1.11 means that any column in Fig. 4.2 also determines the whole square.

Example 4.1.12 If the map $(x, y) \mapsto (\sigma_x(y), \tau_y(x))$ satisfies the YBE, then so does $(x, y) \mapsto (\tau_x(y), \sigma_y(x))$.

Example 4.1.13 Let X be a non-empty set and σ and τ be bijections on X such that $\sigma\tau = \tau\sigma$. Then (X, r), where $r(x, y) = (\sigma(y), \tau(x))$, is a solution to the YBE. This is known as the *permutation solution* associated with permutations σ and τ.

Example 4.1.14 Let G be a group. Then (G, r), where $r(x, y) = (xyx^{-1}, x)$, is a solution to the YBE.

Now we will prove the main theorem of this chapter. The result shows an intriguing connection between group actions and non-degenerate solutions. It was proved by Lu, Yan, and Zhu.

Theorem 4.1.15 *Let G be a group and let $\xi\colon G \times G \to G$, $\xi(x, y) = x \triangleright y$, be a left action of the group G on itself as a set, and let $\eta\colon G \times G \to G$, $\eta(x, y) = x \triangleleft y$, be a right action of the group G on itself as a set. If the compatibility condition*

$$uv = (u \triangleright v)(u \triangleleft v)$$

holds for all $u, v \in G$, then the pair (G, r), where

$$r\colon G \times G \to G \times G, \quad r(u, v) = (u \triangleright v, u \triangleleft v)$$

is a solution to the YBE.

Proof We write $r_1 = r \times \mathrm{id}_G$ and $r_2 = \mathrm{id}_G \times r$. Let

$$r_1r_2r_1(u, v, w) = (u_1, v_1, w_1), \quad r_2r_1r_2(u, v, w) = (u_2, v_2, w_2).$$

The compatibility condition implies that $u_1v_1w_1 = u_2v_2w_2$. So we need to prove that $u_1 = u_2$ and $w_1 = w_2$. We note that

$$u_1 = (u \triangleright v) \triangleright ((u \triangleleft v) \triangleright w), \qquad w_1 = (u \triangleleft v) \triangleleft w,$$
$$u_2 = u \triangleright (v \triangleright w), \qquad w_2 = (u \triangleleft (v \triangleright w)) \triangleleft (v \triangleleft w).$$

Using the compatibility condition and the fact that ξ is a left action,

$$u_1 = ((u \triangleright v)(u \triangleleft v)) \triangleright w = (uv) \triangleright w = u \triangleright (v \triangleright w) = u_2.$$

Similarly, since η is a right action,

$$w_2 = u \triangleleft ((v \triangleright w)(v \triangleleft w)) = u \triangleleft (vw) = (u \triangleleft v) \triangleleft w = w_1.$$

To prove that r is invertible we proceed as follows. Write $r(u, v) = (x, y)$, thus $u \triangleright v = x$, $u \triangleleft v = y$ and $uv = xy$. Since

$$(y \triangleright v^{-1})u = (y \triangleright v^{-1})(y \triangleleft v^{-1}) = yv^{-1} = x^{-1}u,$$

it follows that $y \triangleright v^{-1} = x^{-1}$, i.e., $v^{-1} = y^{-1} \triangleright x^{-1}$. Similarly,

$$v(u^{-1} \triangleleft x) = (u^{-1} \triangleright x)(u^{-1} \triangleleft x) = u^{-1}x = vy^{-1}$$

implies that $u^{-1} = y^{-1} \triangleleft x^{-1}$. Clearly $r^{-1} = \zeta(i \times i)r(i \times i)\zeta$, is the inverse of r, where $\zeta(x, y) = (y, x)$ and $i(x) = x^{-1}$. □

Proposition 4.1.16 *Under the assumptions of Theorem 4.1.15, if $r(x, y) = (u, v)$, then*

$$r(v^{-1}, u^{-1}) = (y^{-1}, x^{-1}), \quad r(x^{-1}, u) = (y, v^{-1}), \quad r(v, y^{-1}) = (u^{-1}, x).$$

Proof In the proof of Theorem 4.1.15 we found that the inverse of the map r is given by $r^{-1} = \zeta(i \times i)r(i \times i)\zeta$, where $\zeta(x, y) = (y, x)$ and $i(x) = x^{-1}$. Hence

$$r^{-1}(y^{-1}, x^{-1}) = \zeta(i \times i)r(i \times i)\zeta(y^{-1}, x^{-1}) = \zeta(i \times i)r(x, y) = (v^{-1}, u^{-1}).$$

It follows that $r(v^{-1}, u^{-1}) = (y^{-1}, x^{-1})$. To prove the equality $r(x^{-1}, u) = (y, v^{-1})$ we proceed as follows. Since $r(x, y) = (u, v)$, it follows that $x \triangleright y = u$. Then $x^{-1} \triangleright u = y$ and hence $r(x^{-1}, u) = (y, z)$ for some $z \in G$. Since $xy = uv$ and $x^{-1}u = yz$, it immediately follows that $yz = yv^{-1}$. Then $z = v^{-1}$. Similarly one proves $r(v, y^{-1}) = (u^{-1}, x)$. □

4.2 Involutive Solutions

In this section we study solutions to the YBE with an intriguing interplay with radical rings.

Definition 4.2.1 A solution (X, r) to the YBE is said to be *involutive* if $r^2 = \mathrm{id}$.

For $n \geq 2$, the *symmetric group* $\mathbb{S}_n$ can be presented as the group with generators $\sigma_1, \dots, \sigma_{n-1}$ and relations

$$\begin{aligned} &\sigma_i\sigma_{i+1}\sigma_i = \sigma_{i+1}\sigma_i\sigma_{i+1} && \text{if } 1 \leq i \leq n-2, \\ &\sigma_i\sigma_j = \sigma_j\sigma_i && \text{if } |i-j| > 1, \\ &\sigma_i^2 = 1 && \text{for all } i \in \{1, \dots, n-1\}. \end{aligned}$$

Let (X, r) be an involutive solution to the YBE. Then the map $\sigma_i \mapsto r_{i,i+1} = \mathrm{id}_{X^{i-1}} \times r \times \mathrm{id}_{X^{n-i-1}}$ extends to an action of $\mathbb{S}_n$ on X^n.

Example 4.2.2 Let X be a non-empty set and σ be a bijection on X. Then (X, r), where $r(x, y) = (\sigma(y), \sigma^{-1}(x))$, is an involutive solution to the YBE.

We now present a very important family of involutive solutions to the YBE. These examples show an intriguing connection between the YBE and the theory of non-commutative rings.

Example 4.2.3 Let p be a prime and let $A = \mathbb{Z}/(p^2) = \mathbb{Z}/p^2\mathbb{Z}$ be the cyclic additive group of order p^2. Then A with a new multiplication $*$ defined by $a * b = pab$ is a radical ring. In this case, $x \circ y = x + y + pxy$, and $x' = -x + px^2$.

Example 4.2.4 Let n be an integer such that $n > 1$. Let $A = \left\{ \frac{nx}{ny+1} : x, y \in \mathbb{Z} \right\} \subseteq \mathbb{Q}$. Note that A is a subring without unity of the field $\mathbb{Q}$. In fact A is a radical ring. A straightforward computation shows that

$$\left(\frac{nx}{ny+1} \right)' = \frac{-nx}{n(x+y)+1}.$$

The following fundamental family of solutions appears in [99]. It turns out to be fundamental in the study of set-theoretic solutions to the YBE.

Proposition 4.2.5 *Let R be a radical ring. Then (R, r), where*

$$r(x, y) = (-x + x \circ y, (-x + x \circ y)' \circ x \circ y)$$

is an involutive solution to the YBE. □

The proposition can be demonstrated using Theorem 4.1.15, see Exercise 4.4.10. We will prove a stronger result in Theorem 8.2.1.

Note that if (X, r) is an involutive solution to the YBE, then

$$(x, y) = r^2(x, y) = r(\sigma_x(y), \tau_y(x)) = (\sigma_{\sigma_x(y)}\tau_y(x), \tau_{\tau_y(x)}\sigma_x(y)).$$

Hence

$$\tau_y(x) = \sigma^{-1}_{\sigma_x(y)}(x), \quad \sigma_x(y) = \tau^{-1}_{\tau_y(x)}(y) \tag{4.1}$$

for all $x, y \in X$. Thus for involutive solutions it is enough to know $\{\sigma_x : x \in X\}$, as from this we obtain the set $\{\tau_x : x \in X\}$.

Proposition 4.2.6 *Let (X, r) be an involutive solution to the YBE. Then the map $T: X \to X$, $x \mapsto \sigma_x^{-1}(x)$, is invertible with inverse $T^{-1}(y) = \tau_y^{-1}(y)$ and*

$$T^{-1}\sigma_x^{-1}T = \tau_x$$

for all $x \in X$.

Proof Let $U(x) = \tau_x^{-1}(x)$. Since r is involutive,

$$(U(x), x) = r^2(U(x), x) = r(\sigma_{U(x)}(x), x) = (\sigma_{\sigma_{U(x)}(x)}(x), \tau_x\sigma_{U(x)}(x)).$$

In particular, $x = \tau_x\sigma_{U(x)}(x)$ and hence $U(x) = \sigma_{U(x)}(x)$. This implies that

$$T(U(x)) = \sigma^{-1}_{U(x)}(U(x)) = x.$$

Similarly one obtains $U(T(x)) = x$.

Since (X, r) is a solution to the YBE, Lemma 4.1.5 implies that

$$\sigma_x\sigma_y = \sigma_{\sigma_x(y)}\sigma_{\tau_y(x)}$$

holds for all $x, y \in X$. Then

$$\sigma_y^{-1}T(x) = \sigma_y^{-1}\sigma_x^{-1}(x) = \sigma^{-1}_{\tau_y(x)}\sigma^{-1}_{\sigma_x(y)}(x) = \sigma^{-1}_{\tau_y(x)}\tau_y(x) = T\tau_y(x)$$

for all $y \in X$, by (4.1). □

Definition 4.2.7 A *cycle set* is a pair $(X, \cdot)$, where X is a non-empty set provided with a binary operation $X \times X \to X$, $(x, y) \mapsto x \cdot y$, such that

$$(x \cdot y) \cdot (x \cdot z) = (y \cdot x) \cdot (y \cdot z) \tag{4.2}$$

holds for all $x, y, z \in X$ and each map $\varphi_x : X \to X$, $y \mapsto x \cdot y$, is bijective. A cycle set $(X, \cdot)$ is said to be *non-degenerate* if the map $X \to X$, $x \mapsto x \cdot x$, is bijective.

Table 4.1 Involutive solutions of size ≤ 10

n	2	3	4	5	6	7	8	9	10
Solutions	2	5	23	88	595	3456	34530	321931	4895272

Definition 4.2.8 Let X and Z be cycle sets. A *homomorphism* between the cycle sets X and Z is a map $f: X \to Z$ such that $f(x \cdot y) = f(x) \cdot f(y)$ for all $x, y \in X$. An *isomorphism* of cycle sets is a bijective homomorphism of cycle sets.

Theorem 4.2.9 *There exists a bijective correspondence between non-isomorphic involutive solutions to the YBE and non-isomorphic non-degenerate cycle sets.* □

For the readers who are not familiar with the above-mentioned result, the bijective correspondence is given by

$$r(x, y) = (x * y, (x * y) \cdot x),$$

where $x * y = \varphi_x^{-1}(y)$. We leave the proof for the reader, see Exercise 4.4.12. However, we will prove a more general result in Theorem 4.3.16.

Theorem 4.2.9 can be used to construct and enumerate small involutive solutions to the YBE [2]. Table 4.1 shows the number of non-isomorphic involutive solutions of size ≤ 10. For size ≤ 7 the numbers of Table 4.1 coincide with those in [56] but differ by two for $n = 8$, as two solutions of size eight are missing in [56].

4.3 Racks

Definition 4.3.1 A *rack* is a pair $(X, \triangleleft)$, where X is a non-empty set and $X \times X \to X$, $(x, y) \mapsto x \triangleleft y$, is a binary operation on X such that the maps $\rho_y : X \to X$, $x \mapsto x \triangleleft y$, are bijective for all $y \in X$, and

$$(x \triangleleft y) \triangleleft z = (x \triangleleft z) \triangleleft (y \triangleleft z) \tag{4.3}$$

for all $x, y, z \in X$.

Racks are used in low-dimensional topology [54], singularities [26] and in the classification of finite-dimensional pointed Hopf algebras [5].

Example 4.3.2 Let X be a set. Then $x \triangleleft y = x$ turns X into a rack. This is the *trivial rack* on X.

Example 4.3.3 Let $X = \mathbb{Z}/(n)$. Then $x \triangleleft y = 2y - x$ turns X into a rack. This is the *dihedral rack* of size n.

Table 4.2 Enumeration of non-isomorphic racks

n	2	3	4	5	6	7	8	9	10	11	12	13
$r(n)$	2	6	19	74	353	2080	16023	159526	2093244	36265070	836395102	25794670618

Example 4.3.4 Let A be an abelian group and $f \in \operatorname{Aut}(A)$. Then

$$x \triangleleft y = (\mathrm{id} - f)(y) + f(x)$$

turns A into a rack. These racks are known as the *Alexander racks.*

Definition 4.3.5 Let X and Z be racks. A *rack homomorphism* between the racks X and Z is a map $f\colon X \to Z$ such that $f(x \triangleleft y) = f(x) \triangleleft f(y)$ for all $x, y \in X$. An *isomorphism* of racks is a bijective rack homomorphism.

For $n \in \mathbb{N}$, let $r(n)$ be the number of isomorphism classes of racks of size n. Some values of $r(n)$ appear in Table 4.2, see for example [112].

Proposition 4.3.6 *Let X be a non-empty set and $X \times X \to X$, $(x, y) \mapsto x \triangleleft y$ be a binary operation on X. Then $r(x, y) = (y, x \triangleleft y)$ is a solution to the YBE if and only if $(X, \triangleleft)$ is a rack.*

Proof The map r satisfies $(r \times \mathrm{id}_X)(\mathrm{id}_X \times r)(r \times \mathrm{id}_X) = (\mathrm{id}_X \times r)(r \times \mathrm{id}_X)(\mathrm{id}_X \times r)$ if and only if (4.3) holds for all $x, y, z \in X$. The map r is bijective if and only if the maps $X \to X$, $x \mapsto x \triangleleft y$, are bijective. Hence the result follows. □

The connection between racks and solutions goes deeper than the phenomenon appearing in Proposition 4.3.6.

Proposition 4.3.7 *Let (X, r) be a solution to the YBE. Then*

$$x \triangleleft y = \sigma_y \tau_{\sigma_x^{-1}(y)}(x) = \sigma_y \widehat{\sigma}_y^{-1}(x) \tag{4.4}$$

turns X into a rack and each σ_x is a rack homomorphism. Moreover, (X, r) is involutive if and only if the rack $(X, \triangleleft)$ is trivial.

Proof Since $r(x, \sigma_x^{-1}(y)) = (y, \tau_{\sigma_x^{-1}(y)}(x))$, it follows that $\widehat{\sigma}_y^{-1}(x) = \tau_{\sigma_x^{-1}(y)}(x)$ for all $x, y \in X$. Hence the second equality of (4.4) holds and the maps $X \to X$, $x \mapsto x \triangleleft y$ are bijective.

Now we show that

$$\sigma_x(y) \triangleleft \sigma_x \sigma_y(z) = \sigma_x(y \triangleleft \sigma_y(z)) \tag{4.5}$$

for all $x, y \in X$. On the one hand, by Lemma 4.1.5,

$$\sigma_x(y) \triangleleft \sigma_x \sigma_y(z) = \sigma_{\sigma_x \sigma_y(z)} \tau_{\sigma_{\tau_y(x)}(z)} \sigma_x(y) = \sigma_{\sigma_x \sigma_y(z)} \sigma_{\tau_{\sigma_y(z)}(x)} \tau_z(y).$$

On the other hand,

$$\sigma_x(y \triangleleft \sigma_y(z)) = \sigma_x\sigma_{\sigma_y(z)}\tau_z(y) = \sigma_{\sigma_x\sigma_y(z)}\sigma_{\tau_{\sigma_y(z)}(x)}\tau_z(y).$$

Therefore (4.5) follows.

By Proposition 4.3.6, to prove that $(X, \triangleleft)$ is a rack it is enough to show that $s(x, y) = (y, x \triangleleft y)$ satisfies the YBE. For that purpose, we demonstrate that the map $J: X^3 \to X^3$, $J(x, y, z) = (x, \sigma_x(y), \sigma_x\sigma_y(z))$ is invertible and satisfies

$$(\mathrm{id}_X \times s)J = J(\mathrm{id}_X \times r), \quad (s \times \mathrm{id}_X)J = J(r \times \mathrm{id}_X).$$

One checks that the map $(x, y, z) \mapsto (x, \sigma_x^{-1}(y), \sigma^{-1}_{\sigma_x^{-1}(y)}\sigma_x^{-1}(z))$ is the inverse of J.

By (4.5),

$$\sigma_x(y) \triangleleft \sigma_x\sigma_y(z) = \sigma_x(y \triangleleft \sigma_y(z)) = \sigma_x\sigma_{\sigma_y(z)}\tau_{\sigma_y^{-1}\sigma_y(z)}(y) = \sigma_x\sigma_{\sigma_y(z)}\tau_z(y).$$

Then it follows that

$$\begin{aligned}(\mathrm{id}_X \times s)J(x, y, z) &= (\mathrm{id}_X \times s)(x, \sigma_x(y), \sigma_x\sigma_y(z))\\ &= (x, \sigma_x\sigma_y(z), \sigma_x(y) \triangleleft \sigma_x\sigma_y(z))\\ &= (x, \sigma_x\sigma_y(z), \sigma_x\sigma_{\sigma_y(z)}\tau_z(y))\\ &= J(x, \sigma_y(z), \tau_z(y))\\ &= J(\mathrm{id}_X \times r)(x, y, z).\end{aligned}$$

Similarly one proves that $(s \times \mathrm{id}_X)J = J(r \times \mathrm{id}_X)$. This implies that (X, s) is a solution to the YBE and hence $(X, \triangleleft)$ is a rack by Proposition 4.3.6.

If (X, r) is involutive, then $x \triangleleft \sigma_x(y) = \sigma_{\sigma_x(y)}\tau_y(x) = x$ by (4.1). Conversely, if $x \triangleleft y = x$ for all $x, y \in X$, then $\sigma_y^{-1}(x) = \widehat{\sigma}_y^{-1}(x)$ for all $x, y \in X$ and r is involutive, as

$$r^2(x, \sigma_x^{-1}(y)) = r(y, \sigma_y^{-1}(x)) = (x, \sigma_x^{-1}(y)).$$ □

Definition 4.3.8 The rack constructed in Proposition 4.3.7 is known as the *derived rack* of (X, r). The solution (X, s) to the YBE, defined by

$$s(x, y) = (y, \sigma_y\widehat{\sigma}_y^{-1}(x))$$

for all $x, y \in X$, is called the *derived solution* of (X, r).

There is a dual version of the derived rack:

Proposition 4.3.9 *Let (X, r) be a solution to the YBE. Then*

$$x \blacktriangleleft y = \tau_y\sigma_{\tau_x^{-1}(y)}(x) = \tau_y\widehat{\tau}_y^{-1}(x)$$

turns X into a rack and each τ_x is a rack homomorphism.

Proof Since (X, r) is a solution, then so is (X, r_0), where $r_0(x, y) = (\tau_x(y), \sigma_y(x))$. Then the claim follows from Proposition 4.3.7 applied to the solution (X, r_0). □

Definition 4.3.10 The rack constructed in Proposition 4.3.9 is known as the *dual derived rack* of (X, r). The solution (X, s') to the YBE, defined by

$$s'(x, y) = (\tau_x \widehat{\tau}_x^{-1}(y), x)$$

for all $x, y \in X$, is called the *dual derived solution* of (X, r).

In general, the racks constructed in Propositions 4.3.7 and 4.3.9 are different:

Example 4.3.11 Let $X = \{1, \dots, 5\}$ and (X, r) be the solution to the YBE given by

$$\begin{aligned} &\sigma_1 = \mathrm{id}, && \sigma_2 = \mathrm{id}, && \sigma_3 = \mathrm{id}, && \sigma_4 = (13)(45), && \sigma_5 = (12)(45), \\ &\tau_1 = \mathrm{id}, && \tau_2 = \mathrm{id}, && \tau_3 = \mathrm{id}, && \tau_4 = (23)(45), && \tau_5 = (23)(45). \end{aligned}$$

On the one hand the derived rack of (X, r) is given by the permutations

$$\sigma_1 \widehat{\sigma}_1^{-1} = \sigma_2 \widehat{\sigma}_2^{-1} = \sigma_3 \widehat{\sigma}_3^{-1} = \mathrm{id}, \qquad \sigma_4 \widehat{\sigma}_4^{-1} = (132), \qquad \sigma_5 \widehat{\sigma}_5^{-1} = (123).$$

On the other hand, the dual derived rack by

$$\tau_1 \widehat{\tau}_1^{-1} = \tau_2 \widehat{\tau}_2^{-1} = \tau_3 \widehat{\tau}_3^{-1} = \mathrm{id}, \qquad \tau_4 \widehat{\tau}_4^{-1} = (123), \qquad \tau_5 \widehat{\tau}_5^{-1} = (132).$$

We now prove that the racks of Propositions 4.3.7 and 4.3.9 are isomorphic. We shall need a lemma.

Lemma 4.3.12 *Let (X, r) be a solution to the YBE. The map $T \colon X \to X$, $x \mapsto \sigma_x^{-1}(x)$, is invertible with inverse $U \colon X \to X$, $x \mapsto \tau_x^{-1}(x \blacktriangleleft x)$.*

Proof Let $x \in X$. Note that $U(x) = \widehat{\tau}_x(x)$ and

$$r^{-1}(U(x), x) = (\widehat{\sigma}_{U(x)}(x), x).$$

By Lemma 4.1.5, $\sigma_{\widehat{\sigma}_{U(x)}(x)} = \sigma_{U(x)}$. Then it follows that

$$U(x) = \sigma_{\widehat{\sigma}_{U(x)}(x)}(x) = \sigma_{U(X)}(x).$$

Therefore $TU(x) = \sigma_{U(x)}^{-1}(U(x)) = x$.

Note also that

$$r(x, T(x)) = (x, \tau_{T(x)}(x)).$$

By Lemma 4.1.5, $\widehat{\tau}_{\tau_{T(x)}(x)} = \widehat{\tau}_{T(x)}$. Hence

$$T(x) = \widehat{\tau}_{\tau_{T(x)}(x)}(x) = \widehat{\tau}_{T(x)}(x).$$

Therefore $UT(x) = \widehat{\tau}_{T(x)}^{-1}(T(x)) = x$, and the result follows. □

There is a version of Proposition 4.2.6 for arbitrary solutions. A similar result appears in Exercise 4.4.8.

Proposition 4.3.13 *Let (X, r) be a solution to the YBE. Then*

$$T: X \to X, \quad x \mapsto \sigma_x^{-1}(x),$$

is a bijective map such that

$$T\tau_y = \widehat{\sigma}_y^{-1}T, \quad T\widehat{\tau}_y = \sigma_y^{-1}T$$

and $T(x \blacktriangleleft y) = T(x) \triangleleft T(y)$ for all $x, y \in X$.

Proof Lemma 4.3.12 proves that T is bijective. We now compute, using Proposition 4.3.7,

$$\begin{aligned} T\tau_y(x) &= \sigma^{-1}_{\tau_y(x)}\tau_y(x) = \sigma^{-1}_{\tau_y(x)}\sigma^{-1}_{\sigma_x(y)}\sigma_{\sigma_x(y)}\tau_y(x) \\ &= \sigma_y^{-1}\sigma_x^{-1}\sigma_{\sigma_x(y)}\tau_y(x) = \sigma_y^{-1}\sigma_x^{-1}(x \triangleleft \sigma_x(y)) = \sigma_y^{-1}(T(x) \triangleleft y) = \widehat{\sigma}_y^{-1}T(x). \end{aligned}$$

Since $\widehat{\tau}_y(x) = \sigma^{-1}_{\widehat{\sigma}_x(y)}(x)$, Lemma 4.1.5 implies that

$$T\widehat{\tau}_y(x) = \sigma^{-1}_{\widehat{\tau}_y(x)}\widehat{\tau}_y(x) = \sigma^{-1}_{\widehat{\tau}_y(x)}\sigma^{-1}_{\widehat{\sigma}_x(y)}(x) = \sigma_y^{-1}\sigma_x^{-1}(x) = \sigma_y^{-1}T(x).$$

These formulas imply that

$$T\tau_y\widehat{\tau}_y^{-1} = \widehat{\sigma}_y^{-1}T\widehat{\tau}_y^{-1} = \widehat{\sigma}_y^{-1}\sigma_y T. \tag{4.6}$$

We evaluate Equality (4.6) on X. On the one hand,

$$T(x \blacktriangleleft y) = T\tau_y\widehat{\tau}_y^{-1}(x).$$

On the other hand, by Proposition 4.3.7,

$$\widehat{\sigma}_y^{-1}\sigma_y T(x) = \sigma_y^{-1}\sigma_y\widehat{\sigma}_y^{-1}\sigma_y T(x) = \sigma_y^{-1}(\sigma_y T(x) \triangleleft y) = T(x) \triangleleft T(y).$$ □

As it happens in the involutive case, there is a nice combinatorial structure that describes a solution to the YBE.

Definition 4.3.14 A *skew cycle set* is a triple $(X, \triangleleft, \cdot)$, where X is a non-empty set, $(X, \triangleleft)$ is a rack and $X \times X \to X$, $(x, y) \mapsto x \cdot y$, is a binary operation on X such that the maps $X \to X$, $y \mapsto x \cdot y$, are rack isomorphisms, and

$$(x \cdot y) \cdot (x \cdot z) = (y \cdot (x \triangleleft y)) \cdot (y \cdot z) \tag{4.7}$$

for all $x, y, z \in X$. A skew cycle set $(X, \triangleleft, \cdot)$ is said to be *non-degenerate* if the map $X \to X, x \mapsto x \cdot x$, is bijective.

If $(X, \triangleleft, \cdot)$ is a skew cycle set and $(X, \triangleleft)$ is a trivial rack, then $(X, \cdot)$ is a cycle set.

Definition 4.3.15 Let X and Z be skew cycle sets. A *homomorphism* between the skew cycle sets X and Z is a map $f\colon X \to Z$ such that $f(x \cdot y) = f(x) \cdot f(y)$ and $f(x \triangleleft y) = f(x) \triangleleft f(y)$ for all $x, y \in X$. An *isomorphism* of skew cycle sets is a bijective homomorphism of skew cycle sets.

Theorem 4.2.9 can be generalized to arbitrary solutions.

Theorem 4.3.16 *There exists a bijective correspondence between solutions to the YBE and non-degenerate skew cycle sets.*

Proof Let (X, r) be a solution to the YBE and $(X, \triangleleft)$ be its derived rack. We will prove that the operation $x \cdot y = \sigma_x^{-1}(y)$ turns $(X, \triangleleft)$ into a skew cycle set. By Proposition 4.3.7, the maps $X \to X, y \mapsto x \cdot y$, are bijective rack homomorphisms.

On the one hand, since $r(x, \sigma_x^{-1}(y)) = (y, \tau_{\sigma_x^{-1}(y)}(x))$,

$$\begin{aligned}(x \cdot y) \cdot (x \cdot z) &= \sigma_x^{-1}(y) \cdot \sigma_x^{-1}(z) = \sigma^{-1}_{\sigma_x^{-1}(y)}\sigma_x^{-1}(z)\\ &= \left(\sigma_x\sigma_{\sigma_x^{-1}(y)}\right)^{-1}(z) = \left(\sigma_y\sigma_{\tau_{\sigma_x^{-1}(y)}(x)}\right)^{-1}(z).\end{aligned}$$

On the other hand,

$$\begin{aligned}(y \cdot (x \triangleleft y)) \cdot (y \cdot z) &= \sigma_y^{-1}(\sigma_y\tau_{\sigma_x^{-1}(y)}(x)) \cdot \sigma_y^{-1}(z)\\ &= \sigma^{-1}_{\tau_{\sigma_x^{-1}(y)}(x)}\sigma_y^{-1}(z) = \left(\sigma_y\sigma_{\tau_{\sigma_x^{-1}(y)}(x)}\right)^{-1}(z).\end{aligned}$$

Therefore $(X, \triangleleft, \cdot)$ is a skew cycle set. Furthermore, by Lemma 4.3.12, this skew cycle set is non-degenerate.

Now we prove the converse statement. Let $(X, \triangleleft, \cdot)$ be a non-degenerate skew cycle set. For $x, y \in X$ let

$$\sigma_x(y) = x * y, \quad \tau_y(x) = \sigma^{-1}_{\sigma_x(y)}(x \triangleleft \sigma_x(y)),$$

where $x * y = z$ if and only if $x \cdot z = y$. Since X is a skew cycle set, each σ_x is bijective. Let us prove that the τ_x are bijective. Equality (4.7) with $y = \sigma_x(t)$ implies that

$$\begin{aligned}\sigma_t^{-1}\sigma_x^{-1} &= \sigma^{-1}_{\sigma_x^{-1}(y)}\sigma_x^{-1}\\ &= \sigma^{-1}_{\sigma_y^{-1}(x \triangleleft y)}\sigma_y^{-1} = \sigma^{-1}_{\sigma^{-1}_{\sigma_x(t)}(x \triangleleft \sigma_x(t))}\sigma^{-1}_{\sigma_x(t)} = \sigma^{-1}_{\tau_t(x)}\sigma^{-1}_{\sigma_x(t)}\end{aligned} \tag{4.8}$$

for all $x, t \in X$. Since each σ_x is a rack homomorphism and $\sigma^{-1}_{\sigma_x(y)}\sigma_x = \sigma_{\tau_y(x)}\sigma^{-1}_y$ holds for all $x, y \in X$, it follows that

$$\tau_y(x) = \sigma^{-1}_{\sigma_x(y)}(x \triangleleft \sigma_x(y)) = \sigma^{-1}_{\sigma_x(y)}\sigma_x(\sigma_x^{-1}(x) \triangleleft y) = \sigma_{\tau_y(x)}\sigma_y^{-1}(\sigma_x^{-1}(x) \triangleleft y).$$

holds for all $x, y \in X$. Therefore $T\tau_y = \sigma_y^{-1}\rho_y T$, where $T\colon X \to X$, $T(x) = x \cdot x$ and $\rho_y\colon X \to X$, $\rho_y(x) = x \triangleleft y$ are bijective maps. In particular, τ_y is bijective for all $y \in X$.

Let $r\colon X \times X \to X \times X$ be the map defined by $r(x, y) = (\sigma_x(y), \tau_y(x))$, for all $x, y \in X$. Now we prove that (X, r) is a solution to the YBE. Let $s\colon X \times X \to X \times X$ be the map defined by $s(x, y) = (y, x \triangleleft y)$. By Proposition 4.3.6, (X, s) is a solution to the YBE. As in the proof of Proposition 4.3.7, the map

$$J\colon X^3 \to X^3, \quad J(x, y, z) = (x, \sigma_x(y), \sigma_x\sigma_y(z))$$

is invertible. Furthermore, by (4.8) we have that

$$\begin{aligned}
(s \times \mathrm{id})J(x, y, z) &= (s \times \mathrm{id})(x, \sigma_x(y), \sigma_x\sigma_y(z)) \\
&= (\sigma_x(y), x \triangleleft \sigma_x(y), \sigma_x\sigma_y(z)) \\
&= (\sigma_x(y), \sigma_{\sigma_x(y)}\tau_y(x), \sigma_{\sigma_x(y)}\sigma_{\tau_y(x)}(z)) \\
&= J(\sigma_x(y), \tau_y(x), z) \\
&= J(r \times \mathrm{id})(x, y, z).
\end{aligned}$$

Similarly one proves that $(id \times s)J = J(id \times r)$. Therefore (X, r) is a solution to the YBE.

Let $(X, \triangleleft, \cdot)$ be a non-degenerate skew cycle set. Let $G(X, \triangleleft, \cdot) = (X, r)$, where $r(x, y) = (x * y, (x * y) \cdot (x \triangleleft (x * y)))$, where $x * y = z$ if and only if $y = x \cdot z$. We have seen that (X, r) is a solution to the YBE.

For every (X, r) we define $F(X, r) = (X, \triangleleft, \cdot)$, where $x \triangleleft y = \sigma_y\tau_{\sigma_x^{-1}(y)}(x)$ and $x \cdot y = \sigma_x^{-1}(y)$. We have seen that $(X, \triangleleft, \cdot)$ is a non-degenerate skew cycle set.

A direct calculation shows that $F(G(X, \triangleleft, \cdot)) = (X, \triangleleft, \cdot)$ for every non-degenerate skew cycle set $(X, \triangleleft, \cdot)$ and $G(F(X, r)) = (X, r)$ for every solution (X, r) to the YBE. □

Theorem 4.3.16 can be used to construct small solutions, see Table 4.3.

Table 4.3 Enumeration of non-involutive solutions

n	2	3	4	5	6	7	8
$s(n)$	2	21	253	3519	100071	4602720	422449480

4.4 Exercises

4.4.1 For $n \geq 2$, the *braid group* $\mathbb{B}_n$ is defined as the group with generators $\sigma_1, \ldots, \sigma_{n-1}$ and relations

$$\sigma_i \sigma_{i+1} \sigma_i = \sigma_{i+1} \sigma_i \sigma_{i+1} \qquad \text{if } 1 \leq i \leq n-2,$$
$$\sigma_i \sigma_j = \sigma_j \sigma_i \qquad \text{if } |i-j| > 1.$$

Let (X, r) be a set-theoretic solution to the YBE. Write $X^n = X \times \cdots \times X$ (n-times). For $i < n$ let $r_{i,i+1} = \mathrm{id}_{X^{i-1}} \times r \times \mathrm{id}_{X^{n-i-1}} : X^n \to X^n$. Then the map $\sigma_i \mapsto r_{i,i+1}$ extends to an action of $\mathbb{B}_n$ on X^n.

4.4.2 Let G be a group. Prove that the following maps satisfy the set-theoretic YBE:

(a) $r(x, y) = (y, x^{-1})$.
(b) $r(x, y) = (y^{-1}, x^{-1})$.
(c) $r(x, y) = (xyx, x^{-1})$.
(d) $r(x, y) = (x^2 y, y^{-1} x^{-1} y)$.

4.4.3 Let G be a group. Prove that the following maps satisfy the set-theoretic YBE:

(a) $r(x, y) = (x^m y x^{-m}, x)$ for every integer m.
(b) $r(x, y) = (x y^{-1} x, x)$.

4.4.4 Let $n \geq 2$ and $X = \mathbb{Z}/(n)$ be the ring of integers modulo n. Prove that the map $r(x, y) = (2x - y, x)$ satisfies the set-theoretic YBE.

4.4.5 Let G be a group and $f \in \mathrm{Aut}(G)$. Prove that the map

$$r(x, y) = (f(y), f(y)^{-1} x y)$$

satisfies the set-theoretic YBE.

4.4.6 Let (X, r) be a solution to the YBE. Let $(X, \triangleleft)$ and $(X, \blacktriangleleft)$ be the derived rack and the dual derived rack of (X, r), respectively. Prove that $x \triangleleft x = x \blacktriangleleft x$ for all $x \in X$.

4.4.7 Let (X, r) be a solution to the YBE and $(X, \triangleleft)$ be its derived rack. Prove that

$$\widehat{\tau}_x(y \triangleleft z) = \widehat{\tau}_x(y) \triangleleft \widehat{\tau}_x(z)$$

for all $x, y, z \in X$.

4.4.8 Let (X, r) be a solution to the YBE and let $(X, \triangleleft)$ be its derived rack. Prove that

$$T \tau_x(y) = \sigma_x^{-1}(T(y) \triangleleft x)$$

for all $x \in X$, where $T : X \to X$, $T(y) = \sigma_y^{-1}(y)$.

4.4.9 Let (X, r) be a solution to the YBE and (X, s) its derived solution, i.e.,

$$s(x, y) = (y, \sigma_y \widehat{\sigma}_y^{-1}(x))$$

for all $x, y \in X$. Let $J_2(x, y) = (x, \sigma_x(y))$ and $J_{n+1} = Q_{n+1} \circ (\mathrm{id} \times J_n)$ for $n \geq 2$, where

$$Q_{n+1}(x_1, \dots, x_{n+1}) = (x_1, \sigma_{x_1}(x_2), \dots, \sigma_{x_1}(x_{n+1})).$$

Prove that

$$J_n \circ r_{i,i+1} = s_{i,i+1} \circ J_n$$

for all $n \geq 2$ and $i \in \{1, \dots, n-1\}$, where

$$r_{i,i+1} = \mathrm{id}^{\times i-1} \times r \times \mathrm{id}^{\times n-i-1}, \quad s_{i,i+1} = \mathrm{id}^{\times i-1} \times s \times \mathrm{id}^{\times n-i-1}.$$

4.4.10 Prove Proposition 4.2.5.

4.4.11 If X is a cycle set, then $x \cdot (y \cdot y) = ((y * x) \cdot y) \cdot ((y * x) \cdot y)$, where $y * x = z$ if and only if $y \cdot z = x$.

4.4.12 Prove Theorem 4.2.9.

4.5 Open Problems

Open problem 4.5.1 Estimate the number of solutions to the YBE of size n for $n \to \infty$.

4.6 Notes

The first papers where set-theoretic solutions to the YBE are studied are those of Etingof, Schedler, and Soloviev [56] and Gateva-Ivanova and Van den Bergh [62]. Both papers deal with non-degenerate involutive solutions, i.e., solutions (X, r) where $r^2 = \mathrm{id}$.

In [53], Drinfeld attributes Example 4.1.13 to Lyubashenko.

Theorem 4.1.15 goes back to Lu, Yan and Zhu, see [87]. Similar results can be found in the work of Etingof, Schedler, and Soloviev [56] for involutive solutions to the YBE and in Soloviev's paper [107].

Rump introduced cycle sets in [98]. The bijective correspondence of Theorem 4.2.9 was also proved by Rump in [98]. A similar result can be found in [56, Proposition 2.2].

The numbers of Table 4.1 were computed in [2] using a combination of [59] and constraint programming techniques. The algorithm is based on an idea of Plemmons [93], originally conceived to construct non-isomorphic semigroups. The construction or enumeration of different families of finite solutions always remains an interesting computational problem. For instance, non-degenerate involutive solutions of size 11 have not yet been constructed. Finding new tools to address this challenge and expanding the database of solutions for concrete computational experiments are important goals.

In [53], Drinfeld attributes Proposition 4.3.6 to Venkov.

A particular family of racks turns out to be useful in combinatorial knot theory. A quandle is a rack $(X, \triangleleft)$ such that $x \triangleleft x = x$ for all $x \in X$.

There are several papers on the enumeration of isomorphic classes of finite racks [6, 23, 67]. Estimations on the number of finite racks of size n appear in [23].

The numbers of Table 4.3 were computed using Theorem 4.3.16 essentially with the same technique used to construct involutive solutions [2]. The construction of non-involutive solutions of size 9 seems to be feasible with these methods. However, it should be noted that a huge number of solutions is expected.

The derived rack of a solution to the YBE was first defined in the work of Soloviev [107]. Most of the properties of the derived racks mentioned in this chapter were proved in [85].

Exercises 4.4.2 and 4.4.3 appear in the work of Wada [113] on representations of braid groups.

Exercises 4.4.6 and 4.4.7 appear in [85].

The map J_n of Exercise 4.4.9 is known as the *guitar map*. It was first considered by Etingof, Schedler, and Soloviev in [56] for involutive solutions. The construction was extended to non-involutive solutions by Soloviev in [107] and Lu, Yan, and Zhu in [87]. In [52] Dehornoy used the inverse of the guitar map to develop his right-cyclic calculus and to obtain short proofs for results on the structure group of involutive solutions. In [5] Andruskiewitsch and Graña use the guitar map to study certain isomorphisms of Nichols algebras. A particular case of the guitar map also appears in the work of Przytycki [95].

Chapter 5
Nilpotent Groups

In this chapter we develop the basic definitions and results on nilpotent groups.

5.1 Central Series

Let G be a group. A subgroup H of G is *central* in G if $H \subseteq Z(G)$ and H is said to be *characteristic* if $f(H) = H$ for all $f \in \operatorname{Aut}(G)$. The center $Z(G)$ of G is a characteristic subgroup of G. Every characteristic subgroup of G is normal in G. If H is a characteristic subgroup of K and K is normal in G, then H is normal in G.

Let X be a non-empty subset of G. The *centralizer* of X in G is

$$C_G(X) = \{y \in G : xy = yx \text{ for all } x \in X\}.$$

Note that $C_G(X)$ is a subgroup of G. In particular, $C_G(G) = Z(G)$.

Let H be a normal subgroup of G. Let $\varphi \colon G \to \operatorname{Aut}(H)$ be the map defined by $\varphi(x) = \varphi_x$ and $\varphi_x(h) = xhx^{-1}$ for all $x \in G$ and $h \in H$. It is easy to check that φ is a homomorphism of groups and $\ker(\varphi) = C_G(H)$. Hence $C_G(H)$ is a normal subgroup of G. We also say that φ is a left action of G on H by conjugation.

If G is a group and $x, y, z \in G$, then the conjugation (as a left action) will be denoted by ${}^{x}y = xyx^{-1}$. The *commutator* between x and y is then

$$[x, y] = xyx^{-1}y^{-1} = ({}^{x}y)y^{-1}.$$

We also write $[x, y, z] = [x, [y, z]]$. It is easy to check the following *commutator identities:*

$$[x, yz] = [x, y]({}^{y}[x, z]), \tag{5.1}$$

$$[xy, z] = ({}^{x}[y, z])[x, z] = [x, y, z][y, z][x, z]. \tag{5.2}$$

F. Cedó and L. Vendramin, *Groups, Radical Rings, and the Yang-Baxter Equation*, Progress in Mathematics 361, https://doi.org/10.1007/978-3-032-01524-2_5

If X, Y and Z are subgroups of G, we write

$$[X, Y] = \langle [x, y] : x \in X,\ y \in Y \rangle$$

and $[X, Y, Z] = [X, [Y, Z]]$. Note that $[X, Y, Z] = [X, Z, Y]$, as $[X, Y] = [Y, X]$. The *commutator subgroup* $[G, G]$ of G is a characteristic subgroup of G.

The proof of the following lemma is an easy exercise.

Lemma 5.1.1 (Hall–Witt identity) *Let G be a group and $x, y, z \in G$. Then*

$$\left({}^{y}[x, y^{-1}, z]\right)\left({}^{z}[y, z^{-1}, x]\right)\left({}^{x}[z, x^{-1}, y]\right) = 1. \tag{5.3}$$

□

Lemma 5.1.2 (Jacobi's identity) *Let G be a group such that $[G, G, G]$ is central and $x, y, z \in G$. Then*

$$[x, y, z][y, z, x][z, x, y] = 1. \tag{5.4}$$

Proof Since $[G, G, G]$ is central, by Hall–Witt's identity, we have that

$$[x, y^{-1}, z][y, z^{-1}, x][z, x^{-1}, y] = 1.$$

Hence it is enough to prove that $[x, y^{-1}, z] = [x, y, z]^{-1}$. Since $[G, G, G]$ is central,

$$[x, y^{-1}, z] = [x, {}^{y^{-1}}[z, y]] = {}^{y^{-1}}[{}^{y}x, [z, y]] = [{}^{y}x, [z, y]].$$

By (5.2),

$$\begin{aligned}[{}^{y}x, [z, y]] &= {}^{y}[xy^{-1}, [z, y]][y, [z, y]] = [xy^{-1}, [z, y]][y, [z, y]] \\ &= {}^{xy^{-1}}[y, [z, y]][xy^{-1}, [z, y]] = [x, [z, y]].\end{aligned}$$

By (5.1),

$$[x, y, z][x, [z, y]] = [x, [y, z]]\left({}^{[y,z]}[x, [z, y]]\right) = [x, [y, z][z, y]] = 1.$$

Hence $[x, y^{-1}, z] = [x, y, z]^{-1}$ and the result follows. □

Lemma 5.1.3 (three-subgroups lemma) *Let N be a normal subgroup of G and let X, Y and Z be subgroups of G. If $[X, Y, Z] \subseteq N$ and $[Y, Z, X] \subseteq N$, then $[Z, X, Y] \subseteq N$.*

Proof We first assume that $N = \{1\}$. Then $[Y, Z] \subseteq C_G(X)$ and $[Z, X] \subseteq C_G(Y)$. It is enough to prove that $[z, x^{-1}, y] = 1$ for all $x \in X$, $y \in Y$ and $z \in Z$. Let $x \in X$, $y \in Y$ and $z \in Z$. Since $[x, y^{-1}, z] = 1$, it follows that ${}^{y}[x, y^{-1}, z] = 1$. Similarly, ${}^{z}[y, z^{-1}, x] = 1$. Hence the Hall–Witt identity yields $[z, x^{-1}, y] = 1$.

We now demonstrate the general case. Let N be a normal subgroup of G and $\pi : G \to G/N$ be the canonical map. Since $[X, Y, Z] \subseteq N$,

$$\begin{aligned}\{1\} &= \pi([X, Y, Z]) = \pi([X, [Y, Z]]) \\ &= [\pi(X), \pi([Y, Z])] = [\pi(X), [\pi(Y), \pi(Z)]] = [\pi(X), \pi(Y), \pi(Z)].\end{aligned}$$

Similarly one proves that $[\pi(Y), \pi(Z), \pi(X)] = \{1\}$. By the previous paragraph, $[\pi(Z), \pi(X), \pi(Y)] = \{1\}$, so $[Z, X, Y] \subseteq N$. □

The *lower central series* of a group G is the sequence $\gamma_k(G)$, for positive integers k, defined recursively as

$$\gamma_1(G) = G, \quad \gamma_{i+1}(G) = [G, \gamma_i(G)] \quad i \geq 1.$$

A group G is said to be *nilpotent* if there exists a non-negative integer c such that $\gamma_{c+1}(G) = \{1\}$. The smallest c such that $\gamma_{c+1}(G) = \{1\}$ is the *nilpotency index* (or *nilpotency class*) of G.

Example 5.1.4 A group is nilpotent of class one if and only if it is non-trivial and abelian.

Example 5.1.5 The group $G = \mathbb{A}_4$ is not nilpotent, as

$$\gamma_1(G) = G, \quad \gamma_j(G) = \{\mathrm{id}, (12)(34), (13)(24), (14)(23)\} \cong C_2 \times C_2$$

for all $j \geq 2$, where C_2 is the (multiplicative) cyclic group of order 2.

Lemma 5.1.6 *Let G be group. Then the following conditions hold.*

(1) Each $\gamma_i(G)$ is a characteristic subgroup of G.
(2) $\gamma_i(G) \supseteq \gamma_{i+1}(G)$ for all $i \geq 1$.
(3) If $f : G \to H$ is a surjective group homomorphism, then $f(\gamma_i(G)) = \gamma_i(H)$ for all $i \geq 1$.

Proof We shall prove the first statement by induction on i. Let $h \in \mathrm{Aut}(G)$. For $i = 1$, the result is clear. Suppose that the result holds for some $i \geq 1$. By the inductive hypothesis,

$$h(\gamma_{i+1}(G)) = h([G, \gamma_i(G)]) = [h(G), h(\gamma_i(G))] = [G, \gamma_i(G)] = \gamma_{i+1}(G).$$

Therefore, the first statement follows by induction.

The second and the third statements can be easily proved by induction on i. □

Theorem 5.1.7 *Let G be a nilpotent group.*

(1) If H is a subgroup of G, then H is nilpotent.
(2) If $f: G \to H$ is a surjective group homomorphism, then H is nilpotent.

Proof For the first statement note that $\gamma_i(H) \subseteq \gamma_i(G)$ for all $i \geq 1$. Let us prove the second claim, if there exists c such that $\gamma_{c+1}(G) = \{1\}$, then, by Lemma 5.1.6,

$$\gamma_{c+1}(H) = f(\gamma_{c+1}(G)) = f(\{1\}) = \{1\}.$$

□

There exist a non-nilpotent group G with a normal subgroup K such that K and G/K are both nilpotent. For example, take $G = \mathbb{S}_3$ and $K = \mathbb{A}_3$.

Theorem 5.1.8 *Let G be a group. Then $[\gamma_i(G), \gamma_j(G)] \subseteq \gamma_{i+j}(G)$ for all $i, j \geq 1$.*

Proof We proceed by induction on i. The case where $i = 1$ is trivial, as by definition one has $[G, \gamma_j(G)] = \gamma_{j+1}(G)$. Assume now that the result holds for some $i \geq 1$ and all $j \geq 1$. We first note that

$$[G, \gamma_i(G), \gamma_j(G)] \subseteq [G, \gamma_{i+j}(G)] = \gamma_{i+j+1}(G)$$

by the inductive hypothesis. Moreover, again using the inductive hypothesis,

$$[\gamma_i(G), \gamma_j(G), G] = [\gamma_i(G), G, \gamma_j(G)] = [\gamma_i(G), \gamma_{j+1}(G)] \subseteq \gamma_{i+j+1}(G).$$

Lemma 5.1.3 implies that $[\gamma_j(G), G, \gamma_i(G)] \subseteq \gamma_{i+j+1}(G)$. Thus

$$[\gamma_{i+1}(G), \gamma_j(G)] = [[G, \gamma_i(G)], \gamma_j(G)] = [\gamma_j(G), G, \gamma_i(G)] \subseteq \gamma_{i+j+1}(G).$$

□

Certainly, we can consider other types of arbitrary commutators, say for example $[[G, G], G]$ and $[G, G, G] = [G, [G, G]]$. This naturally suggests the notion of the weight of a commutator. For example, $[[G, G], G]$ and $[G, G, G] = [G, [G, G]]$ are both commutators of weight three.

Corollary 5.1.9 *Every commutator of weight n is contained in $\gamma_n(G)$.*

Proof We proceed by induction on n. The case $n = 1$ is trivial, so assume that $n \geq 1$ and the result holds for all m such that $n \geq m \geq 1$. Let $[A, B]$ be a commutator, where A is a commutator of weight k, B is a commutator of weight l and $n + 1 = k + l$. Since $k \leq n$ and $l \leq n$, the inductive hypothesis implies that $A \subseteq \gamma_k(G)$ and $B \subseteq \gamma_l(G)$. Thus

$$[A, B] \subseteq [\gamma_k(G), \gamma_l(G)] \subseteq \gamma_{k+l}(G)$$

by the previous theorem. Therefore, the result follows by induction. □

A group G satisfies the *normalizer condition* if each proper subgroup H is properly contained in its normalizer $N_G(H)$, that is $H \subsetneq N_G(H)$.

Lemma 5.1.10 (normalizer condition) *Let G be a nilpotent group. If H is a proper subgroup of G, then $H \subsetneq N_G(H)$.*

Proof Since G is nilpotent, there exists a positive integer c such that

$$G = \gamma_1(G) \supseteq \cdots \supseteq \gamma_{c+1}(G) = \{1\}.$$

Since $\{1\} = \gamma_{c+1}(G) \subseteq H$ and $\gamma_1(G) \nsubseteq H$, let k be the smallest positive integer such that $\gamma_k(G) \subseteq H$. Since

$$[H, \gamma_{k-1}(G)] \subseteq [G, \gamma_{k-1}(G)] = \gamma_k(G) \subseteq H,$$

it follows that $xHx^{-1} \subseteq H$ for all $x \in \gamma_{k-1}(G)$, so $\gamma_{k-1}(G) \subseteq N_G(H)$. If $N_G(H) = H$, then $\gamma_{k-1}(G) \subseteq H$, in contradiction with the minimality of k. Therefore H is a proper subgroup of $N_G(H)$. □

For a group G we define the sequence $\zeta_0(G), \zeta_1(G), \ldots$ recursively as

$$\zeta_0(G) = \{1\}, \quad \zeta_{i+1}(G) = \{g \in G : [x, g] \in \zeta_i(G) \text{ for all } x \in G\}, \quad i \geq 0.$$

In particular, $\zeta_1(G) = Z(G)$.

Lemma 5.1.11 *Let G be a group. Each $\zeta_i(G)$ is a normal subgroup of G.*

Proof We proceed by induction on i. The case $i = 0$ is trivial, as $\zeta_0(G) = \{1\}$. Assume that the result holds for some $i \geq 0$. Thus $\zeta_i(G)$ is a normal subgroup of G. Let $\pi : G \to G/\zeta_i(G)$ be the canonical map. Note that

$$\zeta_{i+1}(G) = \pi^{-1}(Z(G/\zeta_i(g))).$$

Therefore $\zeta_{i+1}(G)$ is a normal subgroup of G, and the result follows by induction. □

For a group G the *upper central series* of G is the sequence

$$\{1\} = \zeta_0(G) \subseteq \zeta_1(G) \subseteq \zeta_2(G) \subseteq \cdots$$

A subgroup K of G *normalizes* a subgroup H if $K \subseteq N_G(H)$. A subgroup K of G *centralizes* a subgroup H if $K \subseteq C_G(H)$, that is if and only if $[H, K] = \{1\}$.

Lemma 5.1.12 *Let G be a group. There exists c such that $\zeta_c(G) = G$ if and only if $\gamma_{c+1}(G) = \{1\}$. In this case,*

$$\gamma_{i+1}(G) \subseteq \zeta_{c-i}(G)$$

for all $i \in \{0, 1, \ldots, c\}$.

Proof For $c = 0$, the result is trivial. Suppose that $c \geq 1$. Assume first that $\zeta_c(G) = G$. To prove that $\gamma_{i+1}(G) \subseteq \zeta_{c-i}(G)$ we proceed by induction. The case where $i = 0$ is trivial, so let us assume the result holds for some $i \geq 0$. Let $x \in G$ and $g \in \gamma_{i+1}(G)$. By the inductive hypothesis $g \in \zeta_{c-i}(G)$. Hence

$$[x, g] \in \zeta_{c-i-1}(G) \text{ and } \gamma_{i+2}(G) \subseteq \zeta_{c-i-1}(G).$$

By induction, this claim follows. In particular, $\gamma_{c+1}(G) \subseteq \zeta_0(G) = \{1\}$.

Assume now that $\gamma_{c+1}(G) = \{1\}$. We prove that $\gamma_{i+1}(G) \subseteq \zeta_{c-i}(G)$ for all i. We proceed by backward induction on i. The case $i = c$ is trivial, so assume the result holds for some $i + 1 \leq c$. Let $g \in \gamma_i(G)$. By the inductive hypothesis,

$$[x, g] \in [G, \gamma_i(G)] = \gamma_{i+1}(G) \subseteq \zeta_{c-i}(G),$$

for all $x \in G$. Thus $g \in \zeta_{c-i+1}(G)$ by definition. Hence $\gamma_i(G) \subseteq \zeta_{c-i+1}(G)$, and the result follows by induction. □

Example 5.1.13 If $G = \mathbb{S}_3$, then $\zeta_j(G) = \{1\}$ for all $j \geq 0$.

A *central series* of a group G is a sequence

$$G = G_0 \supseteq G_1 \supseteq \cdots \supseteq G_n = \{1\}$$

of normal subgroups of G such that for each $i \in \{1, \ldots, n\}$, $\pi_i(G_{i-1})$ is a subgroup of $Z(G/G_i)$, where $\pi_i \colon G \to G/G_i$ is the canonical map.

Proposition 5.1.14 *Let G be a group and let $G = G_0 \supseteq G_1 \supseteq \cdots \supseteq G_n = \{1\}$ be a central series of G. Then $\gamma_{i+1}(G) \subseteq G_i \subseteq \zeta_{n-i}(G)$ for all i.*

Proof We shall prove that $\gamma_{i+1}(G) \subseteq G_i$ by induction on i. The case $i = 0$ is trivial, so we assume that the result holds for some $i \geq 0$. Let $\pi_{i+1} \colon G \to G/G_{i+1}$ be the canonical map. Since $\pi_{i+1}(G_i) \subseteq Z(G/G_{i+1})$, it follows that

$$\pi_{i+1}([G, G_i]) = [\pi_{i+1}(G), \pi_{i+1}(G_i)] = \{1\}.$$

This implies that $[G, G_i] \subseteq \ker \pi_{i+1} = G_{i+1}$. Hence, by the inductive hypothesis,

$$\gamma_{i+2}(G) = [G, \gamma_{i+1}(G)] \subseteq [G, G_i] \subseteq G_{i+1}.$$

Thus the first inclusion follows by induction.

We now prove that $G_i \subseteq \zeta_{n-i}(G)$. We proceed by induction on $n - i$. If $n - i = 0$, then $G_n = \{1\} = \zeta_0(G)$. Suppose that the inclusion we want to prove holds for some $n - i \geq 0$. Since $G_i \subseteq \zeta_{n-i}(G)$, there exists a surjective group homomorphism $f \colon G/G_i \to G/\zeta_{n-i}(G)$, such that $f(gG_i) = g\zeta_{n-i}(G)$ for all $g \in G$. Hence $f(Z(G/G_i)) \subseteq Z(G/\zeta_{n-i}(G))$. Since $f\pi_i \colon G \to G/\zeta_{n-i}(G)$ is the canonical map and $f(\pi_i(G_{i-1})) \subseteq f(Z(G/G_i)) \subseteq Z(G/\zeta_{n-i}(G))$,

$$G_{i-1} \subseteq (f\pi_i)^{-1}(Z(G/\zeta_{n-i}(G))) = \zeta_{n-i+1}(G).$$

Therefore the result follows. □

Theorem 5.1.15 (Hirsch) *Let G be a non-trivial nilpotent group. If H is a non-trivial normal subgroup of G, then $H \cap Z(G) \neq \{1\}$. In particular, $Z(G) \neq \{1\}$.*

Proof Since $\zeta_0(G) = \{1\}$ and there exists $c \geq 1$ such that $\zeta_c(G) = G$, there exists

$$m = \min\{k : H \cap \zeta_k(G) \neq \{1\}\}.$$

Since H is normal,

$$[H \cap \zeta_m(G), G] \subseteq H \cap [\zeta_m(G), G] \subseteq H \cap \zeta_{m-1}(G) = \{1\}.$$

Thus $\{1\} \neq H \cap \zeta_m(G) \subseteq H \cap Z(G)$. If $H = G$, then $Z(G) \neq \{1\}$. □

A subgroup M of G is *maximal normal* if it is maximal among all normal proper subgroups of G.

Corollary 5.1.16 *Let G be a non-abelian nilpotent group and A be a maximal abelian normal subgroup of G. Then $A = C_G(A)$.*

Proof Since A is abelian, $A \subseteq C_G(A)$. Assume that $A \neq C_G(A)$. We know that the centralizer of a normal subgroup is normal. Thus the centralizer $C_G(A)$ is normal in G. Let $\pi : G \to G/A$ be the canonical map. Then $\pi(C_G(A))$ is a non-trivial normal subgroup of $\pi(G)$. Since G is nilpotent, $\pi(G)$ is nilpotent. By Hirsch's theorem, $\pi(C_G(A)) \cap Z(\pi(G)) \neq \{1\}$. Let $x \in C_G(A) \setminus A$ be such that $\pi(x)$ is central in $\pi(G)$. Note that if $g \in G$, then $gxg^{-1} \in xA$. Hence $\langle A, x\rangle$ is an abelian normal subgroup of G such that $A \subsetneq \langle A, x\rangle \subsetneq G$, a contradiction. Therefore $A = C_G(A)$. □

A subgroup M of a group G is said to be *minimal normal* if $M \neq \{1\}$, M is normal in G and the unique normal subgroup of G strictly contained in M is the trivial subgroup. Every finite non-trivial group contains a minimal normal subgroup.

Example 5.1.17 If a non-trivial normal subgroup M is minimal (with respect to the inclusion), then it is minimal normal. The converse statement is not true. The subgroup of $\mathbb{A}_4$ generated by (12)(34), (13)(24) and (14)(23) is minimal normal in $\mathbb{A}_4$ but it is not minimal.

Example 5.1.18 Let $G = \mathbb{D}_6 = \langle r, s : r^6 = s^2 = 1,\ srs = r^{-1}\rangle$ be the dihedral group of order twelve. The subgroups $S = \langle r^2\rangle$ and $T = \langle r^3\rangle$ are minimal normal subgroups.

Theorem 5.1.19 *Let G be a non-trivial nilpotent group.*

(1) Every minimal normal subgroup of G has prime order and it is central.
(2) Every maximal subgroup of G is normal, has prime index and contains $[G, G]$.

Proof We first prove (1). Let N be a minimal normal subgroup of G. Since G is nilpotent, $N \cap Z(G) \neq \{1\}$ by Hirsch's theorem. It follows that $N \cap Z(G)$ is a normal subgroup of G contained in N. Thus $N = N \cap Z(G) \subseteq Z(G)$ by the minimality of N. In particular, N is abelian. Since every subgroup of N is normal in G, N should be cyclic of prime order.

We now prove (2). If M is a maximal subgroup, then M is normal in G by the normalizer condition. The maximality of M implies that G/M contains no non-trivial proper subgroups. It follows that G/M is cyclic of prime order. In particular, G/M is abelian and thus $[G, G] \subseteq M$. □

Theorem 5.1.19 does not prove the existence of maximal subgroups; see, for example, what happens with the additive group of rational numbers.

Proposition 5.1.20 *Let G be a nilpotent group and H be a subgroup of G of index n. If* $g \in G$, *then* $g^n \in H$.

Proof We proceed by induction on n. The case $n = 1$ is trivial. Assume that $n > 1$ and the result holds for all subgroups of index $< n$. If H is a normal subgroup of index n, then clearly $g^n \in H$ for all $g \in G$. Hence we may assume that H is a non-normal subgroup of G of index n. By the normalizer condition, $H \subsetneq N_G(H)$. Hence $(G : N_G(H)), (N_G(H) : H) < n$. By the inductive hypothesis, $g^{(G:N_G(H))} \in N_G(H)$ and thus

$$g^n = g^{(G:H)} = g^{(G:N_G(H))(N_G(H):H)} \in H,$$

for all $g \in G$. The result follows. □

Example 5.1.21 The nilpotency of G is needed in the previous proposition. If $G = \mathbb{S}_3$ and $H = \{\mathrm{id}, (12)\}$, then $(G : H) = 3$. If $g = (13)$, then $g^3 = (13) \notin H$.

The following tool is useful for performing induction on nilpotent groups.

Lemma 5.1.22 *Let G be a nilpotent group of class* $c \geq 2$. *If* $x \in G$, *then the subgroup* $\langle x, [G, G]\rangle$ *is nilpotent of class* $< c$.

Proof Let $H = \langle x, [G, G]\rangle$. Note that

$$H = \{x^n a : n \in \mathbb{Z}, a \in [G, G]\},$$

as $[G, G]$ is normal in G.

It is enough to show that $[H, H] \subseteq \gamma_3(G)$. Let $h = x^n a, k = x^m b \in H$ where $a, b \in [G, G]$. Since by (5.2)

$$[h, x^m] = [x^n, [a, x^m]][a, x^m] \in \gamma_4(G)\gamma_3(G) \subseteq \gamma_3(G),$$

it follows from (5.1) that

$$[h, k] = [h, x^m][x^m, [h, b]][h, b] \in \gamma_3(G)\gamma_4(G)\gamma_3(G) \subseteq \gamma_3(G). \qquad \square$$

Example 5.1.23 Let $G = \mathbb{D}_8 = \langle r, s : r^8 = s^2 = 1, srs = r^{-1}\rangle$ be the dihedral group of order 16. Then G is nilpotent of class three and $[G, G] = \{1, r^2, r^4, r^6\} \cong C_4$. The subgroup $\langle s, [G, G]\rangle \cong \mathbb{D}_4$ is nilpotent of class two.

Now an application of Lemma 5.1.22.

Theorem 5.1.24 *If G is a nilpotent group, then*

$$T(G) = \{g \in G : g^n = 1 \text{ for some positive integer } n\}$$

is a subgroup of G.

Proof We show the result by induction on the nilpotency class c of G. If G is abelian, then the result is clear. Suppose that $c \geq 2$ and that the result holds for all nilpotent groups of class $< c$. Let $a, b \in T(G)$ and let $A = \langle a, [G, G]\rangle$ and $B = \langle b, [G, G]\rangle$. Since A and B are both nilpotent of class $< c$ by the previous lemma, the inductive hypothesis implies that $T(A)$ is a subgroup of A and $T(B)$ is a subgroup of B. Since $T(A)$ is characteristic in A and A is normal in G, it follows that $T(A)$ is normal in G. Similarly, $T(B)$ is normal in G. We claim that every element of $T(A)T(B)$ has finite order. Indeed, if $x \in T(A)T(B)$, say $x = a_1b_1$ with $a_1 \in T(A)$ of order m and $b_1 \in T(B)$, then x has finite order since

$$x^m = (a_1b_1)^m = (a_1b_1a_1^{-1})(a_1^2b_1a_1^{-2}) \cdots (a_1^{m-1}b_1a_1^{-m+1})b_1 \in T(B).$$

In particular, ab^{-1} has finite order. It follows that $T(G)$ is a subgroup of G. $\square$

A group G is said to be *torsion-free* if 1 is the only element of finite order of G. Another application of Lemma 5.1.22.

Theorem 5.1.25 *Let G be a torsion-free nilpotent group and let $a, b \in G$. If $a^n = b^n$ for some $n \neq 0$, then $a = b$.*

Proof We proceed by induction on the nilpotency class c of G. The claim holds if G is abelian. Assume that G is nilpotent of class $c \geq 2$. Since $\langle a, [G, G]\rangle$ is a nilpotent subgroup of G of nilpotency class $< c$ and $bab^{-1} = [b, a]a \in \langle a, [G, G]\rangle$, the inductive hypothesis implies that $ba = ab$, as $a^n = (bab^{-1})^n = b^n$. Thus

$$(ab^{-1})^n = a^nb^{-n} = 1.$$

Since G is torsion-free, it follows that $a = b$. $\square$

Corollary 5.1.26 *Let G be a torsion-free nilpotent group. If $x, y \in G$ are such that $x^ny^m = y^mx^n$ for some $n, m \neq 0$, then $xy = yx$.*

Proof Let $a = x$ and $b = y^m x y^{-m}$. Since $a^n = b^n$, Theorem 5.1.25 implies that $a = b$. Thus $xy^m = y^m x$. By using Theorem 5.1.25 now with $a = y$ and $b = xyx^{-1}$, we conclude that $xy = yx$. □

We need the following lemma.

Lemma 5.1.27 *Let G be a finitely generated group and H be a finite-index subgroup of G. Then H is finitely generated.*

Proof Assume that $G = \langle g_1, \dots, g_m \rangle$. Without loss of generality we may assume that for each i there exists k such that $g_i^{-1} = g_k$. Let $\{1 = t_1, \dots, t_n\}$ be a right transversal of H in G. For $i \in \{1, \dots, n\}$ and $j \in \{1, \dots, m\}$ we write

$$t_i g_j = h(i, j) t_{k(i,j)},$$

where $h(i, j) \in H$. We claim that H is generated by the $h(i, j)$. If $x \in H$, then

$$\begin{aligned} x &= g_{i_1} \cdots g_{i_s} \\ &= (t_1 g_{i_1}) g_{i_2} \cdots g_{i_s} \\ &= h(1, i_1) t_{k_1} g_{i_2} \cdots g_{i_s} \\ &= h(1, i_1) h(k_1, i_2) t_{k_2} g_{i_3} \cdots g_{i_s} \\ &= h(1, i_1) h(k_1, i_2) \cdots h(k_{s-1}, i_s) t_{k_s}, \end{aligned}$$

where $k_1, \dots, k_{s-1} \in \{1, \dots, n\}$. Thus $t_{k_s} \in H$ and hence $t_{k_s} = 1$, which implies the claim. □

A group G is said to be *torsion* or *periodic* if every element of G has finite order.

Theorem 5.1.28 *Let G be a finitely generated torsion nilpotent group. Then G is finite.*

Proof We proceed by induction on the nilpotency class c. The case $c = 1$ is true since G is abelian. Assume that $c \geq 2$ and the result holds for groups of nilpotency class $< c$. Since $G/[G, G]$ is an abelian finitely generated torsion group, it is finite. By Lemma 5.1.27, $[G, G]$ also is finitely generated. Since $[G, G]$ is a finitely generated torsion nilpotent group of class $< c$, the inductive hypothesis implies that $[G, G]$ is finite. Thus G is finite. Therefore the result follows by induction. □

5.2 Finite Nilpotent Groups

Let G be a finite group and p a prime. We denote by $\mathrm{Syl}_p(G)$ the set of the Sylow p-subgroups of G.

Lemma 5.2.1 *Let G be a finite group and let $P \in \mathrm{Syl}_p(G)$. If H is a subgroup of G such that $N_G(P) \subseteq H$, then $N_G(H) = H$.*

Proof Let $x \in N_G(H)$. Since $P \in \mathrm{Syl}_p(H)$ and $xPx^{-1} \subseteq xHx^{-1} = H$, we have that there exists $h \in H$ such that $(hx)P(hx)^{-1} = P$. Hence $hx \in N_G(P) \subseteq H$, and thus $x \in H$. This proves the result. □

Lemma 5.2.2 *Let p be a prime number. Then p-groups are nilpotent.*

Proof We know that every non-trivial p-group has a non-trivial center. Let G be a p-group. Suppose that $|G| = p^n$. We shall prove the result by induction on n. For $n = 1$, G is cyclic of order p, and the result is trivial. Suppose that $n > 1$ and the result holds for p-groups of order $< p^n$. We may assume that G is non-abelian. Since $Z(G)$ is a non-trivial proper subgroup of G, by the inductive hypothesis, $Z(G)$ and $G/Z(G)$ are nilpotent. Let $\pi : G \to G/Z(G)$ be the canonical map. By Lemma 5.1.12, there exists c such that $\zeta_c(G/Z(G)) = G/Z(G)$. Note that

$$\pi^{-1}(\zeta_i(G/Z(G))) = \zeta_{i+1}(G).$$

Hence $\zeta_{c+1}(G) = G$ is nilpotent by Lemma 5.1.12. Therefore the result follows by induction. □

We say that a group G is the *direct product* of its subgroups $H_1, \dots, H_n$ if every subgroup H_i is normal in G, $G = H_1 \cdots H_n$ and $H_{k+1} \cap (H_1 \cdots H_k) = \{1\}$ for all $k = 1, \dots, n-1$. For $i < j$, $h_i \in H_i$ and $h_j \in H_j$, we have that $[h_i, h_j] \in H_i \cap H_j = \{1\}$. Hence the map $f \colon H_1 \times \cdots \times H_n \to G$, defined by $f(h_1, \dots, h_n) = h_1 \cdots h_n$ for all $h_i \in H_i$, is a surjective homomorphism of groups. Let $(a_1, \dots, a_n) \in \ker(f)$. Thus $a_1 \cdots a_n = 1$. We shall prove that $a_1 = \cdots = a_n = 1$ by induction on n. For $n = 1$, $a_1 = 1$. Suppose that $n > 1$ and that $a_1 \cdots a_{n-1} = 1$ implies that $a_1 = \cdots = a_{n-1} = 1$. Since

$$a_n^{-1} = a_1 \cdots a_{n-1} \in H_n \cap (H_1 \cdots H_{n-1}) = \{1\},$$

we have that $a_n = 1 = a_1 \cdots a_{n-1}$. By the inductive hypothesis, $a_1 = \cdots = a_{n-1} = 1$. Hence, by induction, we have proved that $a_1 \cdots a_n = 1$ implies that $a_1 = \cdots = a_n = 1$ for all positive integer n. Therefore $\ker(f) = \{(1, \dots, 1)\}$ and thus f is an isomorphism of groups.

Theorem 5.2.3 *Let G be a finite group. The following conditions are equivalent:*

(1) G is nilpotent.
(2) Every Sylow subgroup of G is normal.
(3) G is the direct product of its Sylow subgroups.

Proof We prove that (1) $\Longrightarrow$ (2). Suppose that G is nilpotent. Let $P \in \mathrm{Syl}_p(G)$. By Lemma 5.2.1, $N_G(N_G(P)) = N_G(P)$. Now $N_G(P) = G$ by the normalizer condition. Hence P is normal in G.

Now we prove that (2) $\Longrightarrow$ (3). Let $p_1, \dots, p_n$ be the distinct prime divisors of $|G|$. For every $i \in \{1, \dots, n\}$, let $P_i \in \mathrm{Syl}_{p_i}(G)$. By hypothesis, each P_i is normal in G. Since $|P_1 \cdots P_n| = |G|$, we have that $G = P_1 \cdots P_n$. Clearly $|P_{k+1}|$, $|P_1 \cdots P_k|$

are coprime for all $k = 1, \ldots, n-1$. Hence $P_{k+1} \cap (P_1 \cdots P_k) = \{1\}$ for all $k = 1, \ldots, n-1$. Thus G is the direct product of its Sylow subgroups.

Finally, to prove that (3) $\implies$ (1), we note that every p-group is nilpotent, and a finite direct product of nilpotent groups is nilpotent. □

Theorem 5.2.4 (Baumslag–Wiegold) *Let G be a finite group such that*

$$o(xy) = o(x)o(y)$$

whenever $x, y \in G$ have coprime order. Then G is nilpotent.

Proof Let $p_1, \ldots, p_n$ be the distinct prime divisors of the order of G. For each $i \in \{1, \ldots, n\}$ let $P_i \in \mathrm{Syl}_{p_i}(G)$. We claim that $G = P_1 \cdots P_n$. The non-trivial inclusion is equivalent to show that the map

$$\psi \colon P_1 \times \cdots \times P_n \to G, \quad (x_1, \ldots, x_n) \mapsto x_1 \cdots x_n$$

is surjective. We first show that ψ is injective. If $\psi(x_1, \ldots, x_n) = \psi(y_1, \ldots, y_n)$, then

$$x_1 \cdots x_n = y_1 \cdots y_n.$$

If $y_n \neq x_n$, then $x_1 \cdots x_{n-1} = (y_1 \cdots y_{n-1})y_n x_n^{-1}$. But $x_1 \cdots x_{n-1}$ has order coprime with p_n and $y_1 \cdots y_{n-1} y_n x_n^{-1}$ has order divisible by p_n, a contradiction. Thus $x_n = y_n$ and hence the same argument proves that ψ is injective. Since $|P_1 \times \cdots \times P_n| = |G|$, we conclude that ψ is bijective. In particular, ψ is surjective and hence $G = P_1 \cdots P_n$.

We now show that each P_j is normal in G. Let $j \in \{1, \ldots, n\}$ and $x_j \in P_j$. Let $g \in G$ and $y_j = g x_j g^{-1}$. Since $y_j \in G$, we write $y_j = z_1 \cdots z_n$ with $z_k \in P_k$ for all k. Since the order of y_j is a power of p_j, it follows that $z_1 \cdots z_n$ has order a power of p_j. Thus $z_k = 1$ for all $k \neq j$ and $y_j = z_j \in P_j$. Since each Sylow subgroup is normal, G is nilpotent, by Theorem 5.2.3. □

5.3 Frattini Subgroup

Let G be a group. If G has maximal subgroups, the *Frattini subgroup* $\Phi(G)$ of G is defined as the intersection of all maximal subgroups of G. Otherwise, $\Phi(G) = G$.

Note that $\Phi(G)$ is a characteristic subgroup of G.

Example 5.3.1 Let $G = \mathbb{S}_3$. The maximal subgroups of G are

$$M_1 = \langle (123) \rangle, \quad M_2 = \langle (12) \rangle, \quad M_3 = \langle (23) \rangle, \quad M_4 = \langle (13) \rangle.$$

Thus $\Phi(G) = \{1\}$.

Example 5.3.2 Let $G = \langle g \rangle \cong C_{12}$ be the cyclic group of order 12. The subgroups of G are

$$\{1\}, \quad \langle g^6 \rangle \cong C_2, \quad \langle g^4 \rangle \cong C_3, \quad \langle g^3 \rangle \cong C_4, \quad \langle g^2 \rangle \cong C_6, \quad G.$$

The maximal subgroups are $\langle g^3 \rangle \cong C_4$ and $\langle g^2 \rangle \cong C_6$. Hence

$$\Phi(G) = \langle g^3 \rangle \cap \langle g^2 \rangle = \langle g^6 \rangle \cong C_2.$$

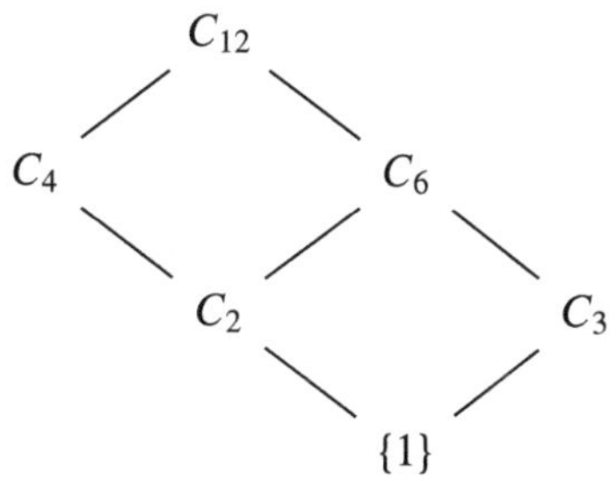

Lemma 5.3.3 (Dedekind) *Let H, K and L be subgroups of G such that $H \subseteq L \subseteq G$. Then $(HK) \cap L = H(K \cap L)$.*

Proof We only need to prove that $(HK) \cap L \subseteq H(K \cap L)$, as the other inclusion is trivial. If $x = hk \in (HK) \cap L$, where $x \in L$, $h \in H$ and $k \in K$, then $k = h^{-1}x \in L \cap K$ since $H \subseteq L$. Thus $x = hk \in H(L \cap K)$. □

Lemma 5.3.4 *Let G be a finite group. If H is a subgroup of G such that $G = H\Phi(G)$, then $H = G$.*

Proof Assume that $H \neq G$. Let M be a maximal subgroup of G such that $H \subseteq M$. Since $\Phi(G) \subseteq M$, it follows that $G = H\Phi(G) \subseteq M$, a contradiction. □

Proposition 5.3.5 *Let G be a finite group and N be a normal subgroup of G. Then $\Phi(N) \subseteq \Phi(G)$.*

Proof Since $\Phi(N)$ is characteristic in N and N is normal in G, it follows that $\Phi(N)$ is normal in G. Let M be a maximal subgroup of G such that $\Phi(N) \nsubseteq M$. Then $\Phi(N)M = G$. By applying Lemma 5.3.3 with $H = \Phi(N)$, $K = M$ and $L = N$,

$$N = G \cap N = (\Phi(N)M) \cap N = \Phi(N)(M \cap N).$$

By Lemma 5.3.4 $N = M \cap N \subseteq M$, a contradiction. Hence every maximal subgroup of G contains $\Phi(N)$ and therefore $\Phi(G) \supseteq \Phi(N)$. □

The Frattini subgroup can be characterized in terms of non-generators.

Lemma 5.3.6 (non-generators) *Let G be a finite group. Then*

$$\Phi(G) = \{x \in G : \textit{if } G = \langle x, Y\rangle \textit{ for some } Y \subseteq G, \textit{ then } G = \langle Y\rangle\}.$$

Proof Let

$$A = \{x \in G : \text{if } G = \langle x, Y\rangle \text{ for some } Y \subseteq G, \text{ then } G = \langle Y\rangle\}.$$

Let $x \in \Phi(G)$ be such that $G = \langle x, Y\rangle$ for some subset Y of G. If $G \neq \langle Y\rangle$, then there exists a maximal subgroup M such that $\langle Y\rangle \subseteq M$. Since $x \in M$, $G = \langle x, Y\rangle \subseteq M$, a contradiction. Hence $x \in A$.

Conversely, let $x \in A$ and M be a maximal subgroup of G. If $x \notin M$, then, since $G = \langle x, M\rangle$, it follows that $G = \langle M\rangle = M$, a contradiction. Thus $x \in M$ for all maximal subgroup M. Hence $x \in \Phi(G)$. □

Lemma 5.3.7 (Frattini's argument) *Let G be a finite group, H a normal subgroup of G and p a prime. Then, for every $P \in \mathrm{Syl}_p(H)$, $G = HN_G(P)$.*

Proof Let $P \in \mathrm{Syl}_p(H)$ and $g \in G$. We have that $gPg^{-1} \subseteq gHg^{-1} = H$. Hence $gPg^{-1} \in \mathrm{Syl}_p(H)$. Therefore, there exists $h \in H$ such that $hPh^{-1} = gPg^{-1}$. Thus $h^{-1}g \in N_G(P)$. Now we have that $g = hh^{-1}g \in HN_G(P)$, and the result follows. □

Theorem 5.3.8 (Frattini) *If G is a finite group, then $\Phi(G)$ is nilpotent.*

Proof Let $P \in \mathrm{Syl}_p(\Phi(G))$ for some prime p. Since $\Phi(G)$ is normal in G, Frattini's argument implies that $G = \Phi(G)N_G(P)$. By Lemma 5.3.4, $G = N_G(P)$. Since Sylow subgroups of $\Phi(G)$ are normal in G, $\Phi(G)$ is nilpotent, by Theorem 5.2.3. □

Theorem 5.3.9 (Gaschütz) *If G is a finite group, then $[G, G] \cap Z(G) \subseteq \Phi(G)$.*

Proof Let $D = [G, G] \cap Z(G)$. Assume that D is not contained in $\Phi(G)$. Hence there exists a maximal subgroup M such that D is not contained in M. Thus $G = MD$. Since $D \subseteq Z(G)$, it is clear that M is normal in G. Since $G/M \cong D/(D \cap M)$ is abelian, $[G, G] \subseteq M$ and hence $D \subseteq [G, G] \subseteq M$, a contradiction. Therefore, the result follows. □

Theorem 5.3.10 (Wielandt) *Let G be a finite group. Then G is nilpotent if and only if $[G, G] \subseteq \Phi(G)$.*

Proof Assume first that $[G, G] \subseteq \Phi(G)$. Let $P \in \mathrm{Syl}_p(G)$. If $N_G(P) \neq G$, then $N_G(P) \subseteq M$ for some maximal subgroup M of G. Let $g \in G$ and $m \in M$. Since

$$gmg^{-1}m^{-1} = [g, m] \in [G, G] \subseteq \Phi(G) \subseteq M,$$

we get that $gmg^{-1} \in M$. Hence M is normal in G, that is $G = N_G(M)$. But, since $N_G(P) \subseteq M$, by Lemma 5.2.1, $N_G(M) = M$, a contradiction. Therefore P is normal in G, and thus G is nilpotent, by Theorem 5.2.3.

The converse follows by Theorem 5.1.19. □

Theorem 5.3.11 *Let G be a finite group. Then G is nilpotent if and only if $G/\Phi(G)$ is nilpotent.*

Proof If G is nilpotent, then clearly $G/\Phi(G)$ also is nilpotent. Suppose that $G/\Phi(G)$ is nilpotent. Let $P \in \mathrm{Syl}_p(G)$. Since $\Phi(G)P/\Phi(G) \in \mathrm{Syl}_p(G/\Phi(G))$ and $G/\Phi(G)$ is nilpotent, $\Phi(G)P/\Phi(G)$ is a normal subgroup of $G/\Phi(G)$. Hence, $\Phi(G)P$ is normal in G. By Frattini's argument,

$$G = \Phi(G)PN_G(P) = \Phi(G)N_G(P).$$

By Lemma 5.3.4, $G = N_G(P)$, and thus P is normal in G. By Theorem 5.2.3, G is nilpotent. □

Theorem 5.3.12 (Hall) *Let G be a finite group and N a normal subgroup of G. If N and $G/[N, N]$ are nilpotent, then G is nilpotent.*

Proof Since N is nilpotent, by Theorem 5.3.10, $[N, N] \subseteq \Phi(N)$. By Proposition 5.3.5, $[N, N] \subseteq \Phi(N) \subseteq \Phi(G)$. Hence, there exists a surjective homomorphism $G/[N, N] \to G/\Phi(G)$. Since $G/[N, N]$ is nilpotent, we get that $G/\Phi(G)$ is nilpotent. Hence G is nilpotent by Theorem 5.3.11. □

Definition 5.3.13 A *minimal generating set* of a group G is a subset X of G such that $G = \langle X \rangle$ and $G \neq \langle Y \rangle$ for every proper subset Y of X.

Note that the minimal number of generators of a group G can be less than the number of elements of a minimal generating set of G.

Example 5.3.14 Let $G = \langle g : g^6 = 1 \rangle$. If $a = g^2$ and $b = g^3$, then $\{a, b\}$ is a minimal generating set of G, but the minimal number of generators of G is 1.

Lemma 5.3.15 *Let p be a prime and G a p-group. Then $G/\Phi(G)$ is a vector space over the field $\mathbb{F}_p$ of p elements.*

Proof Let K be a maximal subgroup of G. Since G is nilpotent, by Theorem 5.1.19, K is normal in G of index p. Hence G/K is a cyclic group of order p. Let $K_1, \ldots, K_m$ be the distinct maximal subgroups of G. Let $f : G \to G/K_1 \times \cdots \times G/K_m$ be the map defined by $f(g) = (gK_1, \ldots, gK_m)$ for all $g \in G$. It is clear that f is a homomorphism of groups and $\ker(f) = \Phi(G)$. Hence $G/\Phi(G)$ is isomorphic to $\mathrm{im}(f)$, a subgroup of the $\mathbb{F}_p$-vector space $G/K_1 \times \cdots \times G/K_m$. Therefore, the result follows. □

Theorem 5.3.16 (Burnside) *Let p be a prime and G a p-group. If X is a minimal generating set of G, then $|X| = \dim_{\mathbb{F}_p}(G/\Phi(G))$.*

Proof By the previous lemma, we know that $G/\Phi(G)$ is an $\mathbb{F}_p$-vector space. Let $\pi : G \to G/\Phi(G)$ be the canonical map and $\{x_1, \dots, x_n\}$ a minimal generating set of G, with $|\{x_1, \dots, x_n\}| = n$. We shall see that $\pi(x_1), \dots, \pi(x_n)$ are linearly independent in $G/\Phi(G)$. Suppose that $\pi(x_1), \dots, \pi(x_n)$ are not linearly independent. Without loss of generality, we may assume that $\pi(x_1) \in \langle \pi(x_2), \dots, \pi(x_n) \rangle$. Then there exists $y \in \langle x_2, \dots, x_n \rangle$ such that $x_1 y^{-1} \in \Phi(G)$. Since

$$G = \langle x_1 y^{-1}, x_2, \dots, x_n \rangle$$

and $x_1 y^{-1} \in \Phi(G)$, Lemma 5.3.6 implies that $G = \langle x_2, \dots, x_n \rangle$, a contradiction. Therefore $n = \dim_{\mathbb{F}_p}(G/\Phi(G))$. □

5.4 Fitting Subgroup

Definition 5.4.1 Let G be a finite group and p a prime. The *p-radical* of G is the subgroup

$$O_p(G) = \bigcap_{P \in \mathrm{Syl}_p(G)} P.$$

The following result is an easy consequence of Sylow's theorem.

Lemma 5.4.2 *Let G be a finite group and p a prime.*

(1) $O_p(G)$ is normal in G.
(2) If N is a normal p-subgroup of G, then $N \subseteq O_p(G)$. □

Definition 5.4.3 Let G be a finite group and let $p_1, \dots, p_k$ be the distinct prime divisors of $|G|$. The *Fitting's subgroup* of G is the subgroup

$$F(G) = O_{p_1}(G) \cdots O_{p_k}(G).$$

Note that $F(G)$ is a characteristic subgroup of G.

Example 5.4.4 Let $G = \mathbb{S}_3$. It is easy to see that $O_2(G) = \{1\}$ and $O_3(G) = \langle (123) \rangle$. Thus $F(G) = \langle (123) \rangle$.

Theorem 5.4.5 (Fitting) *Let G be a finite group. The Fitting subgroup $F(G)$ is nilpotent, and it is normal in G. Furthermore $F(G)$ contains all normal nilpotent subgroups of G.*

Proof Note that for every prime p, $O_p(G)$ is a Sylow p-subgroup of $F(G)$. Since $O_p(G)$ is normal in G, also is normal in $F(G)$. Hence by Theorem 5.2.3, $F(G)$ is nilpotent.

Let N be a normal nilpotent subgroup of G and let $P \in \mathrm{Syl}_p(N)$. Since N is nilpotent, P is normal in N. Then P is the unique Sylow p-subgroup of N. Hence

P is a characteristic subgroup of N and thus P is normal in G. By Lemma 5.4.2, $P \subseteq O_p(G)$. By Theorem 5.2.3, N is the direct product of its Sylow subgroups. Hence $N \subseteq F(G)$. □

Corollary 5.4.6 *Let G be a finite group. Then $Z(G) \subseteq F(G)$.*

Proof Since $Z(G)$ is a normal nilpotent subgroup of G, by Fitting's theorem, $Z(G) \subseteq F(G)$. □

Corollary 5.4.7 (Fitting) *Let K and L be normal nilpotent subgroups of a finite group G. Then KL is nilpotent.*

Proof By Fitting's theorem, K and L are subgroups of $F(G)$. Hence KL also is a subgroup of $F(G)$. By Fitting's theorem, $F(G)$ is nilpotent. Therefore KL also is nilpotent. □

Corollary 5.4.8 *Let G be a finite group and N a normal subgroup of G. Then $N \cap F(G) = F(N)$.*

Proof Since $F(N)$ is a characteristic subgroup of N, $F(N)$ is normal in G. By Fitting's theorem, $F(N)$ is nilpotent and $F(N) \subseteq N \cap F(G)$. On the other hand, since $F(G)$ is normal in G, we have that $F(G) \cap N$ is normal in N. By Fitting's theorem $F(G)$ is nilpotent and thus $F(G) \cap N$ also is nilpotent. Again by Fitting's theorem, $F(G) \cap N \subseteq F(N)$. Therefore the result follows. □

Theorem 5.4.9 *Let G be a finite group. Then $\Phi(G) \subseteq F(G)$ and*

$$F(G)/\Phi(G) \cong F(G/\Phi(G)).$$

Proof By Theorem 5.3.8, $\Phi(G)$ is a normal nilpotent subgroup of G. Hence, by Fitting's theorem, $\Phi(G) \subseteq F(G)$.

Let $\pi : G \to G/\Phi(G)$ be the canonical map. By Fitting's theorem $F(G)$ is nilpotent, thus $\pi(F(G))$ also is nilpotent. Since $\Phi(G) \subseteq F(G)$ and $F(G)$ is normal in G, we have that $\pi(F(G)) = F(G)/\Phi(G)$ is normal in $G/\Phi(G)$. By Fitting's theorem,

$$\pi(F(G)) \subseteq F(G/\Phi(G)).$$

Let $H = \pi^{-1}(F(G/\Phi(G)))$. We know that H is a normal subgroup of G such that $\Phi(G) \subseteq F(G) \subseteq H$. Let $P \in \mathrm{Syl}_p(H)$. Then

$$\pi(P) = (P\Phi(G))/\Phi(G) \cong P/P \cap \Phi(G)$$

is a p-subgroup of $\pi(H) = F(G/\Phi(G))$. Since

$$(\pi(H) : \pi(P)) = \frac{|\pi(H)|}{|\pi(P)|} = \frac{|H/\Phi(G)|}{|P/P \cap \Phi(G)|} = \frac{(H : P)}{(\Phi(G) : P \cap \Phi(G))}$$

is a divisor of $(H : P)$, we have that $\pi(P) \in \mathrm{Syl}_p(\pi(H))$.

By Fitting's theorem $\pi(H)$ is nilpotent. Hence $\pi(P)$ is a characteristic subgroup of $\pi(H)$ and thus $P\Phi(G) = \pi^{-1}(\pi(P))$ is normal in G. Since $P \in \mathrm{Syl}_p(P\Phi(G))$, by Frattini's argument,

$$G = P\Phi(G)N_G(P) = \Phi(G)N_G(P).$$

By Lemma 5.3.4, $G = N_G(P)$, and thus P is normal in G. Hence P is normal in H. By Theorem 5.2.3, H is nilpotent. By Fitting's theorem, $H \subseteq F(G)$. Hence $F(G/\Phi(G)) = \pi(H) \subseteq \pi(F(G))$, and the result follows. □

5.5 Exercises

5.5.1 Let G be a group such that $[G, G, G]$ is central. Prove that $[G, G]$ is abelian.

5.5.2 Let G be a group. Prove that $\zeta_i(G)$ is a characteristic subgroup of G for all i.

5.5.3 Prove that a group is nilpotent if and only if it admits a central series.

5.5.4 Let G be a group. Prove that if K is a central subgroup of G such that G/K is nilpotent, then G is nilpotent.

5.5.5 Let G be a finite group. Prove that the following statements are equivalent:

(1) G is nilpotent.
(2) If $H \subsetneq G$ is a subgroup, then $H \subsetneq N_G(H)$.
(3) Every maximal subgroup of G is normal in G.

5.5.6 Let G be a finite nilpotent group. Prove that if p is a prime dividing the order of G, then there exists a minimal normal subgroup of order p and there exists a maximal subgroup of index p.

5.5.7 Let p be a prime and let G be a non-trivial group of order p^n. Prove the following statements.

(1) G has a normal subgroup of order p.
(2) For each $j \in \{0, \dots, n\}$ there exists a normal subgroup of order p^j.

5.5.8 Let G be a finite group. Prove that the following conditions are equivalent.

(1) G is nilpotent.
(2) Any two elements of coprime order commute.
(3) Every non-trivial quotient of G has non-trivial center.
(4) If d divides $|G|$, there exists a normal subgroup of G of order d.

5.5.9 Let G be a group and $x, y \in G$. Prove the following statements.

(1) If $[x, y] \in C_G(x) \cap C_G(y)$, then $[x, y]^n = [x^n, y] = [x, y^n]$ for all $n \in \mathbb{Z}$.

(2) If G is nilpotent of class two, then $(xy)^n = [y, x]^{n(n-1)/2} x^n y^n$ for all positive integers n.

5.5.10 Let p be an odd prime number and P be a p-group of nilpotency class ≤ 2. Prove the following statements.

(1) If $[y, x]^p = 1$ for all $x, y \in P$, then $P \to P$, $x \mapsto x^p$, is a group homomorphism.
(2) $\{x \in P : x^p = 1\}$ is a subgroup of P.

5.6 Notes

The results in this chapter are classical and can be found, for example, in [96].

Chapter 6
Solvable Groups

In this chapter, we develop the basic definitions and results on solvable groups to prove the theorem of Burnside about groups of order $p^n q^m$ for p and q prime numbers and the classical results of P. Hall.

6.1 Derived Series

For a group G we define

$$G^{(0)} = G, \quad G^{(i+1)} = [G^{(i)}, G^{(i)}] \ \ i \geq 0.$$

The *derived series* of G is the sequence

$$G = G^{(0)} \supseteq G^{(1)} \supseteq G^{(2)} \supseteq \cdots$$

Each $G^{(i)}$ is a characteristic subgroup of G. We say that G is *solvable* if $G^{(n)} = \{1\}$ for some $n \in \mathbb{N}$. Clearly, every abelian group is solvable. A non-trivial group G is said to be *simple* if $\{1\}$ and G are the only normal subgroups of G. Note that a non-abelian simple group cannot be solvable. Nilpotent groups are solvable.

Let p be a prime number. An *elementary abelian* p-group is an abelian p-group P such that $x^p = 1$ for all $x \in P$.

Lemma 6.1.1 *Let G be a non-trivial group. Let M be a minimal normal subgroup of G. If M is solvable and finite, then M is an elementary abelian p-group for some prime number p.*

Proof Since M is solvable, $[M, M] \subsetneq M$. Moreover, $[M, M]$ is normal in G, as $[M, M]$ is characteristic in M and M is normal in G. Since M is minimal normal, it follows that $[M, M] = \{1\}$ and hence M is abelian. Now, since M is finite,

F. Cedó and L. Vendramin, *Groups, Radical Rings, and the Yang-Baxter Equation*, Progress in Mathematics 361, https://doi.org/10.1007/978-3-032-01524-2_6

there exists a prime number p such that $P = \{x \in M : x^p = 1\}$ is a non-trivial subgroup of M. Since P is characteristic in M, the subgroup P is normal in G. Thus $P = M$. □

Theorem 6.1.2 *Let G be a group.*

(1) If G is solvable, then each subgroup H of G is solvable.
(2) Let K be a normal subgroup of G. Then G is solvable if and only if K and G/K are both solvable.

Proof By induction one proves that $H^{(i)} \subseteq G^{(i)}$ for all $i \geq 0$. Let us prove the second claim. Let $Q = G/K$ and let $\pi \colon G \to Q$ be the canonical map. By induction we prove that $\pi(G^{(i)}) = Q^{(i)}$ for all $i \geq 0$. The case $i = 0$ is trivial, as π is surjective. Now assume that the result holds for some $i \geq 0$. Then

$$\pi(G^{(i+1)}) = \pi([G^{(i)}, G^{(i)}]) = [\pi(G^{(i)}), \pi(G^{(i)})] = [Q^{(i)}, Q^{(i)}] = Q^{(i+1)}.$$

Assume that Q and K are both solvable. Since Q is solvable, there exists n such that $Q^{(n)} = \{1\}$. Since $\pi(G^{(n)}) = Q^{(n)} = \{1\}$, it follows that $G^{(n)} \subseteq K$. Since K is solvable, there exists m such that

$$G^{(n+m)} = (G^{(n)})^{(m)} \subseteq K^{(m)} = \{1\},$$

and hence G is solvable.

Let us now assume that G is solvable. There exists $n \in \mathbb{N}$ such that $G^{(n)} = \{1\}$. Thus Q is solvable, as $Q^{(n)} = \pi(G^{(n)}) = \pi(\{1\}) = \{1\}$. The group K also is solvable, as it is a subgroup of G. □

Theorem 6.1.3 *Let G be a finite non-trivial solvable group and let N be a minimal normal subgroup of G. Then $N \subseteq Z(F(G))$ and thus $F(G) \subseteq C_G(N)$.*

Proof Let N be a minimal normal subgroup of G. By Theorem 6.1.2, N is solvable. By Lemma 6.1.1, N is an elementary abelian p-group for some prime number p. In particular, N is a normal nilpotent subgroup of G. By Fitting's theorem, $N \subseteq F(G)$ and $F(G)$ is a normal nilpotent subgroup of G. By Hirsch's theorem $N \cap Z(F(G)) \neq \{1\}$. Since $Z(F(G))$ is a characteristic subgroup of $F(G)$ and $F(G)$ is normal in G, we have that $Z(F(G))$ is a normal subgroup of G. Hence $N \cap Z(F(G))$ also is a normal subgroup of G. Since N is a minimal normal subgroup of G and $N \cap Z(F(G))$ is non-trivial, we have that $N = N \cap Z(F(G))$. Therefore the result follows. □

6.2 Burnside's Theorem

In this section, we shall prove an important theorem of Burnside, which says that every finite group of order $p^n q^m$, for some primes p, q, and positive integers m, n, is solvable. For its proof, we shall introduce some results of representation theory and integral extensions.

Let K be a field. A K-*algebra* is a K-vector space R joint with a multiplication on R such that $(R, +, \cdot)$ is a ring and $\alpha(ab) = (\alpha a)b = a(\alpha b)$, for all $\alpha \in K$ and $a, b \in R$.

Let G be a group and K a field. The *group algebra* $K[G]$ of the group G over the field K is a K-vector space with basis G with a multiplication on $K[G]$ defined by

$$\left(\sum_{i=1}^{n} a_i x_i\right)\left(\sum_{j=1}^{m} b_j y_j\right) = \sum_{i=1}^{n}\sum_{j=1}^{m}(a_i b_j)(x_i y_j),$$

for all $a_1, \ldots, a_n, b_1, \ldots, b_m \in K$ and $x_1, \ldots, x_n, y_1, \ldots, y_m \in G$. We will write the elements of $a \in K[G]$ in the form

$$a = \sum_{x \in G} a_x x,$$

with $a_x \in K$, assuming that $\{x \in G : a_x \neq 0\}$ is finite. With this notation, we have that

$$\left(\sum_{x \in G} a_x x\right)\left(\sum_{x \in G} b_x x\right) = \sum_{x \in G} c_x x,$$

where

$$c_x = \sum_{y \in G} a_y b_{y^{-1}x},$$

for all $x \in G$. One can easily check that the K-vector space $K[G]$ with this multiplication is a K-algebra. Note that the unit-element 1_G of G is the unit-element of the ring $(K[G], +, \cdot)$. The map $K \to K[G] : \alpha \mapsto \alpha 1_G$ is an injective ring homomorphism, and we will identify the elements $\alpha \in K$ with $\alpha 1_G \in K[G]$, thus $K \subseteq K[G]$. We will denote the unit element of $K[G]$ by 1.

Definition 6.2.1 A *(linear) representation* of a group G over a field K is a group homomorphism

$$\rho \colon G \to \mathrm{Aut}(V),$$

where V is a K-vector space. If $\dim(V) = n < \infty$, then we say that ρ has degree n. It is said that ρ is *faithful* if $\ker(\rho) = \{1\}$.

Let $\rho \colon G \to \mathrm{Aut}(V)$ be a representation of a group G over a field K of finite degree n. Fixing a basis $\mathcal{B}$ of V, we have a group isomorphism $\psi_{\mathcal{B}} \colon \mathrm{Aut}(V) \to \mathbf{GL}_n(K)$, where

$$\mathbf{GL}_n(K) = \{A \in M_n(K) : A \text{ is invertible}\}$$

is the *general linear group* of degree n over K, defined by $\psi_{\mathcal{B}}(f) = M(f, \mathcal{B})$, the matrix of f with respect the basis $\mathcal{B}$, for all $f \in \mathrm{Aut}(V)$.

We say that the composition

$$G \xrightarrow{\rho} \mathrm{Aut}(V) \xrightarrow{\psi_{\mathcal{B}}} \mathbf{GL}_n(K)$$

is a matrix representation of G over K. The map

$$\chi : G \to K, \quad \chi(x) = \mathrm{tr}(\psi_{\mathcal{B}}(\rho(x))),$$

the trace of the matrix $\psi_{\mathcal{B}}(\rho(x))$, for all $x \in G$, is called the *character* of ρ. It is easy to see that χ is independent of the choice of the basis $\mathcal{B}$ of V.

Example 6.2.2 Let G be a finite group of order n. Let K be a field. Then $K[G]$ is a K-vector space of dimension n. The map

$$\rho : G \to \mathrm{Aut}_K(K[G]), \quad \rho(x)(a) = xa,$$

is a representation of G over K of degree n. It is called the *regular representation* of G over K. If χ is the character of ρ, then $\chi(1) = n$ and $\chi(x) = 0$ for all $x \in G \setminus \{1\}$.

Given a representation $\rho : G \to \mathrm{Aut}(V)$ of a group G over a field K, we define

$$\widehat{\rho} : K[G] \to \mathrm{End}_K(V), \quad \widehat{\rho}\left(\sum_{x \in G} a_x x\right) = \sum_{x \in G} a_x \rho(x).$$

Note that $\widehat{\rho}$ is a ring homomorphism. We define on V a multiplication by elements of $K[G]$ by

$$\left(\sum_{x \in G} a_x x\right) v := \sum_{x \in G} a_x \rho(x)(v)$$

for all $\sum_{x \in G} a_x x \in K[G]$ and $v \in V$. One can check that V with the sum and this product by elements of $K[G]$ is a $K[G]$-module. We say that this $K[G]$-module corresponds to the representation ρ.

Conversely, if W is a $K[G]$-module, then W also is a K-vector space, and the map $\rho : G \to \mathrm{Aut}_K(W)$, defined by $\rho(x)(w) = xw$, for all $x \in G$ and $w \in W$, is a representation of G over K. We say that ρ is the representation of G over K corresponding to the $K[G]$-module W.

One can quickly check that this gives a bijective correspondence between the representations of G over K and the $K[G]$-modules.

Lemma 6.2.3 *Let G be a group and K a field. Let M_1, M_2 be isomorphic $K[G]$-modules of finite dimension over K. Let ρ_1, ρ_2 be the representations over K corresponding to M_1, M_2 respectively. Let χ_1, χ_2 be the characters of ρ_1, ρ_2 respectively. Then $\chi_1(g) = \chi_2(g)$, for all $g \in G$.*

Proof Let $f : M_1 \to M_2$ be an isomorphism of $K[G]$-modules. Then

$$\rho_2(g)(f(m)) = gf(m) = f(gm) = f(\rho_1(g)(m)),$$

for all $g \in G$ and $m \in M_1$. Hence

$$\rho_2(g) = f\rho_1(g)f^{-1}.$$

Therefore $\chi_2(g) = \chi_1(g)$, and the result follows. □

Definition 6.2.4 A representation of a group G over a field K is said to be *irreducible* if its corresponding $K[G]$-module is simple.

Theorem 6.2.5 (Maschke's theorem) *Let G be a finite group. Let K be a field of characteristic $p \geq 0$. If p is not a divisor of $|G|$, then $J(K[G]) = \{0\}$ and therefore $K[G]$ is semisimple.*

Proof Since $K[G]$ is finite-dimensional, it is artinian. By Wedderburn–Artin theorem, it is enough to show that $J(K[G]) = \{0\}$. Let W be a vector subspace of $K[G]$ such that $K[G] = J(K[G]) \oplus W$. Let

$$\pi : K[G] \to K[G], \quad \pi(a+b) = a,$$

for all $a \in J(K[G])$ and $b \in W$. It is clear that π is a K-linear map. We define

$$\pi^* : K[G] \to K[G], \quad \pi^*(c) = \frac{1}{|G|}\sum_{x\in G} x^{-1}\pi(xc).$$

Note that π^* is a K-linear map. Let $c \in K[G]$ and $y \in G$. Then

$$\pi^*(yc) = \frac{1}{|G|}\sum_{x\in G} x^{-1}\pi(xyc) = \frac{y}{|G|}\sum_{x\in G} y^{-1}x^{-1}\pi(xyc) = y\pi^*(c).$$

Hence π^* is a homomorphism of $K[G]$-modules.

We need to prove that $J(K[G]) = \{0\}$. We first prove that $\pi^*(K[G]) = J(K[G])$. Clearly, $\pi^*(K[G]) \subseteq J(K[G])$. Now let $a \in J(K[G])$. Since $\pi(xa) = xa$, for all $x \in G$, we have that $\pi^*(a) = a$. Hence $\pi^*(K[G]) = J(K[G])$ and $(\pi^*)^2 = \pi^*$. Thus $\ker(\pi^*) = \operatorname{im}(\mathrm{id} - \pi^*)$ and $K[G] = J(K[G]) \oplus \ker(\pi^*)$. Now there exist $a \in J(K[G])$ and $b \in \ker(\pi^*)$ such that $1 = a + b$. Hence $a = a^2 + ab$. Since

$$ab \in J(K[G]) \cap \ker(\pi^*) = \{0\},$$

we get that $a = a^2$. Hence $(1-a)a = 0$. Since $1 - a$ is invertible, we have that $a = 0$. Hence $1 = b \in \ker(\pi^*)$, and thus $\ker(\pi^*) = K[G]$. Therefore $J(K[G]) = \{0\}$. □

The *center* of a ring R is

$$Z(R) = \{a \in R : ab = ba, \text{ for all } b \in R\}.$$

Let G be a group. The *conjugacy class* of an element $x \in G$ is

$$C_x = \{yxy^{-1} : y \in G\}.$$

Note that if C_x is finite, then $|C_x| = (G : C_G(x))$. In fact, if T is a left transversal of $C_G(x)$ in G, then $C_x = \{yxy^{-1} : y \in T\}$, and if $yxy^{-1} = zxz^{-1}$, for $y, z \in T$, then $z^{-1}y \in C_G(x)$, and thus $y = z$.

Lemma 6.2.6 *Let G be a finite group. Let K be a field. Let $C_1, \ldots, C_r$ be the distinct conjugacy classes of G. Let*

$$\alpha_i = \sum_{x \in C_i} x.$$

Then $\alpha_1, \ldots, \alpha_r$ is a K-basis of $Z(K[G])$.

Proof It is clear that $\alpha_1, \ldots, \alpha_r$ are K-linearly independent. Since

$$y\alpha_i y^{-1} = \sum_{x \in C_i} yxy^{-1} = \sum_{x \in C_i} x = \alpha_i$$

for all $y \in G$, we have that $\alpha_i \in Z(K[G])$. Let $b = \sum_{x \in G} b_x x \in Z(K[G])$. For every $y \in G$,

$$b = yby^{-1} = \sum_{x \in G} b_x yxy^{-1}.$$

Hence $b_x = b_{yxy^{-1}}$ for all $x, y \in G$. For every $i = 1, \ldots, r$, choose an element $x_i \in C_i$. Now

$$b = \sum_{i=1}^{r} b_{x_i} \alpha_i,$$

and the result follows. □

Definition 6.2.7 Let R be a subring of a commutative ring S. An element $a \in S$ is said to be *integral* over R if it is a root of a monic polynomial

$$a_0 + a_1 X + \cdots + a_{n-1} X^{n-1} + X^n \in R[X]$$

for some positive integer n.

Proposition 6.2.8 *Let S be a commutative integral domain. Let R be a subring of S. For any $a \in S$, the following conditions are equivalent.*

(1) a is integral over R.

(2) $R[a]$ is a finitely generated R-module.
(3) There is a nonzero finitely generated R-submodule M of S such that $aM \subseteq M$.

Proof (1) $\Longrightarrow$ (2). Suppose that a is integral over R. Then there exist a positive integer n and $a_0, \ldots, a_{n-1} \in R$ such that

$$a_0 + a_1 a + \cdots + a_{n-1}a^{n-1} + a^n = 0.$$

We shall prove by induction on m that $a^m \in R + Ra + \cdots + a^{n-1}R$. If $m < n$, then it is clear that $a^m \in R + Ra + \cdots + a^{n-1}R$. Assume that $m \geq n$ and that

$$a^k \in R + Ra + \cdots + a^{n-1}R$$

for all $k < m$. Then

$$a^m = -a_0 a^{m-n} - a_1 a^{m-n+1} - \cdots - a_{n-1}a^{m-1}.$$

Hence, by the inductive hypothesis, $a^m \in R + Ra + \cdots + a^{n-1}R$. Thus, by induction, $R[a] = R + Ra + \cdots + Ra^{n-1}$.

(2) $\Longrightarrow$ (3) follows from the fact that $aR[a] \subseteq R[a]$.

(3) $\Longrightarrow$ (1). Suppose that there exists a nonzero finitely generated R-submodule M of S such that $aM \subseteq M$. Let $s_1, \ldots, s_m \in S$ be elements such that $M = Rs_1 + \cdots + Rs_m$. Hence, there exist elements $a_{i,j} \in R$ such that

$$as_i = \sum_{j=1}^{m} a_{i,j}s_j,$$

that is

$$\sum_{j=1}^{m} (a\delta_{i,j} - a_{i,j})s_j = 0$$

for all $i = 1, \ldots, m$. Since $M \neq \{0\}$, some $s_j \neq 0$. Therefore

$$\det \begin{pmatrix} a - a_{1,1} & -a_{1,2} & \cdots & -a_{1,m} \\ -a_{2,1} & a - a_{2,2} & \ddots & \vdots \\ \vdots & \ddots & \ddots & -a_{m-1,m} \\ -a_{m,1} & \cdots & -a_{m,m-1} & a - a_{m,m} \end{pmatrix} = 0.$$

Therefore a is integral over R. □

Corollary 6.2.9 *Let R be a subring of a commutative integral domain S. Then*

$$\{a \in S : a \text{ is integral over } R\}$$

is a subring of S containing R.

Proof Let $T = \{a \in S : a \text{ is integral over } R\}$. It is clear that $R \subseteq T$. Let $a, b \in T$. Since a is integral over R, by Proposition 6.2.8, $R[a]$ is a finitely generated R-submodule of S. Since b is integral over R, b also is integral over $R[a]$. Hence, by Proposition 6.2.8, $R[a, b]$ is a finitely generated $R[a]$-submodule of S. Since $R[a]$ is finitely generated as R-module, we have that $R[a, b]$ also is finitely generated as R-module. Since

$$(a - b)R[a, b] \subseteq R[a, b] \quad \text{and} \quad abR[a, b] \subseteq R[a, b],$$

$a - b \in T$ and $ab \in T$, by Proposition 6.2.8. Therefore the result follows. □

Example 6.2.10 We shall see that $\mathbb{Z} = \{a \in \mathbb{Q} : a \text{ is integral over } \mathbb{Z}\}$. Let $a \in \mathbb{Q}$ be integral over $\mathbb{Z}$. We shall see that $a \in \mathbb{Z}$. We may assume that $a = b/c$, where b, c are nonzero coprime integers. Since a is integral over $\mathbb{Z}$, there exist a positive integer n and $a_0, \ldots, a_{n-1} \in \mathbb{Z}$ such that

$$a_0 + a_1 \frac{b}{c} + \cdots + a_{n-1} \frac{b^{n-1}}{c^{n-1}} + \frac{b^n}{c^n} = 0.$$

Hence

$$-b^n = c(a_0 c^{n-1} + \cdots + a_{n-1} b^{n-1}).$$

Since b, c are coprime, we get that $c = \pm 1$, and thus $a \in \mathbb{Z}$.

A field K is said to be *algebraically closed* if every polynomial $p(X) \in K[X]$ of positive degree has a root in K.

Proposition 6.2.11 *Let K be an algebraically closed field. Let R be a semiprimitive finite-dimensional K-algebra. Then*

$$R \cong M_{n_1}(K) \times \cdots \times M_{n_r}(K),$$

for some positive integers $n_1, \ldots, n_r$.

Proof Since $\dim_K(R) < \infty$, R is artinian. By the Wedderburn–Artin theorem there exist positive integers $n_1, \ldots, n_r$ and division rings $D_1, \ldots, D_r$ such that

$$R \cong M_{n_1}(D_1) \times \cdots \times M_{n_r}(D_r).$$

Furthermore, each D_i is a finite-dimensional K-algebra. We may assume that $K \subseteq D_i$. Let $a \in D_i$. Since $\dim_K(D_i) < \infty$, a is algebraic over K. Hence $K[a]$ is an algebraic field extension of K. Since K is algebraically closed, $K[a] = K$. Therefore $D_i = K$, and the result follows. □

Proposition 6.2.12 *Let K be an algebraically closed field of characteristic zero. Let G be a finite group. Let ρ be an irreducible representation of G over K of finite degree n. Let χ be the character of ρ. If $g \in G$ has l conjugate elements in G, then*

$$\frac{l\chi(g)}{n}$$

is integral over $\mathbb{Z}$.

Proof Let $C_1, \ldots, C_r$ be the distinct conjugacy classes of G. We may assume that $C_1 = \{1\}$. Let $\alpha_i = \sum_{x \in C_i} x$. By Lemma 6.2.6, $\alpha_1, \ldots, \alpha_r$ is a K-basis of $Z(K[G])$. For each i, fix an element $x_i \in C_i$. Since $\alpha_i \alpha_j \in Z(K[G])$,

$$\alpha_i \alpha_j = \sum_{k=1}^{r} m_{i,j,k} \alpha_k, \tag{6.1}$$

where

$$m_{i,j,k} = |\{(x, y) \in C_i \times C_j : xy = x_k\}|.$$

Let M be the $K[G]$-module corresponding to ρ. Since ρ is irreducible, M is simple. By Schur's lemma, $\mathrm{End}_{K[G]}(M)$ is a division ring. Since $\dim_K(M) = n$, we have that $\dim_K(\mathrm{End}_{K[G]}(M)) \leq n^2$. Thus $\mathrm{End}_{K[G]}(M)$ is a finite-dimensional K-algebra. As we have seen in the proof of Proposition 6.2.11, we have that $\mathrm{End}_{K[G]}(M) \cong K$. Since $\alpha_i \in Z(K[G])$,

$$\widehat{\rho}(\alpha_i) = \sum_{x \in C_i} \rho(x) \in \mathrm{End}_{K[G]}(M).$$

Hence there exists $\lambda_i \in K$ such that

$$\widehat{\rho}(\alpha_i)(m) = \lambda_i m$$

for all $m \in M$. Now we have that

$$|C_i| \chi(x_i) = \sum_{x \in C_i} \chi(x) = \mathrm{tr}(\widehat{\rho}(\alpha_i)) = n \lambda_i.$$

Hence

$$\lambda_i = \frac{|C_i| \chi(x_i)}{n}.$$

Applying $\widehat{\rho}$ to (6.1), we get $\lambda_i \lambda_j = \sum_{k=1}^{r} m_{i,j,k} \lambda_k$. Hence

$$\sum_{k=1}^{r} (\lambda_i \delta_{i,j} - m_{i,j,k}) \lambda_k = 0$$

for all $j = 1, \ldots, r$. Since $\alpha_1 = 1$, we have that $\lambda_1 = 1 \neq 0$. Hence

$$\det \begin{pmatrix} \lambda_i - m_{i,1,1} & -m_{i,1,2} & \cdots & -m_{i,1,r} \\ -m_{i,2,1} & \lambda_i - m_{i,2,2} & \ddots & \vdots \\ \vdots & \ddots & \ddots & -m_{i,r-1,r} \\ -m_{i,r,1} & \cdots & -m_{i,r,r-1} & \lambda_i - m_{i,r,r} \end{pmatrix} = 0.$$

Therefore λ_i is integral over $\mathbb{Z}$. Note that if $g \in C_i$, then there exists $y \in G$ such that $g = yx_iy^{-1}$, and thus

$$\lambda_i = \frac{|C_i|\chi(x_i)}{n} = \frac{l\chi(g)}{n}.$$

Therefore the result follows. □

Note that if S is a non-empty set and V is a vector space over a field F, then the set $\mathrm{Map}(S, V)$ of all maps $f : S \to V$ with the sum and the product by elements of F defined by

$$(f + g)(x) = f(x) + g(x), \quad (\lambda f)(x) = \lambda f(x)$$

for all $f, g \in \mathrm{Map}(S, V)$, $\lambda \in F$ and $x \in S$, is an F-vector space.

Lemma 6.2.13 (Dedekind's lemma) *Any family of distinct homomorphisms of a field K into another field F is F-linearly independent.*

Proof Suppose that the result is not true. Then there exist a smallest integer $n > 1$, distinct homomorphisms $f_1, \ldots, f_n$ of K into F and nonzero $a_1, \ldots, a_n \in F$, such that

$$a_1 f_1 + \cdots + a_n f_n = 0.$$

Hence

$$a_1 f_1(x) + \cdots + a_n f_n(x) = 0 \tag{6.2}$$

for all $x \in K$. Let $x, y \in K$. Then

$$0 = a_1 f_1(yx) + \cdots + a_n f_n(yx) = a_1 f_1(y) f_1(x) + \cdots + a_n f_n(y) f_n(x).$$

Multiplying (6.2) by $f_1(y)$, we get

$$0 = a_1 f_1(y) f_1(x) + \cdots + a_n f_1(y) f_n(x).$$

Hence, subtracting the last two equalities, we get

$$0 = a_2(f_2(y) - f_1(y)) f_2(x) + \cdots + a_n(f_n(y) - f_1(y)) f_n(x).$$

Therefore

$$a_2(f_2(y) - f_1(y))f_2 + \cdots + a_n(f_n(y) - f_1(y))f_n = 0$$

for all $y \in K$. By the minimality of n, we have that $f_2, \ldots, f_n$ are F-linearly independent. Hence $a_2(f_2(y) - f_1(y)) = 0$ and, since $a_2 \neq 0$, we get that $f_2(y) = f_1(y)$ for all $y \in K$, a contradiction, because $f_1 \neq f_2$. Therefore the result follows. □

Lemma 6.2.14 *Let $\xi_1, \ldots, \xi_n \in \mathbb{C}$ be roots of unity. Let $F = \mathbb{Q}(\xi_1, \ldots, \xi_n)$ be the subfield of $\mathbb{C}$ generated by $\xi_1, \ldots, \xi_n$. Let G be the group of automorphisms of F as $\mathbb{Q}$-algebra. Then*

$$\mathbb{Q} = \{a \in F : \alpha(a) = a \text{ for all } \alpha \in G\}.$$

Proof Note that the multiplicative subgroup of $\mathbb{C} \setminus \{0\}$ generated by $\xi_1, \ldots, \xi_n$ is a finite cyclic group generated by some root of unity ξ. Hence $F = \mathbb{Q}[\xi]$. Let $p(X) \in \mathbb{Q}[X]$ be a monic irreducible polynomial such that $p(\xi) = 0$. Let r be the degree of $p(X)$. Note that if the multiplicative order of ξ is m, then $p(X)$ is a divisor of $X^m - 1$. Hence all roots of $p(X)$ are roots of $X^m - 1$, and thus they belong to F. Since $p(X)$ is irreducible over $\mathbb{Q}$, all its roots are simple. Let $\eta_1, \ldots \eta_r \in F$ be the distinct roots of $p(X)$. Every automorphism $\alpha \in G$ is determined by $\alpha(\xi)$. Note that $\alpha(\xi) \in \{\eta_1, \ldots, \eta_r\}$. Furthermore, for $i = 1, \ldots, r$, the map $\alpha_i : F \to F$, defined by $\alpha_i(q(\xi)) = q(\eta_i)$ for all $q(X) \in \mathbb{Q}[X]$ of degree $< r$, is an automorphism of $F \cong \mathbb{Q}[X]/(p(X))$. Hence $|G| = r$. Let

$$L = \{a \in F : \alpha(a) = a \text{ for all } \alpha \in G\}.$$

Then L is a subfield of F and $\mathbb{Q} \subseteq L \subseteq F$. We have that $\dim_{\mathbb{Q}} F = r$. Suppose that $\dim_L F = s < r$. Let $a_1, \ldots a_s \in F$ be an L-basis of F. Since $s < r$, the system of s equations in r unknowns x_i:

$$\sum_{i=1}^{r} \alpha_i(a_j)x_i = 0 \quad (j = 1, \ldots, s),$$

has a nonzero solution $x_i = b_i$ in F. Let $a \in F$. There exist $c_1, \ldots, c_s \in L$ such that $a = \sum_{j=1}^{s} c_j a_j$. Hence

$$\sum_{i=1}^{r} \alpha_i(a)b_i = \sum_{i=1}^{r}\sum_{j=1}^{s} c_j\alpha_i(a_j)b_i = \sum_{j=1}^{s} c_j \sum_{i=1}^{r} \alpha_i(a_j)b_i = 0.$$

Thus

$$\sum_{i=1}^{r} b_i\alpha_i = 0,$$

a contradiction to Dedekind's lemma. Therefore $\dim_L F = r$ and thus $L = \mathbb{Q}$. □

Lemma 6.2.15 *Let ρ be an irreducible representation of a finite group G over $\mathbb{C}$ of finite degree n. Let χ be the character of ρ. Assume that $g \in G$ has l conjugate* *elements in G and that l and n are coprime. Then either $\chi(g) = 0$ or $\rho(g)$ acts by scalar multiplication.*

Proof By Proposition 6.2.12, $\frac{l\chi(g)}{n}$ is integral over $\mathbb{Z}$. Since l and n are coprime, there are integers r, s such as $1 = rl + sn$. Hence

$$\frac{\chi(g)}{n} = \frac{rl\chi(g)}{n} + s\chi(g).$$

Since G is finite, there exists a positive integer m such that $g^m = 1$. Hence $\chi(g)$ is a sum of roots of unity in $\mathbb{C}$. Hence, by Corollary 6.2.9, $\frac{\chi(g)}{n}$ is integral over $\mathbb{Z}$. Let $\xi_1, \ldots, \xi_n \in \mathbb{C}$ roots of unity such that $\chi(g) = \sum_{i=1}^{n} \xi_i$. Hence

$$\left|\frac{\chi(g)}{n}\right| = \frac{\left|\sum_{i=1}^{n} \xi_i\right|}{n} \leq 1.$$

Let M be the $\mathbb{C}[G]$-module corresponding to ρ. If $\xi_1 = \cdots = \xi_n$, then we may assume that there exists a $\mathbb{C}$-basis $\mathcal{B}$ of M such that the matrix of $\rho(g)$ with respect to $\mathcal{B}$ is

$$A = \begin{pmatrix} \xi_1 & 0 & \ldots & \ldots & 0 \\ \varepsilon_1 & \xi_1 & \ddots & & \vdots \\ 0 & \varepsilon_2 & \ddots & \ddots & \vdots \\ \vdots & \ddots & \ddots & \ddots & 0 \\ 0 & \ldots & 0 & \varepsilon_{n-1} & \xi_1 \end{pmatrix},$$

where $\varepsilon_i \in \{0, 1\}$. Since $g^m = 1$, we have that $\rho(g)^m = \mathrm{id}$. Let $I \in M_n(\mathbb{C})$ be the identity matrix and $N = A - \xi_1 I$. Note that N is nilpotent; in fact, $N^n = 0$. We have

$$I = A^m = (\xi_1 I + N)^m = \xi_1^m I + \sum_{i=1}^{m} \binom{m}{i} \xi_1^{m-i} N^i.$$

Hence $\xi_1^m = 1$ and

$$0 = \sum_{i=1}^{m} \binom{m}{i} \xi_1^{m-i} N^i = N \left(m\xi_1^{-1} I + \sum_{i=2}^{m} \binom{m}{i} \xi_1^{-i} N^{i-1} \right).$$

Since $m\xi_1^{-1} I + \sum_{i=2}^{m} \binom{m}{i} \xi_1^{-i} N^{i-1}$ is invertible, we get that $N = 0$. Therefore $A = \xi_1 I$, and thus, in this case, $\rho(g)$ acts as a scalar multiplication.

Suppose that $\xi_1, \ldots, \xi_n$ are not equal. In this case

$$\frac{|\chi(g)|}{n} < 1.$$

Let $F = \mathbb{Q}(\xi_1, \dots, \xi_n)$ be the subfield of $\mathbb{C}$ generated by $\xi_1, \dots, \xi_n$. Let H be the group of automorphisms of F. Then, for every $\alpha \in H, \alpha(\xi_i)$ is a root of unity. Hence

$$\frac{|\alpha(\chi(g))|}{n} = \frac{|\alpha(\xi_1) + \cdots + \alpha(\xi_n)|}{n} < 1.$$

By Lemma 6.2.14,

$$\prod_{\alpha \in H} \alpha\left(\frac{\chi(g)}{n}\right) \in \mathbb{Q}.$$

Since $\frac{\chi(g)}{n}$ is integral over $\mathbb{Z}$, $\alpha\left(\frac{\chi(g)}{n}\right)$ also is integral over $\mathbb{Z}$. By Corollary 6.2.9,

$$\prod_{\alpha \in H} \alpha\left(\frac{\chi(g)}{n}\right)$$

is integral over $\mathbb{Z}$. By Example 6.2.10,

$$\prod_{\alpha \in H} \alpha\left(\frac{\chi(g)}{n}\right) \in \mathbb{Z}.$$

Since

$$\left|\prod_{\alpha \in H} \alpha\left(\frac{\chi(g)}{n}\right)\right| < 1,$$

we have that $\chi(g) = 0$ and the result follows. □

Theorem 6.2.16 (Burnside) *Let G be a finite group with a conjugacy class with p^m elements, for some prime p and positive integer m. Then G is not simple.*

Proof Suppose that G is simple. It is clear that G is not abelian. Let $g \in G$ be an element with p^m conjugate elements in G. Let ρ be an irreducible non-trivial representation of G over $\mathbb{C}$ of degree n. Let χ be the character of ρ. Suppose that $\chi(g) \neq 0$ and that p is not a divisor of n. By Lemma 6.2.15, $\rho(g)$ is scalar and thus it is central in $\rho(G)$. Since G is simple and ρ is non-trivial, we have that $\ker(\rho) = \{1\}$ and $\rho(G) \cong G$ is simple. Hence $\rho(g) = \mathrm{id}$, and thus $g = 1$, in contradiction with the fact that g has p^m conjugate elements. Therefore, if p is not a divisor of n, then $\chi(g) = 0$.

Let σ be the regular representation of G over $\mathbb{C}$, that is,

$$\sigma\colon G \to \mathrm{Aut}_{\mathbb{C}}(\mathbb{C}[G]), \quad \sigma(h)(a) = ha.$$

Let ψ be the character of σ. By Maschke's theorem $J(\mathbb{C}[G]) = \{0\}$. Since $\mathbb{C}[G]$ is artinian, by Proposition 6.2.11, there exist positive integers $n_1, \dots, n_s$ such that $\mathbb{C}[G] \cong M_{n_1}(\mathbb{C}) \times \cdots \times M_{n_s}(\mathbb{C})$. Let $f\colon \mathbb{C}[G] \to M_{n_1}(\mathbb{C}) \times \cdots \times M_{n_s}(\mathbb{C})$ be an isomorphism. Let

$$I_{i,j} = f^{-1}(\{0\} \times \cdots \times \{0\} \times M_{n_i}(\mathbb{C})e_{j,j} \times \{0\} \times \cdots \times \{0\}),$$

where $e_{j,j} \in M_{n_i}(\mathbb{C})$ is the matrix with 1 in the (j, j) entry and 0 elsewhere. Note that $I_{i,1}, \ldots, I_{i,n_i}$ are minimal isomorphic left ideals of $\mathbb{C}[G]$ of dimension n_i over $\mathbb{C}$. Furthermore

$$\mathbb{C}[G] = \bigoplus_{i=1}^{s} \bigoplus_{j=1}^{n_i} I_{i,j}.$$

Let $\sigma_{i,j} : G \to \mathrm{Aut}_{\mathbb{C}}(I_{i,j})$ be the map defined by $\sigma_{i,j}(h)(a) = ha$, for all $h \in G$ and $a \in I_{i,j}$. Let $\psi_{i,j}$ be the character of the representation $\sigma_{i,j}$. Then it is easy to see that $\psi(h) = \sum_{i=1}^{s} \sum_{j=1}^{n_i} \psi_{i,j}(h)$, for all $h \in G$. By Lemma 6.2.3, $\psi_{i,1}(h) = \psi_{i,j}(h)$, for all i and all $j = 1, \ldots, n_i$. Hence

$$\psi(h) = \sum_{i=1}^{s} n_i \psi_{i,1}(h),$$

for all $h \in G$. We may assume that $n_1 = 1$ and

$$I_{1,1} = \mathbb{C}[G]\frac{1}{|G|}\sum_{x \in G} x = \mathbb{C}\frac{1}{|G|}\sum_{x \in G} x,$$

which corresponds to the trivial representation $\sigma_{1,1}$ of degree 1. Note that $\sigma_{i,1}$ is irreducible and non-trivial, for all $i > 1$. Since n_i is the degree of $\sigma_{i,1}$ and p^m is the number of conjugate elements of g, we have seen that if p is not a divisor of n_i and $i > 1$, then $\psi_{i,1}(g) = 0$. Hence

$$0 = \psi(g) = \psi_{1,1}(g) + \sum_{i=2}^{s} n_i \psi_{i,1}(g) = 1 + p \sum_{i=2}^{s} r_i \psi_{i,1}(g),$$

where

$$r_i = \begin{cases} \frac{n_i}{p} & \text{if } p \text{ is a divisor of } n_i, \\ 0 & \text{otherwise.} \end{cases}$$

By Corollary 6.2.9,

$$-\frac{1}{p} = \sum_{i=2}^{s} r_i \psi_{i,1}(g)$$

is integral over $\mathbb{Z}$, because $\psi_{i,1}$ is a sum of roots of unity, a contradiction to Example 6.2.10. Therefore the result follows. □

Theorem 6.2.17 (Burnside's p-q-theorem) *Let G be a finite group of order $p^n q^m$ for some prime numbers p, q and non-negative integers m, n. Then G is solvable.*

Proof Let G be a counterexample of minimal order, that is G is a finite non-solvable group, $|G| = p^n q^m$ and every group H of order $|H| = p_1^r q_1^s < |G|$, for some prime numbers p_1, q_1 and non-negative integers r, s, is solvable. Let S be a Sylow q-subgroup of G. Since finite q-groups are nilpotent and thus solvable, we have that S is non-trivial. Since S is a non-trivial q-subgroup, we know that $Z(S)$ is non-trivial. Let $g \in Z(S) \setminus \{1\}$. Then the number of conjugate elements of g in G is $(G : C_G(g))$. Since $S \subseteq C_G(g)$ and $p^n = (G : S) = (G : C_G(g))(C_g(g) : S)$, we have that either $g \in Z(G)$ or g has $p^r > 1$ conjugate elements. By Theorem 6.2.16, in both cases, G is not simple. Let N be a non-trivial proper normal subgroup of G. Since $|N|, |G/N| < |G|$ and $|N|, |G/N|$ are divisors of $|G|$, we have that N and G/N are solvable. By Theorem 6.1.2, G is solvable, a contradiction. □

A non-character-theoretic proof of this theorem is known but it is harder, see [68]. Burnside's theorem has several generalizations.

Theorem 6.2.18 (Kegel–Wielandt) *Let G be a finite group such that $G = AB$ for nilpotent subgroups A and B of G. Then G is solvable.* □

For the proof we refer to [3, Theorem 2.4.3].

Theorem 6.2.19 (Feit–Thompson) *Every finite group of odd order is solvable.* □

The proof of the theorem is extremely hard. It occupies a full volume of *Pacific Journal of Mathematics*, see [58].

6.3 Hall Subgroups

In this section we shall prove a characterization of finite solvable groups due to P. Hall. For this we will use Burnside's p-q-theorem and Wielandt's theorem. To prove Wielandt's theorem on solvable groups we need the following lemma.

Lemma 6.3.1 *Let G be a finite group. If H and K are subgroups of G of coprime indices, then $G = HK$ and $(H : H \cap K) = (G : K)$.*

Proof Let $D = H \cap K$. Since

$$(G : D) = \frac{|G|}{|H \cap K|} = (G : H)(H : H \cap K),$$

$(G : H)$ divides $(G : D)$. Similarly, $(G : K)$ divides $(G : D)$. Since $(G : H)$ and $(G : K)$ are coprime, $(G : H)(G : K)$ divides $(G : D)$. In particular,

$$\frac{|G|}{|H|}\frac{|G|}{|K|} = (G : H)(G : K) \leq (G : D) = \frac{|G|}{|H \cap K|}$$

and hence $|G| = |HK|$. Since

$$|G| = |HK| = |H||K|/|H \cap K|,$$

it follows that $(G : K) = (H : H \cap K)$. □

The *normal closure* H^G of a subgroup H of G is the subgroup

$$H^G = \langle xHx^{-1} : x \in G \rangle$$

generated by all conjugates of H. The subgroup H^G is the smallest normal subgroup of G containing H.

Example 6.3.2 Let $G = \mathbb{A}_4$ and $H = \{\text{id}, (12)(34)\}$. Then

$$H^G = \{\text{id}, (12)(34), (13)(24), (14)(23)\} \cong C_2 \times C_2.$$

Theorem 6.3.3 (Wielandt) *Let G be a finite group and H, K and L be subgroups of G with pair-wise coprime indices. If H, K and L are solvable, then G is solvable.*

Proof Assume the theorem is not valid and let G be a minimal counterexample. Then G is not trivial. Let N be a minimal normal subgroup of G and $\pi : G \to G/N$, $g \mapsto gN$, be the canonical map. Since by definition N is non-trivial, it follows that $|G/N| < |G|$. The subgroups $\pi(H) = \pi(HN)$, $\pi(K) = \pi(KN)$ and $\pi(L) = \pi(LN)$ of $\pi(G) = G/N$ are solvable. The correspondence theorem implies that the indices of $\pi(H)$, $\pi(K)$, and $\pi(L)$ in $\pi(G)$ are pair-wise coprime. By the minimality of G, the group $\pi(G)$ is solvable. If $H = \{1\}$, then $|G| = (G : H)$ is coprime with $(G : K)$ and hence $G = K$ is solvable. So we may assume that $H \neq \{1\}$. Let M be a minimal normal subgroup of H. By Lemma 6.1.1, M is a p-group for some prime number p. We may assume that p does not divide $(G : K)$ (if p divides $(G : K)$, then p does no divide $(G : L)$ and hence it is enough to replace K by L). There exists $P \in \text{Syl}_p(G)$ such that $P \subseteq K$. By Sylow's theorem, there exists $g \in G$ such that $M \subseteq gKg^{-1}$. Since $(G : gKg^{-1}) = (G : K)$ and $(G : H)$ are coprime, Lemma 6.3.1 implies that $G = (gKg^{-1})H$.

We claim that all conjugates of M are included in gKg^{-1}. If $x \in G$, then $x = uv$ for some $u \in gKg^{-1}$ and $v \in H$. Since M is normal in H,

$$xMx^{-1} = (uv)M(uv)^{-1} = uMu^{-1} \subseteq gKg^{-1}.$$

In particular, $\{1\} \neq M^G \subseteq gKg^{-1}$ is solvable, as gKg^{-1} is solvable. The minimality of G implies that G/M^G is solvable. Hence G is solvable by Theorem 6.1.2. □

Let G be a finite group of order $p^\alpha m$ with p a prime number coprime with m. A subgroup H of G is a *p-complement* if $|H| = m$.

Example 6.3.4 Let $G = \mathbb{S}_3$. Then $H = \langle (123) \rangle$ is a 2-complement and $K = \langle (12) \rangle$ is a 3-complement.

Theorem 6.3.5 (Hall) *Let G be a finite group that admits a p-complement for all primes p dividing the order of G. Then G is solvable.*

Proof Let $|G| = p_1^{\alpha_1} \cdots p_k^{\alpha_k}$ with the p_j being distinct primes. We proceed by induction on k. If $k = 1$, then G is a p_1-group and the result is clear. If $k = 2$, then Burnside's p-q-theorem implies the claim. Assume now that $k \geq 3$. For each $j \in \{1, 2, 3\}$ let H_j be p_j-complement in G. Since $|H_j| = |G|/p_j^{\alpha_j}$, the subgroups H_j have pair-wise coprime indices.

We claim that H_1 is solvable. Note that $|H_1| = p_2^{\alpha_2} \cdots p_k^{\alpha_k}$. Let p be a prime number that divides $|H_1|$ and let Q be a p-complement in G. Since $(G : H_1)$ and $(G : Q)$ are coprime, Lemma 6.3.1 implies that

$$(H_1 : H_1 \cap Q) = (G : Q).$$

Thus $H_1 \cap Q$ is a p-complement in H_1. Hence H_1 is solvable by the inductive hypothesis. Similarly, H_2 and H_3 are both solvable.

Since H_1, H_2, and H_3 are solvable and have pair-wise coprime indices, by Wielandt's theorem, G also is solvable. Therefore the result follows by induction. □

Definition 6.3.6 Let π be a non-empty set of primes. A finite group G is a *π-group* if every prime divisor p of $|G|$ belongs to π.

Definition 6.3.7 Let π be a non-empty set of primes. A π-subgroup H of a finite group G is a *Hall π-subgroup* of G if $(G : H)$ and p are coprime, for all $p \in \pi$.

Theorem 6.3.8 (Hall) *Let G be a finite solvable group. Then the following conditions hold.*

(1) For every non-empty set π of primes, there exists a Hall π-subgroup of G.
(2) Let π be a non-empty set of primes. Let H be a Hall π-subgroup of G. Then for every π-subgroup L of G there exists $g \in G$ such that $L \subseteq gHg^{-1}$.

Proof Let G be a counterexample to (1) of the smallest order. Hence G is a finite solvable group and there exists a non-empty set π of prime divisors of $|G|$ such that G has no Hall π-subgroups. Furthermore, every finite solvable group K such that $|K| < |G|$ has Hall τ-subgroups for all non-empty sets of primes τ. Let N be a minimal normal subgroup of G. By Theorem 6.1.2, N and G/N are solvable. By Lemma 6.1.1, N is an elementary abelian p-subgroup of G, for some prime p. Suppose that $p \in \pi$. In this case, since $|G/N| < |G|$, there exists a subgroup H of G such that $N \subseteq H$ and H/N is a Hall π-subgroup of G/N. Since $|H| = |H/N| \cdot |N|$, we have that H is a π-subgroup of G. Since $(G : H) = (G/N : H/N)$, we get that H is a Hall π-subgroup of G, a contradiction. Therefore, $p \notin \pi$. Since $|G/N| < |G|$, there exists a subgroup H of G such that $N \subseteq H$ and H/N is a Hall π-subgroup of G/N. Since $|H| = |H/N| \cdot |N|$, we have that H is a $\pi \cup \{p\}$-subgroup of G. Note that every Hall π-subgroup of H also is a Hall π-subgroup of G. Hence $H = G$. Note that N is a Sylow p-subgroup of G. Let M be a normal subgroup of G such that $N \subseteq M$ and M/N is a minimal normal subgroup of G/N. Since M/N is solvable, by Lemma 6.1.1, M/N is an elementary abelian q-group, for some prime q. Since N is a Sylow p-subgroup of G, we have that $q \neq p$. Hence $q \in \pi$. Let S be a Sylow q-subgroup of M. By Frattini's argument, $G = MN_G(S) = NSN_G(S) = NN_G(S)$.

Note that, by the above argument, S is not normal in G. Hence $|N_G(S)| < |G|$. Since $N_G(S)$ is solvable, there exists a Hall π-subgroup H_1 of $N_G(S)$. Since $(G : H_1) = (G : N_G(S))(N_G(S) : H_1)$ is a divisor of $|N|(N_G(S) : H_1)$, we have that H_1 also is a Hall π-subgroup of G, a contradiction. Therefore (1) is proved.

Assume now that (2) does not hold. Let G be a counterexample to (2) of the smallest order. Hence G is a finite solvable group and there exists a non-empty set of primes π, a Hall π-subgroup H of G and a π-subgroup L of G such that, for every $g \in G, L \nsubseteq gHg^{-1}$. Furthermore, if K is a finite solvable group such that $|K| < |G|$, then for every non-empty set of primes τ, every Hall τ-subgroup U and every τ-subgroup V of K, there exists $x \in K$ such that $V \subseteq xUx^{-1}$. Let N be a minimal normal subgroup of G. As above, we know that N is an elementary abelian p-group. Then NH/N and NL/N are π-subgroups of G/N. Since H is a Hall π-subgroup of G, we have that NH/N is a Hall π-subgroup of G/N. Suppose that $p \in \pi$. In this case, NH is a π-subgroup. Since H is a Hall π-subgroup of G, we have that $H = NH$. Since $|G/N| < |G|$, there exists $x \in G$ such that $NL \subseteq (xN)NH(x^{-1}N) = xNHx^{-1} = xHx^{-1}$, and thus $L \subseteq xHx^{-1}$, a contradiction. Hence $p \notin \pi$, and $L \subseteq xNHx^{-1}$. If $|xNHx^{-1}| < |G|$, then, since xHx^{-1} is a Hall π-subgroup of $xNHx^{-1}$, there exists $y \in xNHx^{-1}$ such that $L \subseteq yxHx^{-1}y^{-1}$, a contradiction. Hence $|NH| = |xNHx^{-1}| = |G|$, and thus $G = NH$. Suppose that $|L| < |H|$. In this case, $|NL| = |N| \cdot |L| < |N| \cdot |H| = |G|$. Note that L is a Hall π-subgroup of NL and

$$|(NL) \cap H| = \frac{|NL| \cdot |H|}{|(NL)H|} = \frac{|N| \cdot |L| \cdot |H|}{|NH|} = \frac{|N| \cdot |L| \cdot |H|}{|N| \cdot |H|} = |L|.$$

Hence there exists $z \in NL$ such that $L \subseteq z((NL) \cap H)z^{-1} \subseteq zHz^{-1}$, a contradiction. Hence $|L| = |H|$. Let M be a normal subgroup of G such that $N \subseteq M$ and M/N is a minimal normal subgroup of G/N. As above, we know that M/N is an elementary abelian q-subgroup, for some prime $q \in \pi$. Note that

$$|H \cap M| = \frac{|H| \cdot |M|}{|MH|} = \frac{|H| \cdot |M|}{|NH|} = \frac{|M|}{|N|}.$$

Hence $H \cap M$ is a Sylow q-subgroup of M. Since $L \cap M$ is a q-subgroup of M, by Sylow's theorem, there exists $z \in M$ such that $L \cap M \subseteq z(H \cap M)z^{-1}$. Since $|L| = |H|$, we have that $G = NH = NL = MH = ML$ and

$$|L \cap M| = \frac{|L| \cdot |M|}{|ML|} = \frac{|H| \cdot |M|}{|MH|} = |H \cap M|.$$

Hence $L \cap M = z(H \cap M)z^{-1}$ and thus

$$L \subseteq N_G(L \cap M) = N_G(z(H \cap M)z^{-1}) = zN_G(H \cap M)z^{-1}.$$

Note that $H \cap M \neq \{1\}$. If $G = N_G(H \cap M)$, then $L \cap M = z(H \cap M)z^{-1} = H \cap M$. Since $|G/(H \cap M)| < |G|$, there exists $t \in G$ such that

$$\frac{L}{L \cap M} = \frac{L}{H \cap M} = \frac{tHt^{-1}}{H \cap M}.$$

Hence $L = tHt^{-1}$, a contradiction. Thus we have that $G \neq N_G(H \cap M)$. Hence $|zN_G(H \cap M)z^{-1}| = |N_G(H \cap M)| < |G|$. Since zHz^{-1} is a Hall π-subgroup of $zN_G(H \cap M)z^{-1}$, there exists $t \in zN_G(H \cap M)z^{-1}$ such that $L \subseteq tzHz^{-1}t^{-1}$, a contradiction. Therefore (2) follows. □

Definition 6.3.9 Let G be a finite group. Let $p_1, \ldots, p_r$ be the distinct prime divisors of $|G|$. A *Sylow system* of G is a set $\{S_1, \ldots, S_r\}$, where S_i is a Sylow p_i-subgroup of G, such that $S_i S_j = S_j S_i$, for all i, j.

Theorem 6.3.10 (Hall) *Let G be a finite group. Then G is solvable if and only if G has a Sylow system.*

Proof Suppose that G has a Sylow system $\{S_1, \ldots, S_r\}$, where S_i is a Sylow p_i-subgroup of G. Since $S_i S_j = S_j S_i$, for all i, j, $C_i = S_1 \cdots S_{i-1} S_{i+1} \cdots S_r$ is a subgroup of G. Furthermore

$$|C_i| = |S_1| \cdots |S_{i-1}| \cdot |S_{i+1}| \cdots |S_r|$$

and$(G : C_i) = |S_i|$. Hence C_i is a p_i-complement of G. By Theorem 6.3.5, G is solvable.

Conversely, suppose that G is solvable. By Theorem 6.3.8, for every prime divisor p_i of $|G|$, G has a p_i-complement H_i. Let $p_1, \ldots, p_r$ be the distinct prime divisors of $|G|$. Clearly, to prove that G has a Sylow system, we may assume that $r > 2$. Let I be a non-empty subset of $\{1, \ldots, r\}$. Let $H_I = \bigcap_{i \in I} H_i$. We shall show that

$$(G : H_I) = \prod_{i \in I} (G : H_i) \tag{6.3}$$

by induction on $|I|$. For $|I| = 1$, it is clear. Assume that $|I| > 1$ and that (6.3) holds for every non-empty subset J of $\{1, \ldots, r\}$ such that $|J| < |I|$. Let $i \in I$ and $I_1 = I \setminus \{i\}$. By the inductive hypothesis, we have that

$$(G : H_{I_1}) = \prod_{k \in I_1} (G : H_k).$$

Hence $(G : H_i)$ and $(G : H_{I_1})$ are coprime. By Lemma 6.3.1, $G = H_i H_{I_1}$. Now we have that

$$\begin{aligned}
\prod_{k \in I} (G : H_k) &= (G : H_i)(G : H_{I_1}) = \frac{|G|^2}{|H_i| \cdot |H_{I_1}|} \\
&= \frac{|G| \cdot |H_i H_{I_1}|}{|H_i| \cdot |H_{I_1}|} = \frac{|G|}{|H_i \cap H_{I_1}|} = (G : H_i \cap H_{I_1}) = (G : H_I).
\end{aligned}$$

Hence (6.3) follows by induction. Let I, J be non-empty subsets of $\{1, \ldots, r\}$. We shall prove that

$$H_I H_J = H_J H_I \tag{6.4}$$

by induction on $|I| \cdot |J|$. For $|I| = |J| = 1$, (6.4) follows from Lemma 6.3.1. Suppose that $|I| \cdot |J| > 1$ and that (6.4) holds for all non-empty subsets I_1, J_1 of $\{1, \ldots r\}$, such that $|I_1| \cdot |J_1| < |I| \cdot |J|$. If $I \subseteq J$, then $H_J \subseteq H_I$ and $H_I H_J = H_I = H_J H_I$. Hence we may assume that there exist $i \in I \setminus J$. Let $I_1 = I \setminus \{i\}$. By the inductive hypothesis, $H_{I_1} H_J = H_J H_{I_1}$ is a subgroup of G. Note that H_I, H_J are subgroups of the group $H_{I_1} H_J$. Since $H_I = H_{I_1} \cap H_i$,

$$|H_I H_J| = \frac{|H_I| \cdot |H_J|}{|H_I \cap H_J|} = \frac{|H_{I_1}| \cdot |H_i| \cdot |H_J|}{|H_{I_1} H_i| \cdot |H_{I_1} \cap H_i \cap H_J|} = \frac{|H_{I_1} H_J| \cdot |H_i| \cdot |H_{I_1} \cap H_J|}{|H_{I_1} H_i| \cdot |H_{I_1} \cap H_i \cap H_J|}.$$

Note that $(G : H_i)$ and $(G : H_{I_1})$ are coprime. Similarly, $(G : H_i)$ and $(G : H_J)$ are coprime. By Lemma 6.3.1, $G = H_i H_{I_1} = H_i H_J = H_i(H_{I_1} \cap H_J)$. Hence

$$|H_I H_J| = \frac{|H_{I_1} H_J| \cdot |H_i| \cdot |H_{I_1} \cap H_J|}{|H_i H_{I_1}| \cdot |H_{I_1} \cap H_i \cap H_J|} = \frac{|H_{I_1} H_J| \cdot |H_i(H_{I-1} \cap H_J)|}{|H_i H_{I_1}|} = |H_{I_1} H_J|.$$

Hence

$$H_I H_J = H_{I_1} H_J = H_J H_I.$$

Therefore (6.4) follows by induction. Let $i \in \{1, \ldots, r\}$, $J_i = \{1, \ldots, r\} \setminus \{i\}$ and $S_i = H_{J_i}$. By (6.3), S_i is a Sylow p_i-subgroup of G. By (6.4), $S_i S_j = S_j S_i$. Hence $\{S_1, \ldots, S_r\}$ is a Sylow system of G, and the result follows. □

6.4 Perfect Groups

A group G is said to be *perfect* if $[G, G] = G$. Note that G is perfect if and only if $G/[G, G]$ is trivial. Every non-abelian simple group is perfect.

Let K be a field. The *special linear* group of degree n over K is

$$\mathbf{SL}_n(K) = \{A \in \mathbf{GL}_n(K) : \det(A) = 1\}.$$

If $K = \mathbb{F}_q$ is the finite field of q elements, then we will write $\mathbf{SL}_n(q) = \mathbf{SL}_n(\mathbb{F}_q)$.

Example 6.4.1 The groups $\mathbf{SL}_2(2)$ and $\mathbf{SL}_2(3)$ are not perfect.

Let p be a prime number and $q = p^m$, for some positive integer m. The groups $\mathbf{SL}_n(q)$ are perfect except the cases $\mathbf{SL}_2(2)$ and $\mathbf{SL}_2(3)$. As an example, let us prove that $\mathbf{SL}_2(q)$ is perfect if $q > 3$. We first claim that $\mathbf{SL}_2(q)$ is generated by matrices $X_{ij}(\lambda) = I + \lambda E_{ij}$, where I denotes the identity matrix, E_{ij} is the matrix with a one at position (i, j) and zero in all other entries, $i, j \in \{1, 2\}$ are distinct and $\lambda \in \mathbb{F}_q \setminus \{0\}$. First note that a matrix $\begin{pmatrix} 1 & b \\ c & d \end{pmatrix} \in \mathbf{SL}_2(q)$ is a product of some of the $X_{ij}(\lambda)$, as

$$\begin{pmatrix} 1 & b \\ c & d \end{pmatrix} = \begin{pmatrix} 1 & 0 \\ c & 1 \end{pmatrix} \begin{pmatrix} 1 & b \\ 0 & 1 \end{pmatrix} = X_{21}(c)X_{12}(b).$$

This implies that a matrix $\begin{pmatrix} a & b \\ c & d \end{pmatrix} \in \mathbf{SL}_2(q)$, with $c \neq 0$, also is a product of some $X_{ij}(\lambda)$. Indeed, if λ is such that $a = 1 - \lambda c$, then

$$\begin{pmatrix} a & b \\ c & d \end{pmatrix} = \begin{pmatrix} 1 & -\lambda \\ 0 & 1 \end{pmatrix} \begin{pmatrix} 1 & b + \lambda d \\ c & d \end{pmatrix} = X_{12}(-\lambda)X_{21}(c)X_{12}(b + \lambda d).$$

Finally,

$$\begin{pmatrix} a & b \\ 0 & a^{-1} \end{pmatrix} = \begin{pmatrix} 1 & 0 \\ -1 & 1 \end{pmatrix} \begin{pmatrix} a & b \\ a & b + a^{-1} \end{pmatrix}$$

and therefore $\begin{pmatrix} a & b \\ 0 & a^{-1} \end{pmatrix}$ is a product of some $X_{ij}(\lambda)$ since $\begin{pmatrix} a & b \\ a & b + a^{-1} \end{pmatrix}$ is a product of some $X_{ij}(\lambda)$. To prove that $[\mathbf{SL}_2(q), \mathbf{SL}_2(q)] = \mathbf{SL}_2(q)$ we first note that

$$\left[\begin{pmatrix} a & 0 \\ 0 & a^{-1} \end{pmatrix}, \begin{pmatrix} 1 & b \\ 0 & 1 \end{pmatrix} \right] = \begin{pmatrix} 1 & (a^2 - 1)b \\ 0 & 1 \end{pmatrix}.$$

Since $q > 3$, given $\lambda \in \mathbb{F}_q$ and $a \in \mathbb{F}_q \setminus \{-1, 0, 1\}$, there exists $b \in \mathbb{F}_q$ such that $\lambda = (a^2 - 1)b$. This implies that each $X_{ij}(\lambda)$ belongs to the commutator subgroup of $\mathbf{SL}_2(q)$.

Theorem 6.4.2 (Grün) *If G is a perfect group, then $Z(G/Z(G)) = \{1\}$.*

Proof The three-subgroups lemma with $X = Y = G$, $Z = \zeta_2(G)$ and $N = \{1\}$ yields

$$\{1\} = [\zeta_2(G), G, G] = [\zeta_2(G), [G, G]] = [\zeta_2(G), G].$$

Thus $\zeta_2(G) \subseteq Z(G)$. Now we prove that $Z(G/Z(G))$ is trivial. Let $\pi \colon G \to G/Z(G)$ be the canonical map and $x \in G$ be such that $\pi(x)$ is a central element. Then

$$[\pi(x), \pi(y)] = \pi([x, y]) = 1$$

for all $y \in G$. In particular, $[x, y] \in Z(G) = \zeta_1(G)$ for all $y \in G$. This means that $x \in \zeta_2(G) \subseteq Z(G)$. □

6.5 Exercises

6.5.1 Prove that the group $\mathbf{SL}_2(3)$ is solvable.

6.5.2 Compute the Frattini subgroup of $\mathbf{SL}_2(3)$.

6.5.3 Let p be a prime. Prove that if G is an elementary abelian p-group, then $\Phi(G) = \{1\}$.

6.5.4 Prove that the only minimal normal subgroup of $\mathbf{SL}_2(3)$ is $Z(\mathbf{SL}_2(3)) \cong C_2$.

6.5.5 Let $q \geq 5$. Prove that $\mathbf{SL}_n(q)$ is perfect.

6.5.6 Let G be a perfect group and N be a normal subgroup of G. Then G/N is perfect.

6.6 Notes

The results in this chapter are classical. The proof of Burnside's p-q-theorem can be found in [66]. The classical results of P. Hall can be found, for example, in [96].

Chapter 7
Complements

In this chapter, we explain the relation between 1-cocycles and complements. The main result is the theorem of Schur–Zassenhaus.

7.1 Extensions and 1-Cocycles

Let K, Q be groups. An *extension* of K by Q is a short exact sequence of group homomorphisms

$$\{1\} \longrightarrow K \xrightarrow{f} G \xrightarrow{g} Q \longrightarrow \{1\}$$

This means that f is injective, g is surjective, and $\ker g = \operatorname{im} f$. Note that in this case, K is isomorphic to $f(K)$, which is a normal subgroup of G and $G/f(K) \cong Q$. We also say that G is an extension of K by Q.

Example 7.1.1 C_6 and $\mathbb{S}_3$ are both extensions of C_3 by C_2.

Example 7.1.2 C_6 is an extension of C_2 by C_3.

Example 7.1.3 The direct product $K \times Q$ of the groups K and Q is an extension of K by Q and an extension of Q by K.

Example 7.1.4 Let G be an extension of K by Q. If L is a subgroup of G containing K, then L is an extension of K by L/K.

Let

$$E: \ \{1\} \longrightarrow K \longrightarrow G \xrightarrow{p} Q \longrightarrow \{1\}$$

F. Cedó and L. Vendramin, *Groups, Radical Rings, and the Yang-Baxter Equation*, Progress in Mathematics 361, https://doi.org/10.1007/978-3-032-01524-2_7

be an extension of groups. A *lifting* of E is a map $\ell\colon Q \to G$ such that $p(\ell(x)) = x$, for all $x \in Q$.

An extension E *splits* if there is a lifting of E that it is a group homomorphism.

Let Q and K be groups. Assume that Q acts by automorphisms on K, that is, there is a group homomorphism $\alpha\colon Q \to \operatorname{Aut}(K)$. We write $\alpha_x = \alpha(x)$, for all $x \in Q$. A map $\varphi\colon Q \to K$ is said to be a 1-*cocycle* (or a *derivation*) if

$$\varphi(xy) = \varphi(x)\alpha_x(\varphi(y))$$

for all $x, y \in Q$. The set of 1-cocycles $Q \to K$ is defined as

$$\operatorname{Der}(Q, K) = Z^1(Q, K) = \{\delta\colon Q \to K : \delta \text{ is a 1-cocycle}\}.$$

Example 7.1.5 Let Q and K be groups. Let $\alpha\colon Q \to \operatorname{Aut}(K)$ be a group homomorphism. For each $k \in K$, the map $\delta_k\colon Q \to K, x \mapsto \delta_k(x) = [k, x] = k\alpha_x(k)^{-1}$, is a derivation.

A subgroup K of a group G admits a *complement* Q if G admits an exact factorization through K and Q, i.e., $G = KQ$ with $K \cap Q = \{1\}$. A classical example is the (inner) semidirect product $G = K \rtimes Q$, where K is a normal subgroup of G and Q is a subgroup of G such that $K \cap Q = \{1\}$.

Let Q and K be groups. Let $\alpha\colon Q \to \operatorname{Aut}(K)$, $x \mapsto \alpha_x$ be a group homomorphism. Consider the set $K \times Q$. We define a multiplication on this set by

$$(k_1, q_1)(k_2, q_2) = (k_1\alpha_{q_1}(k_2), q_1q_2),$$

for all $k_1, k_2 \in K$ and $q_1, q_2 \in Q$. Then $K \times Q$ with this multiplication is a group, called the *semidirect product* of K by Q via α, and denoted by $K \rtimes_\alpha Q$. Note that $K \times \{1\}$ is a normal subgroup of $K \rtimes_\alpha Q$, and $\{1\} \times Q$ is a subgroup of $K \rtimes_\alpha Q$. Thus $K \rtimes_\alpha Q$ is the inner semidirect product of $K \times \{1\}$ by $\{1\} \times Q$. Since $K \cong K \times \{1\}$ and $Q \cong \{1\} \times Q$, we identify these groups, i.e., $k = (k, 1)$ and $q = (1, q)$, for all $k \in K$ and $q \in Q$. Note that

$$qkq^{-1} = (1, q)(k, 1)(1, q^{-1}) = (\alpha_q(k), q)(1, q^{-1}) = (\alpha_q(k), 1) = \alpha_q(k),$$

for all $k \in K$ and $q \in Q$. This justifies the notation in Example 7.1.5.

Proposition 7.1.6 *Let*

$$E : \{1\} \longrightarrow K \xrightarrow{f} G \xrightarrow{g} Q \longrightarrow \{1\}$$

be an extension of groups. Then E splits if and only if G is the inner semidirect product of $f(K)$ and a complement C of $f(K)$ in G such that $g(C) = Q$.

Proof Suppose that E splits. Let $h\colon Q \to G$ be a group homomorphism such that $gh = \mathrm{id}_Q$. Let $C = h(Q)$. Note that for every $x \in G$, $xhg(x)^{-1} \in \ker(g) = f(K)$.

Hence $G = f(K)C$. Let $y \in C \cap f(K)$. There exists $q \in Q$ such that $h(q) = y$. Hence $1 = g(y) = gh(q) = q$ and thus $y = h(q) = h(1) = 1$. Therefore G is the inner semidirect product of $f(K)$ and C.

Conversely, suppose that G is the inner semidirect product of $f(K)$ and a complement C of $f(K)$ in G such that $g(C) = Q$. Since $\ker(g) \cap C = f(K) \cap C = \{1\}$, the restriction $g|_C$ of g to C is an isomorphism. Let $h' \colon Q \to C$ be the inverse of $g|_C$. Let $h = \iota h' \colon Q \to G$, where $\iota \colon C \to G$ is the inclusion map. We have that $gh(q) = gh'(q) = q$ for all $q \in Q$. Hence E splits. □

Theorem 7.1.7 *Let Q, K be groups and let $\alpha \colon Q \to \mathrm{Aut}(K)$ be a group homomorphism. Then there exists a bijective correspondence between the set $\mathcal{C}$ of complements of K in $K \rtimes Q$ and the set $\mathrm{Der}(Q, K)$ of 1-cocycles $Q \to K$.*

Proof Since Q acts by conjugation on K, it follows that $\delta \in \mathrm{Der}(Q, K)$ if and only if $\delta(xy) = \delta(x)x\delta(y)x^{-1}$ for all $x, y \in Q$. In this case, one obtains that $\delta(1) = 1$ and $\delta(x^{-1}) = x^{-1}\delta(x)^{-1}x$.

Let $C \in \mathcal{C}$. If $x \in Q$, then there exist unique elements $k \in K$ and $c \in C$ such that $x = k^{-1}c$. Hence the map $\delta_C \colon Q \to K, x \mapsto k$, is well defined and $\delta_C(x)x = c \in C$.

We claim that $\delta_C \in \mathrm{Der}(Q, K)$. If $x, x_1 \in Q$, we write $x = k^{-1}c$ and $x_1 = k_1^{-1}c_1$ for $k, k_1 \in K$ and $c, c_1 \in C$. Since K is a normal subgroup of the semidirect product $K \rtimes Q$, we can write xx_1 as $xx_1 = k_2c_2$, where $k_2 = k^{-1}(ck_1^{-1}c^{-1}) \in K$, $c_2 = cc_1 \in C$. Thus $\delta_C(xx_1)xx_1 = cc_1 = \delta_C(x)x\delta_C(x_1)x_1$ implies that $\delta_C(xx_1) = \delta_C(x)x\delta_C(x_1)x^{-1}$. So there is a map $F \colon \mathcal{C} \to \mathrm{Der}(Q, K)$, defined by $F(C) = \delta_C$ for all $C \in \mathcal{C}$.

We now construct a map $G \colon \mathrm{Der}(Q, K) \to \mathcal{C}$. For each $\delta \in \mathrm{Der}(Q, K)$ we find a complement Δ of K in $K \rtimes Q$. Let $\Delta = \{\delta(x)x : x \in Q\}$. We claim that Δ is a subgroup of $K \rtimes Q$. Since $\delta(1) = 1$, $1 \in \Delta$. If $x, y \in Q$, then

$$\delta(x)x\delta(y)y = \delta(x)x\delta(y)x^{-1}xy = \delta(xy)xy \in \Delta.$$

Finally, if $x \in Q$, then

$$(\delta(x)x)^{-1} = x^{-1}\delta(x)^{-1}xx^{-1} = \delta(x^{-1})x^{-1} \in \Delta.$$

Thus Δ is a subgroup of $K \rtimes Q$. We claim that $\Delta \cap K = \{1\}$. If $x \in Q$ is such that $\delta(x)x \in K$, then since $\delta(x) \in K$, it follows that $x \in K \cap Q = \{1\}$. If $g \in K \rtimes Q$, then there are unique $k \in K$ and $x \in Q$ such that $g = kx$. We write $g = k\delta(x)^{-1}\delta(x)x$. Since $k\delta(x)^{-1} \in K$ and $\delta(x)x \in \Delta$, we conclude that $K \rtimes Q = K\Delta$. Thus there is a well-defined map $G \colon \mathrm{Der}(Q, K) \to \mathcal{C}, \delta \mapsto \Delta = \{\delta(x)x : x \in Q\}$.

We claim that $GF = \mathrm{id}_{\mathcal{C}}$. Let $C \in \mathcal{C}$. Then

$$G(F(C)) = G(\delta_C) = \{\delta_C(x)x : x \in Q\} = C,$$

by construction. (We know that $\delta_C(x)x \in C$. Conversely, if $c \in C$, we write $c = kx$ for unique elements $k \in K$ and $x \in Q$. Thus $x = k^{-1}c$ and hence $c = \delta_C(x)x$.)

Finally, we prove that $FG = \mathrm{id}_{\mathrm{Der}(Q,K)}$. Let $\delta \in \mathrm{Der}(Q, K)$. Then

$$F(G(\delta)) = F(\Delta) = \delta_\Delta.$$

Finally, we need to show that $\delta_\Delta = \delta$. Let $x \in Q$. There exists $\delta(y)y \in \Delta$ for some $y \in Q$ such that $x = k^{-1}\delta(y)y$. Thus $\delta_\Delta(x)x = \delta(y)y$ and hence $x = y$ and $\delta_\Delta(x) = \delta(y)$ by the uniqueness. Therefore, $\delta_\Delta = \delta$, and the result follows. □

Let K, Q be groups and let $\alpha\colon Q \to \mathrm{Aut}(K)$ be a group homomorphism. A derivation $\delta \in \mathrm{Der}(Q, K)$ is said to be *inner* if there exists $k \in K$ such that $\delta(x) = [k, x]$ for all $x \in Q$. The set of *inner derivations* will be denoted by

$$\mathrm{Inn}(Q, K) = B^1(Q, K) = \{\delta \in \mathrm{Der}(Q, K) : \delta \text{is inner}\}.$$

An inner derivation is also called a 1-*coboundary*.

Theorem 7.1.8 (Sysak) *Let K, Q be groups and let $\alpha\colon Q \to \mathrm{Aut}(K)$ be a group homomorphism. Let $\delta \in \mathrm{Der}(Q, K)$.*

(1) $\Delta = \{\delta(x)x : x \in Q\}$ *is a complement of K in $K \rtimes Q$.*
(2) $\delta \in \mathrm{Inn}(Q, K)$ *if and only if $\Delta = kQk^{-1}$ for some $k \in K$.*
(3) $\ker \delta = Q \cap \Delta$.
(4) *δ is surjective if and only if $K \rtimes Q = \Delta Q$.*

Proof In the proof of Theorem 7.1.7 we found that Δ is a complement of K in $K \rtimes Q$.

Let us prove the second statement. If δ is inner, then there exists $k \in K$ such that $\delta(x) = [k, x] = kxk^{-1}x^{-1}$ for all $x \in Q$. Since $\delta(x)x = kxk^{-1}$ for all $x \in Q$, $\Delta = kQk^{-1}$. Conversely, if there exists $k \in K$ such that $\Delta = kQk^{-1}$, for each $x \in Q$ there exists $y \in Q$ such that $\delta(x)x = kyk^{-1}$. Since $[k, y] = kyk^{-1}y^{-1} \in K$, $\delta(x) \in K$ and $\delta(x)x = [k, y]y \in KQ$, we conclude that $x = y$ and hence $\delta(x) = [k, x]$.

Let us prove the third statement. If $x \in Q$ is such that $\delta(x)x = y \in Q$, then

$$\delta(x) = yx^{-1} \in K \cap Q = \{1\}.$$

Conversely, if $x \in Q$ is such that $\delta(x) = 1$, then $x = \delta(x)x \in Q \cap \Delta$.

Finally, we prove the fourth statement. If δ is surjective, then for each $k \in K$ there exits $y \in Q$ such that $\delta(y) = k$. Thus $K \rtimes Q \subseteq \Delta Q$, as

$$kx = \delta(y)x = (\delta(y)y)y^{-1}x \in \Delta Q$$

for all $x \in Q$. Since Δ and Q are subgroups of $K \rtimes Q$, we have that $\Delta Q \subseteq K \rtimes Q$, and therefore $\Delta Q = K \rtimes Q$. Conversely, if $k \in K$ and $x \in Q$ there exist $y, z \in Q$ such that $kx = \delta(y)yz$. Then it follows that $k = \delta(y)$. □

A group G admits a *triple factorization* if there are subgroups A, B, and M such that $G = MA = MB = AB$ and $A \cap M = B \cap M = \{1\}$. The following result is an immediate consequence of Sysak's theorem.

Corollary 7.1.9 *If the group Q acts by automorphisms on K and $\delta \in \mathrm{Der}(Q, K)$ is surjective, then $G = K \rtimes Q$ admits a triple factorization.* □

7.2 Schur–Zassenhaus Theorem

In this section, we shall prove Schur–Zassenhaus theorem about complements of a normal subgroup N of a finite group G.

Lemma 7.2.1 *Let G be a group and N be a normal subgroup of G. If G acts on N by conjugation and $\varphi\colon G \to N$ is a 1-cocycle with kernel K, then $\varphi(x) = \varphi(y)$ if and only if $xK = yK$. In particular, $(G : K) = |\varphi(G)|$.*

Proof If $\varphi(x) = \varphi(y)$, then, since

$$\varphi(x^{-1}y) = \varphi(x^{-1})x^{-1}\varphi(y)x = \varphi(x^{-1})x^{-1}\varphi(x)x = \varphi(x^{-1}x) = \varphi(1) = 1,$$

we obtain that $xK = yK$. Conversely, if $x^{-1}y \in K$, then, since

$$\varphi(x^{-1})x^{-1}\varphi(x)x = \varphi(x^{-1}x) = 1 = \varphi(x^{-1}y) = \varphi(x^{-1})x^{-1}\varphi(y)x,$$

we conclude that $\varphi(x) = \varphi(y)$.

The second claim is now trivial, as φ is constant in each coset of K and there are $(G : K)$ different possible values. □

Lemma 7.2.2 *Let G be a finite group; N be a normal abelian subgroup of G; and S, T, and U be transversals of N in G. Let*

$$d(S, T) = \prod st^{-1} \in N,$$

where the product is taken over all $s \in S$ and $t \in T$ such that $sN = tN$. The following statements hold:

(1) $d(S, T)d(T, U) = d(S, U)$.
(2) $d(gS, gT) = gd(S, T)g^{-1}$ *for all* $g \in G$.
(3) $d(nS, S) = n^{(G:N)}$ *for all* $n \in N$.

Proof If $s \in S$, $t \in T$ and $u \in U$ are such that $sN = tN = uN$, then, since N is abelian and $(st^{-1})(tu^{-1}) = su^{-1}$,

$$d(S, T)d(T, U) = \prod (st^{-1})(tu^{-1}) = \prod su^{-1} = d(S, U).$$

Since $sN = tN$ if and only if $gsN = gtN$ for all $g \in G$,

$$g\left(\prod st^{-1}\right)g^{-1} = \prod gst^{-1}g^{-1} = \prod (gs)(gt)^{-1} = d(gS, gT).$$

Finally, since N is normal, $nsN = sN$ for all $n \in N$. Thus

$$d(nS, S) = \prod (ns)s^{-1} = n^{(G:N)}.$$

□

We now prove the first version of Schur–Zassenhaus theorem.

Theorem 7.2.3 (Schur–Zassenhaus) *Let G be a finite group and let N be an abelian normal subgroup of G. If $|N|$ and $(G : N)$ are coprime, then N admits a complement in G. In this case, all complements of N are conjugate in G.*

Proof Let T be a transversal of N in G. Let $\theta\colon G \to N$, $\theta(g) = d(gT, T)$. Since N is abelian, Lemma 7.2.2 implies that θ is a 1-cocycle, where G acts on N by conjugation:

$$\begin{aligned}\theta(xy) &= d(xyT, T) = d(xyT, xT)d(xT, T)\\ &= (xd(yT, T)x^{-1})d(xT, T) = (x\theta(y)x^{-1})\theta(x).\end{aligned}$$

Claim. $\theta|_N\colon N \to N$ is surjective.

If $n \in N$, then $\theta(n) = d(nT, T) = n^{(G:N)}$ by Lemma 7.2.2. Since $|N|$ and $(G : N)$ are coprime, there exist $r, s \in \mathbb{Z}$ such that $r|N| + s(G : N) = 1$. Thus

$$n = n^{r|N|+s(G:N)} = (n^s)^{(G:N)} = \theta(n^s).$$

Let $H = \ker\theta$. We claim that H is a complement for N. We know that H is a subgroup of G. Since, by Lemma 7.2.1,

$$|N| = |\theta(G)| = (G : H) = \frac{|G|}{|H|},$$

it follows that $N \cap H = \{1\}$ because $|N|$ and $(G : N) = |H|$ are coprime. Since $|NH| = |N||H| = |G|$, we conclude that $G = NH$ and hence H is a complement of N in G.

We now prove that two complements of N in G are conjugate. Let K be a complement of N in G. Since $NK = G$ and $N \cap K = \{1\}$, it follows that K is a transversal of N in G. Let $m = d(T, K) \in N$. Since $\theta|_N$ is surjective, there exists $n \in N$ such that $\theta(n) = m$. By Lemma 7.2.2, for each $k \in K$,

$$kmk^{-1} = kd(T, K)k^{-1} = d(kT, kK) = d(kT, K) = d(kT, T)d(T, K) = \theta(k)m$$

holds. Since N is abelian, $\theta(n^{-1}) = m^{-1}$ and hence

$$\begin{aligned}\theta(nkn^{-1}) &= \theta(n)n\theta(kn^{-1})n^{-1} = m\theta(kn^{-1})\\ &= m\theta(k)k\theta(n^{-1})k^{-1} = m\theta(k)km^{-1}k^{-1} = 1.\end{aligned}$$

Therefore $nKn^{-1} \subseteq H = \ker\theta$. Since $|K| = (G : N) = |H|$, $nKn^{-1} = H$. □

The general version of Schur–Zassenhaus theorem does not need N to be abelian.

Theorem 7.2.4 (Schur–Zassenhaus) *Let G be a finite group and let N be a normal subgroup of G. If $|N|$ and $(G : N)$ are coprime, then N admits a complement in G.*

Proof We proceed by induction on $|G|$. Suppose that $|G| > 1$ and the result holds for all finite groups of order $< |G|$. If there exists a proper subgroup K of G such that $NK = G$, then, since $(K : K \cap N) = (G : N)$ is coprime with $|N|$, it is coprime with $|K \cap N|$. Moreover, $K \cap N$ is normal in K. By the inductive hypothesis, $K \cap N$ admits a complement in K. Hence there exists a subgroup H of K such that

$$|H| = (K : K \cap N) = (G : N).$$

Therefore H is a complement of N in G.

Hence we may assume that G has no proper subgroups K such that $NK = G$. We may also assume that $N \neq \{1\}$, otherwise G is a complement of $\{1\}$ in G. Note that $N \subseteq \Phi(G)$, because if M is a maximal subgroup of G such that $N \not\subseteq M$, then $NM = G$, in contradiction with our assumption. By Frattini's Theorem 5.3.8, $\Phi(G)$ is nilpotent and thus so is N. Since N is a non-trivial nilpotent group, we have that $Z(N) \neq \{1\}$. Since N is normal in G and $Z(N)$ is characteristic in N, $Z(N)$ is normal in G. Let $\pi : G \to G/Z(N)$ the canonical map. We have that

$$(\pi(G) : \pi(N)) = \frac{|\pi(G)|}{|\pi(N)|} = \frac{|G/Z(N)|}{|N/Z(N)|} = (G : N)$$

is coprime with $|N|$, and thus it is coprime with $|\pi(N)|$. By the inductive hypothesis, $\pi(N)$ admits a complement in $G/Z(N)$. Let K be a subgroup of G such that $Z(N) \subseteq K$ and $\pi(K)$ is a complement of $\pi(N)$ in $\pi(G)$. Since $\pi(G) = \pi(N)\pi(K) = \pi(NK)$, we have that $NK = G$. By assumption, we get that $K = G$ and

$$\pi(Z(N)) = \pi(N) \cap \pi(K) = \pi(N) \cap \pi(G) = \pi(N).$$

Hence $N = Z(N)$ is abelian. By Theorem 7.2.3, we get a contradiction. Hence N admits a complement in G, and the result follows by induction. □

Theorem 7.2.5 *Let G be a finite group and N be a normal subgroup of G such that $|N|$ and $(G : N)$ are coprime. If either N or G/N is solvable, then all complements of N in G are conjugate.*

Proof Let G be a minimal counterexample, so there are complements K_1 and K_2 of N in G such that K_1 and K_2 are not conjugate and $|G|$ is minimal with this property.

Claim. Each subgroup U of G satisfies the hypotheses of the theorem with respect to the normal subgroup $U \cap N$.

Since N is normal in G, the subgroup $U \cap N$ is normal in U. Moreover, $|U \cap N|$ and $(U : U \cap N)$ are coprime, as $|U \cap N|$ divides $|N|$ and $(U : U \cap N) = (UN : N)$ divides $(G : N)$. If G/N is solvable, then $U/U \cap N$ is solvable since $U/U \cap N$ is isomorphic to a subgroup of G/N. If N is solvable, the subgroup $U \cap N$ is solvable. This proves the claim.

Claim. Let L be a normal subgroup of G and let $\pi : G \to G/L$ be the canonical map. Then $\pi(G)$ satisfies the hypotheses of the theorem with respect to $\pi(N) = \pi(NL)$. In this case, if H is a complement of N in G, then $\pi(H)$ is a complement of $\pi(N)$ in $\pi(G)$.

If N is solvable, then $\pi(N)$ is solvable. If G/N is solvable, then the group $\pi(G)/\pi(N) \cong G/NL$ is solvable. Moreover, $(\pi(G) : \pi(N)) = \frac{|G/L|}{|NL/L|} = (G : NL)$ divides the index $(G : N)$ of N in G.

If H is a complement of N in G, $|\pi(H)|$ and $|\pi(N)|$ are coprime. Thus $\pi(H)$ is a complement of $\pi(N)$, as $\pi(G) = \pi(NH) = \pi(N)\pi(H)$ and $\pi(N) \cap \pi(H) = \{1\}$. Thus the claim follows.

Claim. N is minimal normal in G.

Let $M \neq \{1\}$ be a normal subgroup of G such that $M \subseteq N$. Let $\pi : G \to G/M$ be the canonical map. The group $\pi(G)$ satisfies the hypotheses of the theorem with respect to the normal subgroup $\pi(N)$. By the minimality of $|G|$, there exists $x \in G$ such that $\pi(xK_1x^{-1}) = \pi(K_2)$. The subgroup $U = MK_2$ satisfies the hypotheses of the theorem with respect to the normal subgroup $U \cap N$. Since $xK_1x^{-1} \cup K_2 \subseteq U$, we conclude that xK_1x^{-1} and K_2 are complements of $U \cap N$ in U. Thus $U = MK_2 = G$, as xK_1x^{-1} and K_2 are not conjugate and $|G|$ is minimal. Therefore $M = N$, as

$$|K_2| = \frac{|K_2|}{|M \cap K_2|} = (MK_2 : M) = (G : M) = \frac{|NK_2|}{|M|} = (N : M)|K_2|.$$

Hence N is a minimal normal subgroup of G.

Claim. N is not solvable and G/N is solvable.

Otherwise, by Lemma 6.1.1, N is minimal normal and hence abelian. This yields a contradiction with Theorem 7.2.3.

Let $\pi : G \to G/N$ be the canonical map and let S be a subgroup of G such that $N \subseteq S$ and $\pi(S)$ is minimal normal in $\pi(G) = G/N$. By Lemma 6.1.1, $\pi(S)$ is a p-group for some prime number p. Since $G = NK_1 = NK_2$ and $N \subseteq S$, Lemma 5.3.3 implies that

$$S = N(S \cap K_1) = N(S \cap K_2).$$

Thus $S \cap K_1$ and $S \cap K_2$ are complements of N in S. Since

$$\pi(S) = \pi(S \cap K_1) = \pi(S \cap K_2)$$

is a p-group, p divides $|S|$. The group S satisfies the hypotheses of the theorem with respect to the normal subgroup N, so $|N|$ and $(S : N)$ are coprime. If $p \mid |N|$, then $p \nmid (S : N) = |\pi(S)|$, a contradiction. Therefore $p \nmid |N|$. This implies that $S \cap K_1$ and $S \cap K_2$ are Sylow p-subgroups of S, as

$$|S \cap K_1| = (S : N) = |S \cap K_2|.$$

By Sylow's theorem, there exists $s \in S$ such that

$$S \cap sK_1s^{-1} = S \cap K_2.$$

In particular, $S \neq G$. Let

$$L = S \cap K_2 = S \cap sK_1s^{-1} \neq \{1\}.$$

Since S is normal in G, it follows that $sK_1s^{-1} \cup K_2 \subseteq N_G(L)$ (because L is normal both in sK_1s^{-1} and in K_2). The subgroups $sK_1s^{-1} \subseteq N_G(L)$ and $K_2 \subseteq N_G(L)$ are complements of $N \cap N_G(L)$ in $N_G(L)$. Thus $N_G(L) = G$ by the minimality of $|G|$. Therefore L is normal in G. Let $\pi_L : G \to G/L$ be the canonical map. Since both $\pi_L(K_1)$ and $\pi_L(K_2)$ are complements of $\pi_L(N)$ in $\pi_L(G)$, the minimality of $|G|$ implies that there exists $g \in G$ such that $\pi_L(gK_1g^{-1}) = \pi_L(K_2)$, so there exists $g \in G$ such that $(gK_1g^{-1})L = K_2L$. Thus $gK_1g^{-1} \cup K_2 \subseteq \langle K_2, L\rangle = K_2$, as $L \subseteq K_2$. In conclusion, $gK_1g^{-1} = K_2$, a contradiction. Therefore the result follows. □

By Feit–Thompson theorem (Theorem 6.2.19), we do not need to assume that N or G/N is solvable. Indeed, since every group of odd order is solvable and $|N|$ and $(G : N)$ are coprime, it follows that either $|N|$ or $(G : N)$ is odd.

Corollary 7.2.6 *Let G be a finite group and let N be a normal subgroup of G of order n. Suppose that either N or G/N is solvable. If $(G : N) = m$ is coprime with n and m_1 is a divisor of m, then every subgroup of G of order m_1 is contained in some subgroup of order m of G.*

Proof By Schur–Zassenhaus theorem, there exists a complement H of N in G. Hence $|H| = m$. Let H_1 be a subgroup of G such that $|H_1| = m_1$. Since n and m are coprime,

$$\frac{|H||N||H_1|}{|H \cap (NH_1)|} = \frac{|H||NH_1|}{|H \cap (NH_1)|} = |H(NH_1)| = |G| = |NH| = |N||H|.$$

Hence $m_1 = |H_1| = |H \cap (NH_1)|$, and thus H_1 and $H \cap (NH_1)$ are complements of N in NH_1. By Theorem 7.2.5, there exists $g \in NH_1$ such that

$$H_1 = g(H \cap NH_1)g^{-1} \subseteq gHg^{-1},$$

and clearly $|gHg^{-1}| = |H| = m$. □

7.3 Exercises

7.3.1 Let $E: \ 1 \longrightarrow K \longrightarrow G \xrightarrow{p} Q \longrightarrow 1$ be an extension. Prove the following statements:

(1) If $\ell\colon Q \to G$ is a lifting, then $\ell(Q)$ is a transversal of $\ker p$ in G.
(2) Each transversal of $\ker p$ in G induces a lifting $\ell\colon Q \to G$.
(3) If $\ell\colon Q \to G$ is a lifting, then $\ell(xy)\ker p = \ell(x)\ell(y)\ker p$.

7.3.2 Let $\varphi\colon Q \to K$ be a 1-cocycle with respect to the action $\alpha\colon Q \to \mathrm{Aut}(K)$. Prove the following statements:

(1) $\varphi(1) = 1$.
(2) $\varphi(y^{-1}) = (\alpha_{y^{-1}}(\varphi(y)))^{-1} = \alpha_{y^{-1}}(\varphi(y)^{-1})$, where $\alpha_x = \alpha(x)$ for all $x, y \in Q$.
(3) The set $\ker\varphi = \{x \in Q : \varphi(x) = 1\}$ is a subgroup of Q.

7.3.3 Let $\delta \in \mathrm{Der}(Q, K)$.

(1) Prove that δ is injective if and only if $\ker\delta = \{1\}$.
(2) Prove that if δ is bijective, then K admits a complement Δ in $K \rtimes Q$ such that $K \rtimes Q = K \rtimes \Delta = \Delta Q$ and $Q \cap \Delta = \{1\}$.

7.4 Notes

The results in this chapter are classical and can be found for example in [97]. Our presentation of Schur–Zassenhaus theorem is based on that of [68].

Chapter 8
Skew Braces

In this chapter, we develop the basic definitions and tools for studying skew left braces and their relation with bijective 1-cocycles of groups. We also study the solution to the YBE associated with a skew left brace.

8.1 Basic Definitions

By convention, an additive group A will be a (not necessarily abelian) group with a binary operation $(a, b) \mapsto a + b$. As usual in additive notation, the neutral element of A will be denoted by 0, and the additive inverse of an element a will be denoted by $-a$.

Definition 8.1.1 A *skew left brace* is a triple $(A, +, \circ)$, where $(A, +)$ and $(A, \circ)$ are (not necessarily abelian) groups and

$$a \circ (b + c) = (a \circ b) - a + (a \circ c) \tag{8.1}$$

holds for all $a, b, c \in A$. The groups $(A, +)$ and $(A, \circ)$ are, respectively, the *additive* and *multiplicative* group of the skew left brace A.

As usual, multiplication is granted higher precedence than addition. This means that one can write (8.1) simply as $a \circ (b + c) = a \circ b - a + a \circ c$. We write a' to denote the inverse of a for the circle operation $\circ$.

Skew right braces are defined similarly; one needs to replace (8.1) by

$$(a + b) \circ c = a \circ c - c + b \circ c.$$

F. Cedó and L. Vendramin, *Groups, Radical Rings, and the Yang-Baxter Equation*, Progress in Mathematics 361, https://doi.org/10.1007/978-3-032-01524-2_8

There is a bijective correspondence between skew left braces and skew right braces; see Exercise 8.6.1. For that reason, a skew brace will always mean a skew left brace.

Definition 8.1.2 Let $\mathcal{X}$ be a family of groups. A skew brace A is said to be of $\mathcal{X}$-type if its additive group belongs to $\mathcal{X}$.

One exciting family of skew braces is the family of *skew braces of abelian type*, that is skew braces with abelian additive group. Skew braces of abelian type were introduced by Rump in [99] to study involutive solutions to the YBE. In the literature, skew braces of abelian type are called *left braces*.

Example 8.1.3 Let A be an additive group. Then A is a skew brace with $a \circ b = a + b$ for all $a, b \in A$. A skew brace $(A, +, \circ)$ such that $a \circ b = a + b$ for all $a, b \in A$ is said to be *trivial*. Similarly, the operation $a \circ b = b + a$ turns A into a skew brace.

Example 8.1.4 Let A and B be skew braces. Then $A \times B$ with

$$(a, b) + (a_1, b_1) = (a + a_1, b + b_1), \quad (a, b) \circ (a_1, b_1) = (a \circ a_1, b \circ b_1)$$

is a skew brace. This is the *direct product* of the skew braces A and B.

Example 8.1.5 Let A and M be additive groups and $\alpha\colon A \to \mathrm{Aut}(M)$ be a group homomorphism. We write $\alpha(a) = \alpha_a$ for all $a \in A$. Then $M \times A$ with

$$(x, a) + (y, b) = (x + y, a + b), \quad (x, a) \circ (y, b) = (x + \alpha_a(y), a + b)$$

is a skew brace. Similarly, $M \times A$ with

$$(x, a) + (y, b) = (x + \alpha_a(y), a + b), \quad (x, a) \circ (y, b) = (x + y, b + a)$$

is a skew brace.

Example 8.1.6 Let A be the symmetric group $\mathbb{S}_3$ written additively. For example, $(12) + (13) = (132)$. Let $\lambda\colon A \to \mathrm{Aut}(A)$ be the map given by

$$\begin{aligned} \lambda_{\mathrm{id}} &= \lambda_{(123)} = \lambda_{(132)} = \mathrm{id}, \\ \lambda_{(12)} &= \lambda_{(23)} = \lambda_{(13)} = \text{conjugation by } (23), \end{aligned}$$

where $\lambda_a = \lambda(a)$ for all $a \in A$. The operation $a \circ b = a + \lambda_a(b)$ turns A into a skew brace with additive group $\mathbb{S}_3$ and multiplicative group a cyclic group of order six. The multiplicative group of A is generated by (12).

Example 8.1.7 Let A be an additive group and B and C be subgroups of A such that $B \cap C = \{0\}$ and $A = B + C$. In this case, one says that A admits an *exact factorization* through the subgroups B and C. Thus each $a \in A$ can be written uniquely as $a = b + c$, for some $b \in B$ and $c \in C$. The map

$$f: B \times C^{\mathrm{op}} \to A, \quad (b, c) \mapsto b + c$$

is bijective. Since f is bijective, A is a group with the operation

$$a \circ a_1 = f\left(f^{-1}(a) f^{-1}(a_1)\right). \tag{8.2}$$

To compute (8.2) write $a = b + c$ and $a_1 = b_1 + c_1$ for $b, b_1 \in B$ and $c, c_1 \in C$. Then

$$\begin{aligned} a \circ a_1 &= f\left((b, c)(b_1, c_1)\right) \\ &= f\left((b + b_1, c_1 + c)\right) = b + (b_1 + c_1) + c = b + a_1 + c. \end{aligned}$$

It follows that $(A, \circ)$ is a group isomorphic to $B \times C^{\mathrm{op}} \cong B \times C$. Moreover, if $x, y \in A$, then

$$a \circ x - a + a \circ y = b + x + c - (b + c) + b + y + c = b + x + y + c = a \circ (x + y),$$

and therefore $(A, +, \circ)$ is a skew brace.

We now give some concrete examples of the previous construction.

Example 8.1.8 Let n be a positive integer. For

$$A = \begin{pmatrix} a_{1,1} & a_{1,2} & \cdots & a_{1,n} \\ a_{2,1} & a_{2,2} & \cdots & a_{2,n} \\ \vdots & \vdots & \ddots & \vdots \\ a_{n,1} & a_{n,2} & \cdots & a_{n,n} \end{pmatrix} \in \mathbf{GL}_n(\mathbb{C}),$$

let

$$A^* = \bar{A}^t = \begin{pmatrix} \bar{a}_{1,1} & \bar{a}_{2,1} & \cdots & \bar{a}_{n,1} \\ \bar{a}_{1,2} & \bar{a}_{2,2} & \cdots & \bar{a}_{n,2} \\ \vdots & \vdots & \ddots & \vdots \\ \bar{a}_{n,1} & \bar{a}_{2,n} & \cdots & \bar{a}_{n,n} \end{pmatrix},$$

where $\overline{a + bi} = a - bi$ for all $a, b \in \mathbb{R}$. The group $\mathbf{GL}_n(\mathbb{C})$ admits an exact factorization through the subgroups $U(n)$ and $T(n)$, where $U(n) = \{A \in \mathbf{GL}_n(\mathbb{C}) : AA^* = I\}$ is the unitary group and $T(n)$ is the group of upper triangular matrices with positive diagonal entries. Therefore there exists a skew brace with additive group isomorphic to $\mathbf{GL}_n(\mathbb{C})$ and multiplicative group isomorphic to $U(n) \times T(n)$.

The following examples appeared in the theory of Hopf–Galois structures, see [29, Corollary 1.1].

Example 8.1.9 The alternating simple group $\mathbb{A}_5$ admits an exact factorization through the subgroups $A = \langle(123), (12)(34)\rangle \cong \mathbb{A}_4$ and $B = \langle(12345)\rangle \cong C_5$. There exists a skew brace with additive group isomorphic to $\mathbb{A}_5$ and multiplicative group isomorphic to $\mathbb{A}_4 \times C_5$.

Example 8.1.10 The simple group $\mathbf{PSL}_2(7) = \mathbf{SL}_2(7)/Z(\mathbf{SL}_2(7))$ admits an exact factorization through the subgroups

$$A = \left\langle \begin{pmatrix} 0 & 1 \\ -1 & -1 \end{pmatrix} Z(\mathbf{SL}_2(7)), \begin{pmatrix} 1 & -2 \\ 1 & -1 \end{pmatrix} Z(\mathbf{SL}_2(7)) \right\rangle \cong \mathbb{S}_4$$

and

$$B = \left\langle \begin{pmatrix} 1 & 1 \\ 0 & 1 \end{pmatrix} Z(\mathbf{SL}_2(7)) \right\rangle \cong C_7.$$

Hence there exists a skew brace with additive group isomorphic to $\mathbf{PSL}_2(7)$ and multiplicative group isomorphic to $\mathbb{S}_4 \times C_7$.

Lemma 8.1.11 *Let A be a skew brace. Then the following properties hold:*

(1) $0 = 1$.
(2) $a \circ (-b + c) = a - (a \circ b) + (a \circ c)$ *for all* $a, b, c \in A$.
(3) $a \circ (b - c) = (a \circ b) - (a \circ c) + a$ *for all* $a, b, c \in A$.

Proof The first claim follows from the compatibility condition (8.1) with $c = 1$. To prove the second claim, let $d = b + c$. Then (8.1) becomes

$$a \circ d = a \circ b - a + a \circ (-b + d)$$

and the claim follows. The third claim is proved similarly. □

Proposition 8.1.12 *Let A be a skew brace. For each $a \in A$, the map*

$$\lambda_a \colon A \to A, \quad b \mapsto -a + (a \circ b)$$

is an automorphism of $(A, +)$. Moreover, the map $\lambda \colon (A, \circ) \to \operatorname{Aut}(A, +)$, $a \mapsto \lambda_a$ is a group homomorphism.

Proof The inverse of λ_a is given by $\lambda_a^{-1} \colon A \to A$, $b \mapsto a' \circ (a + b)$. To prove that $\lambda_a \in \operatorname{Aut}(A, +)$ we note that

$$\lambda_a(b + c) = -a + a \circ (b + c) = -a + a \circ b - a + a \circ c = \lambda_a(b) + \lambda_a(c).$$

Note that $\lambda_a(b) = -a + a \circ b = a \circ (a' + b)$ for all $a, b \in A$. Hence

$$\begin{aligned} \lambda_a(\lambda_b(c)) &= a \circ (a' + b \circ (b' + c)) = -a + a \circ b \circ (b' + c) \\ &= -a + a - a \circ b + a \circ b \circ c = -a \circ b + a \circ b \circ c = \lambda_{a \circ b}(c). \end{aligned}$$

□

If A is a skew brace, then the map λ in the previous proposition yields a left action from $(A, \circ)$ on $(A, +)$ by automorphisms.

The following formulas

$$a \circ b = a + \lambda_a(b), \qquad a + b = a \circ \lambda_a^{-1}(b), \qquad \lambda_a(a') = -a \tag{8.3}$$

hold for $a, b \in A$. Moreover, if

$$a * b = \lambda_a(b) - b = -a + a \circ b - b,$$

then the following identities are easily verified:

$$a * (b + c) = a * b + b + a * c - b, \tag{8.4}$$

$$(a \circ b) * c = (a * (b * c)) + b * c + a * c. \tag{8.5}$$

These last two identities are similar to the usual commutator identities (5.1) and (5.2).

There is also a right action of $(A, \circ)$ on the set A.

Proposition 8.1.13 *Let A be a skew brace. For each $a \in A$, the map*

$$\mu_a \colon A \to A, \quad b \mapsto \lambda_b(a)' \circ b \circ a,$$

is bijective. Moreover, the map $\mu \colon (A, \circ) \to \mathbb{S}_A$, $a \mapsto \mu_a$ satisfies $\mu_b \mu_a = \mu_{a \circ b}$ for all $a, b \in A$.

Proof Note that

$$\mu_a(b) = \lambda_b(a)' \circ b \circ a = (b \circ (b' + a))' \circ b \circ a = (b' + a)' \circ a$$

for all $a, b \in A$. Hence μ_a is bijective and

$$\mu_a^{-1}(b) = ((b \circ a')' - a)' = (a \circ b' - a)' = (b' + a')' \circ a',$$

for all $a, b \in A$. Now we have

$$\begin{aligned} \mu_b(\mu_a(c)) &= \mu_b((c' + a)' \circ a) = (a' \circ (c' + a) + b)' \circ b \\ &= (a' \circ c' - a' + b)' \circ b = (a' \circ (c' + a \circ b))' \circ b \\ &= (c' + a \circ b)' \circ a \circ b = \mu_{a \circ b}(c) \end{aligned}$$

for all $a, b, c \in A$. Therefore the result follows. □

Definition 8.1.14 A map $f \colon A \to B$ between two skew braces A and B is a *homomorphism of skew braces* if $f(x \circ y) = f(x) \circ f(y)$ and $f(x + y) = f(x) + f(y)$ for all $x, y \in A$. The *kernel* of f is

$$\ker f = \{a \in A : f(a) = 0\}.$$

We write $\mathrm{Hom}_b(A, B)$ to denote the set of skew brace homomorphisms $A \to B$. A bijective homomorphism of skew braces is an *isomorphism*. An *automorphism* of a skew brace A is an isomorphism from the skew brace A to itself. We write $\mathrm{Aut}_b(A)$ to denote the group of skew brace automorphisms of A. Two skew braces A and B are isomorphic if there exists an isomorphism $f\colon A \to B$. We write $A \cong B$ when the skew braces A and B are isomorphic.

Proposition 8.1.15 *Let A, B be skew braces. Let $\alpha\colon (B, \circ) \to \mathrm{Aut}_b(A)$ be a homomorphism of groups. Define two operations on $A \times B$ by*

$$(a_1, b_1) + (a_2, b_2) = (a_1 + a_2, b_1 + b_2) \text{ and } (a_1, b_1) \circ (a_2, b_2) = (a_1 \circ \alpha_{b_1}(a_2), b_1 \circ b_2),$$

where $\alpha_{b_1} = \alpha(b_1)$ for all $a_1, a_2 \in A$ and $b_1, b_2 \in B$. Then $(A \times B, +, \circ)$ is a skew brace. This skew brace is the semidirect product *of the skew brace A by B via α, and it is denoted by $A \rtimes_\alpha B$.*

Proof Note that $(A \times B, +)$ is the direct product of the groups $(A, +)$ and $(B, +)$, and $(A \times B, \circ)$ is the semidirect product of the group $(A, \circ)$ by $(B, \circ)$ via α. Hence it is enough to check the compatibility condition. Let $a_1, a_2, a_3 \in A$ and $b_1, b_2, b_3 \in B$. We have

$$\begin{aligned}
(a_1, b_1)\circ&((a_2, b_2) + (a_3, b_3))\\
&= (a_1, b_1) \circ (a_2 + a_3, b_2 + b_3)\\
&= (a_1 \circ \alpha_{b_1}(a_2 + a_3), b_1 \circ (b_2 + b_3))\\
&= (a_1 \circ (\alpha_{b_1}(a_2) + \alpha_{b_1}(a_3)), b_1 \circ b_2 - b_1 + b_1 \circ b_3)\\
&= (a_1 \circ \alpha_{b_1}(a_2) - a_1 + a_1 \circ \alpha_{b_1}(a_3)), b_1 \circ b_2 - b_1 + b_1 \circ b_3)\\
&= (a_1 \circ \alpha_{b_1}(a_2), b_1 \circ b_2) - (a_1, b_1) + (a_1 \circ \alpha_{b_1}(a_3)), b_1 \circ b_3)\\
&= (a_1, b_1) \circ (a_2, b_2) - (a_1, b_1) + (a_1, b_1) \circ (a_2, b_3).
\end{aligned}$$

Therefore the result follows. □

Lemma 8.1.16 *Let A and B be skew braces. A map $f\colon A \to B$ is a homomorphism of skew braces if and only if $f(a \circ b) = f(a) \circ f(b)$ and $f(\lambda_a(b)) = \lambda_{f(a)}(f(b))$ for all $a, b \in A$.*

Proof Let $a, b \in A$. Suppose that f is a homomorphism of skew braces. Then

$$f(\lambda_a(b)) = f(-a + a \circ b) = -f(a) + f(a) \circ f(b) = \lambda_{f(a)}(f(b)).$$

Conversely, suppose that $f(x \circ y) = f(x) \circ f(y)$ and $f(\lambda_x(y)) = \lambda_{f(x)}(f(y))$ for all $x, y \in A$. We have that

$$f(a + b) = f(a \circ \lambda_{a'}(b)) = f(a) \circ f(\lambda_{a'}(b)) = f(a) \circ \lambda_{f(a)'}(f(b)) = f(a) + f(b).$$

Therefore the result follows. □

Definition 8.1.17 A skew brace A is said to be a *two-sided skew brace* if

$$(a+b)\circ c = a\circ c - c + b\circ c \tag{8.6}$$

holds for all $a, b, c \in A$.

If A is a two-sided skew brace, then

$$a\circ(-b) = a - a\circ b + a, \qquad (-a)\circ b = b - a\circ b + b \tag{8.7}$$

hold for all $a, b \in A$. The first equality holds for every skew brace and follows from Lemma 8.1.11. The second equality follows from (8.6).

Example 8.1.18 Any skew brace with abelian multiplicative group is two sided.

Example 8.1.19 Let $n \in \mathbb{N}$ be such that $n = p_1^{a_1}\cdots p_k^{a_k}$, where the p_j are distinct primes, all $a_j \in \{1,2\}$ and $p_i^m \not\equiv 1 \bmod p_j$ for all i, j, m with $1 \le m \le a_i$. Then every skew brace of size n is a two-sided skew brace of abelian type, since every group of order n is abelian, see for example [91].

Theorem 8.1.20 *A skew brace of abelian type* $(A, +, \circ)$ *is two sided if and only if* $(A, +, *)$ *is a radical ring.*

Proof Assume first that A is a two-sided skew brace of abelian type. Then $(A, +)$ is an abelian group. Let us prove that the operation

$$a * b = -a + a\circ b - b$$

turns A into a radical ring. Left distributivity follows from the compatibility condition and the fact that the addition is commutative:

$$a*(b+c) = -a + a\circ(b+c) - (b+c) = -a + a\circ b - a + a\circ c - c - b = a*b + a*c.$$

Similarly, since A is two sided, one proves $(a+b)*c = a*c + b*c$. It remains to show that the operation $*$ is associative. On the one hand, using the first equality of (8.7) and the compatibility condition, we write

$$\begin{aligned} a*(b*c) &= a*(-b + b\circ c - c) \\ &= -a + a\circ(-b + b\circ c - c) - (-b + b\circ c - c) \\ &= -a + a\circ(-b) - a + a\circ(b\circ c) - a + a\circ(-c) + c - b\circ c + b \\ &= a\circ(b\circ c) - a\circ b - a\circ c - b\circ c + a + b + c, \end{aligned}$$

since the group $(A, +)$ is abelian. On the other hand, the second equality of (8.7) and Equality (8.6) imply that

$$(a*b)*c = (-a + a\circ b - b)*c = -(-a + a\circ b - b) + (-a + a\circ b - b)\circ c - c$$

$$= b - a \circ b + a + (-a) \circ c - c + (a \circ b) \circ c - c + (-b) \circ c - c$$
$$= (a \circ b) \circ c - a \circ b - a \circ c - b \circ c + a + b + c.$$

It then follows that the operation $*$ is associative.

Conversely, if A is a radical ring, say with ring multiplication $(a, b) \mapsto ab$, then $a \circ b = a + ab + b$ turns A into a two-sided skew brace of abelian type. In fact, since A is a radical ring, then $(A, +)$ is an abelian group and $(A, \circ)$ is a group. Moreover,

$$a \circ (b + c) = a + a(b + c) + (b + c) = a + ab + ac + b + c = a \circ b - a + a \circ c.$$

Similarly one proves $(a + b) \circ c = a \circ c - c + b \circ c$. □

A skew brace is said to be *associative* if the operation $(x, y) \mapsto x * y = \lambda_x(y) - y$ is associative. In [37, Question 2.1(2)], Cedó, Gateva-Ivanova and Smoktunowicz asked whether associative skew braces of abelian type are always radical rings. To answer this question, we need some lemmas.

Lemma 8.1.21 *If A is an associative skew brace of abelian type, then*

$$(-a) * b = -(a * b)$$

holds for all $a, b \in A$. In particular, $(-a) \circ b = 2b - (a \circ b)$ for all $a, b \in A$.

Proof Since A is of abelian type, Identity (8.5) implies that

$$\begin{aligned}(a * (-a)) * b &= (a * (-a) + a + (-a)) * b \\ &= (a \circ (-a)) * b \\ &= a * ((-a) * b) + (-a) * b + a * b.\end{aligned}$$

The associativity of A then implies that $(-a) * b = -(a * b)$. □

If A is a skew brace of abelian type, then one proves by induction that

$$a \circ \left(\sum_{i=1}^{n} b_i - \sum_{j=1}^{m} c_j \right) = \sum_{i=1}^{n} a \circ b_i - \sum_{j=1}^{m} a \circ c_j + (m - n + 1)a \qquad (8.8)$$

holds for all $a, b, c \in A$, see Exercise 8.6.8.

Theorem 8.1.22 *If $(A, +, \circ)$ is an associative skew brace of abelian type, then $(A, +, *)$ is a radical ring.*

Proof We need to prove that the right compatibility condition holds. Since A is associative, $(a * b) * c = a * (b * c)$ for all $a, b, c \in A$. Write the associativity condition between $a, b, c \in A$ as

$$(a \circ b - a - b) \circ c - (a \circ b - a - b) - c = a \circ (b \circ c - b - c) - a - (b \circ c - b - c),$$

which is equivalent to

$$a' \circ ((a \circ b - a - b) \circ c - a \circ b) = a' \circ (a \circ (b \circ c - b - c) - a - a - b \circ c + 2c).$$

By Lemma 8.1.21, this is equivalent to

$$a' \circ ((a \circ b - a - b) \circ c - a \circ b) = a' \circ (a \circ (b \circ c - b - c) - a - a + (-b) \circ c).$$

By using the formula (8.8), this is equivalent to

$$(b + a' \circ (-b)) \circ c - b + a' = b \circ c - b - c + a' \circ (-b) \circ c + a'.$$

Hence the associativity of $(A, *)$ is equivalent to

$$(b + a' \circ (-b)) \circ c + c = b \circ c + a' \circ (-b) \circ c. \tag{8.9}$$

Let $b, c \in A$. If $d \in A$, then there exists $a \in A$ such that $d = a' \circ (-b)$. Equality (8.9) implies that

$$(b + d) \circ c + c = b \circ c + d \circ c.$$

□

In the database of [64], one finds associative skew braces (of size 16) that are not two sided, see for example [80].

8.2 Associated Solution to the YBE

In Proposition 4.2.5 we used radical rings to produce examples of solutions of the YBE. A natural question arises: Does one need radical rings? Surprisingly, radical rings are just the tip of the iceberg.

Theorem 8.2.1 *Let A be a skew brace. Then (A, r_A), where*

$$r_A : A \times A \to A \times A, \quad r_A(x, y) = (-x + x \circ y, (-x + x \circ y)' \circ x \circ y)$$

is a solution to the YBE.

Proof By Theorem 4.1.15, since $x \circ y = (-x + x \circ y) \circ ((-x + x \circ y)' \circ x \circ y)$ for all $x, y \in A$, we only need to check that $x \triangleright y = \lambda_x(y) = -x + x \circ y$ is a left action of $(A, \circ)$ on the set A and that $x \triangleleft y = \mu_y(x) = (-x + x \circ y)' \circ x \circ y$ is a right action of $(A, \circ)$ on the set A. For the left action, we use Proposition 8.1.12 and for the right action we use Proposition 8.1.13. □

We say that (A, r_A) is the solution to the YBE associated with the skew brace A.

In Theorem 8.2.1, it is possible to prove that the solution is involutive if and only if the additive group of the skew brace is abelian. The next result generalizes this fact. We shall need a lemma.

Lemma 8.2.2 *Let A be a skew brace and (A, r_A) be its associated solution to the YBE. Then*

$$r_A^{2n}(a,b) = (-n(a\circ b)+a+n(a\circ b),\quad (-n(a\circ b)+a+n(a\circ b))'\circ a\circ b), \tag{8.10}$$

$$r_A^{2n+1}(a,b) = (-n(a\circ b)-a+(n+1)(a\circ b),\quad (-n(a\circ b)-a+(n+1)(a\circ b))'\circ a\circ b), \tag{8.11}$$

for all $n \geq 0$. Moreover, the following statements hold:

(1) $r_A^{2n} = \mathrm{id}$ *if and only if $a + nb = nb + a$ for all $a, b \in A$.*
(2) $r_A^{2n+1} = \mathrm{id}$ *if and only if $\lambda_a(b) = n(a\circ b) + a - n(a\circ b)$ for all $a, b \in A$.*

Proof First we shall prove (8.10) and (8.11) by induction on n. The case $n = 0$ is trivial for (8.10) and (8.11). Assume that the claim holds for some $n \geq 0$. By applying the map r_A to Equation (8.11) we obtain that

$$\begin{aligned} r_A^{2(n+1)}(a,b) &= r_A\big(-n(a\circ b)-a+(n+1)(a\circ b),\\ &\qquad (-n(a\circ b)-a+(n+1)(a\circ b))'\circ a\circ b\big)\\ &= \big(-(n+1)(a\circ b)+a+(n+1)(a\circ b),\\ &\qquad (-(n+1)(a\circ b)+a+(n+1)(a\circ b))'\circ a\circ b\big). \end{aligned}$$

By applying r_A again to this equality, we get

$$\begin{aligned} r_A^{2(n+1)+1}(a,b) &= r_A\big(-(n+1)(a\circ b)+a+(n+1)(a\circ b),\\ &\qquad (-(n+1)(a\circ b)+a+(n+1)(a\circ b))'\circ a\circ b\big)\\ &= \big(-(n+1)(a\circ b)-a+(n+2)(a\circ b),\\ &\qquad (-(n+1)(a\circ b)-a+(n+2)(a\circ b))'\circ a\circ b\big). \end{aligned}$$

Thus Equations (8.10) and (8.11) hold by induction. The other claims follow easily from Equations (8.10) and (8.11). □

Recall that the (minimal) *exponent* $\exp(G)$ of a finite group G is the least positive integer n such that $g^n = 1$ for all $g \in G$.

Theorem 8.2.3 *Let A be a finite skew brace with more than one element and let G be the additive group of A. If (A, r_A) is the solution to the YBE associated with A, then r_A has order $2\exp(G/Z(G))$.*

Proof Assume that r has order $2n + 1$. Since $r^{2n+1} = \mathrm{id}$, by applying Lemma 8.2.2, one obtains that $-a + (n+1)(a\circ b) = n(a\circ b) + a$ for all $a, b \in A$. In particular, for $b = 0$, we get $a = 0$ for all $a \in A$, a contradiction. Therefore we may assume that the order of the permutation r is $2n$, where

$$n = \min\{k \in \mathbb{Z} : k > 0 \text{ and } kb + a = a + kb \text{ for all } a, b \in A\}.$$

Now,

$$\begin{aligned} n &= \min\{k \in \mathbb{Z} : k > 0 \text{ and } kb \in Z(G) \text{ for all } b \in A\} \\ &= \min\{k \in \mathbb{Z} : k > 0 \text{ and } k(b + Z(G)) = Z(G) \text{ for all } b \in A\} = \exp(G/Z(G)). \end{aligned}$$

□

An immediate consequence is the following result.

Corollary 8.2.4 *Let A be a finite skew brace and (A, r_A) be its associated solution to the YBE. Then (A, r_A) is involutive if and only if A is of abelian type.* □

8.3 Bijective 1-Cocycles

Let A be an additive group and G be a group and let $\alpha\colon G \to \mathrm{Aut}(A)$ be a group homomorphism. We write $\alpha_x = \alpha(x)$ for all $x \in G$. A *bijective* 1*-cocycle* is a bijective map $\pi : G \to A$ such that

$$\pi(gh) = \pi(g) + \alpha_g(\pi(h)) \tag{8.12}$$

for all $g, h \in G$. We now prove the equivalence between skew braces and bijective 1-cocycles.

Theorem 8.3.1 *Over any additive group A the following data are equivalent:*

(1) A group G and a bijective 1-cocycle $\pi : G \to A$.
(2) A skew brace structure over A.

Proof Suppose that G is a group, $\alpha\colon G \to \mathrm{Aut}(A)$ is a group homomorphism, and $\pi : G \to A$ is bijective 1-cocycle. Consider on A a second group structure given by

$$a \circ b = \pi(\pi^{-1}(a)\pi^{-1}(b)) = a + \alpha_{\pi^{-1}(a)}(b)$$

for all $a, b \in A$. Since G acts on A by automorphisms,

$$\begin{aligned} a \circ (b + c) &= \pi(\pi^{-1}(a)\pi^{-1}(b + c)) \\ &= a + \alpha_{\pi^{-1}(a)}(b + c) \\ &= a + \alpha_{\pi^{-1}(a)}(b) + \alpha_{\pi^{-1}(a)}(c) \\ &= a \circ b - a + a \circ c \end{aligned}$$

holds for all $a, b, c \in A$.

Conversely, assume that the additive group A has a skew brace structure. Let G be the multiplicative group of A and $\pi = \mathrm{id}$. By Proposition 8.1.12, $a \mapsto \lambda_a$ is a group homomorphism from G to $\mathrm{Aut}(A, +)$. Since $a \circ b = a + \lambda_a(b)$ for all $a, b \in A$, π is a bijective 1-cocycle. □

Example 8.3.2 Let

$$D_4 = \langle r, s : r^4 = s^2 = 1, srs = r^{-1} \rangle$$

be the dihedral group of eight elements and let

$$Q_8 = \{1, -1, i, -i, j, -j, k, -k\}$$

be the quaternion group of eight elements. Let $\alpha \colon Q_8 \to \mathrm{Aut}(D_4)$ be the map defined by

$$\alpha(1) = \alpha(-1) = \alpha(j) = \alpha(-j) = \mathrm{id} \quad \text{and} \quad \alpha(i) = \alpha(-i) = \alpha(k) = \alpha(-k) = \alpha_i,$$

where $\alpha_i(r) = r$ and $\alpha_i(s) = sr^2$. It is easy to check that α is a group homomorphism. Let $\pi : Q_8 \to D_4$ be given by

$$\begin{aligned} &1 \mapsto 1, && -1 \mapsto r^2, && -k \mapsto r^3 s, && k \mapsto rs, \\ &i \mapsto s, && -i \mapsto r^2 s, && j \mapsto r^3, && -j \mapsto r. \end{aligned}$$

One can check that π is a bijective 1-cocycle. A straightforward calculation shows that D_4 with

$$x + y = xy, \quad x \circ y = \pi(\pi^{-1}(x)\pi^{-1}(y))$$

is a two-sided skew brace with additive group isomorphic to D_4 and multiplicative group isomorphic to Q_8.

8.4 Subbraces and Ideals

Definition 8.4.1 Let A be a skew brace. A *subbrace* of A is a subset B of A such that $(B, +)$ is a subgroup of $(A, +)$ and $(B, \circ)$ is a subgroup of $(A, \circ)$.

Definition 8.4.2 Let A be a skew brace. A *left ideal* of A is a subgroup $(I, +)$ of $(A, +)$ such that $\lambda_a(I) \subseteq I$ for all $a \in A$, i.e., $\lambda_a(x) \in I$ for all $a \in A$ and $x \in I$. A *strong left ideal* of A is a left ideal I of A such that $(I, +)$ is a normal subgroup of $(A, +)$.

Proposition 8.4.3 *A left ideal I of a skew brace A is a subbrace of A.*

Proof We need to prove that $(I, \circ)$ is a subgroup of $(A, \circ)$. Clearly I is non-empty, as it is an additive subgroup of A. If $x, y \in I$, then $x \circ y = x + \lambda_x(y) \in I + I \subseteq I$ and $x' = -\lambda_{x'}(x) \in I$. □

Example 8.4.4 Let A be a skew brace. Then

$$\mathrm{Fix}(A) = \{a \in A : \lambda_x(a) = a \text{ for all } x \in A\}$$

is a left ideal of A.

Definition 8.4.5 An *ideal* of a skew brace A is a strong left ideal I of A such that $(I, \circ)$ is a normal subgroup of $(A, \circ)$.

In general, left ideals, strong left ideals, and ideals are different notions, see Exercise 8.6.13.

Example 8.4.6 Consider the semidirect product $A = \mathbb{Z}/(3) \rtimes \mathbb{Z}/(2)$ of the trivial skew braces $\mathbb{Z}/(3)$ and $\mathbb{Z}/(2)$ via the non-trivial action of $\mathbb{Z}/(2)$ over $\mathbb{Z}/(3)$. Then

$$\lambda_{(x,y)}(a, b) = -(x, y) + (x, y) \circ (a, b) = -(x, y) + (x + (-1)^y a, y + b) = ((-1)^y a, b)$$

and hence $\mathrm{Fix}(A) = \{(0, 0), (0, 1)\}$ is not a normal subgroup of $(A, \circ)$. In particular, $\mathrm{Fix}(A)$ is not an ideal of A.

Example 8.4.7 Let $f : A \to B$ be a homomorphism of skew braces. Then $\ker f$ is an ideal of A.

If X and Y are subsets of a brace A, $X * Y$ is defined as the subgroup of $(A, +)$ generated by elements of the form $x * y$, $x \in X$ and $y \in Y$, i.e.:

$$X * Y = \langle x * y : x \in X, y \in Y \rangle_+.$$

Proposition 8.4.8 *Let A be a skew brace. A subgroup I of $(A, +)$ is a left ideal of A if and only if $A * I \subseteq I$.*

Proof Let $a \in A$ and $x \in I$. If I is a left ideal, then $a * x = \lambda_a(x) - x \in I$. Conversely, if $A * I \subseteq I$, then $\lambda_a(x) = a * x + x \in I$. □

Proposition 8.4.9 *Let A be a skew brace. A normal subgroup I of $(A, +)$ is an ideal of A if and only $\lambda_a(I) \subseteq I$ for all $a \in A$, and $I * A \subseteq I$.*

Proof Let $x \in I$ and $a \in A$. Assume first that I is invariant under the action of λ and that $I * A \subseteq I$. Then

$$
\begin{aligned}
a \circ x \circ a' &= a + \lambda_a(x \circ a') \\
&= a + \lambda_a(x + \lambda_x(a')) = a + \lambda_a(x) + \lambda_a\lambda_x(a') + a - a \qquad (8.13) \\
&= a + \lambda_a(x + \lambda_x(a') - a') - a = a + \lambda_a(x + x * a') - a \in I,
\end{aligned}
$$

and hence, by Proposition 8.4.3, I is an ideal.

Conversely, assume that I is an ideal. Then $I * A \subseteq I$ since

$$
\begin{aligned}
x * a &= -x + x \circ a - a \\
&= -x + a \circ (a' \circ x \circ a) - a = -x + a + \lambda_a(a' \circ x \circ a) - a \in I
\end{aligned}
$$

for all $x \in I$ and $a \in A$. □

Let I and J be ideals of a skew brace A. Then $I \cap J$ is an ideal of A, see Exercise 8.6.11. The sum $I + J$ of I and J is defined as the additive subgroup of A generated by all the elements of the form $u + v$, $u \in I$, and $v \in J$.

Proposition 8.4.10 *Let A be a skew brace and I and J be ideals of A. Then $I + J$ is an ideal of A.*

Proof Since I and J are normal subgroups of $(A, +)$, we have that

$$
I + J = \{u + v : u \in I,\ v \in J\}
$$

is a normal subgroup of $(A, +)$. Let $a \in A$, $u \in I$ and $v \in J$. Then

$$
\lambda_a(u + v) = \lambda_a(u) + \lambda_a(v) \in I + J
$$

and hence it follows that $\lambda_a(I + J) \subseteq I + J$. Moreover, by Propositions 8.4.8 and 8.4.9,

$$
(u + v) * a = (u \circ \lambda_u^{-1}(v)) * a = u * (\lambda_u^{-1}(v) * a) + \lambda_u^{-1}(v) * a + u * a \in I + J.
$$

Hence $(I + J) * A \subseteq I + J$. Therefore the result follows by Proposition 8.4.9. □

Definition 8.4.11 Let A be a skew brace. The subset $\mathrm{Soc}(A) = \ker \lambda \cap Z(A, +)$ is the *socle* of A.

Lemma 8.4.12 *Let A be a skew brace and $x \in Soc(A)$. Then*

$$
x \circ a = x + a, \quad a + a \circ x = a \circ x + a \quad \textit{and} \quad \lambda_a(x) = a \circ x \circ a'
$$

hold for all $a \in A$. In particular, $x' = -x$ for all $x \in Soc(A)$.

Proof Let $a \in A$ and $x \in \mathrm{Soc}(A)$. Note that $-x + x \circ a = \lambda_x(a) = a$. Hence the first claim follows. Since

$$
a' \circ (a \circ x + a) = x - a' \quad \text{and} \quad a' \circ (a + a \circ x) = -a' + x,
$$

the second claim follows since $x \in Z(A, +)$. Now we prove the third claim:

$$a \circ x \circ a' = a \circ (x \circ a') = a \circ (x + a') = a \circ x - a = -a + a \circ x = \lambda_a(x).$$

□

Proposition 8.4.13 *Let A be a skew brace. Then $Soc(A)$ is an ideal of A.*

Proof We first prove that $(\mathrm{Soc}(A), +)$ is a normal subgroup of $(A, +)$. Clearly $0 \in \mathrm{Soc}(A)$, since λ is a group homomorphism. Let $x, y \in \mathrm{Soc}(A)$ and $a \in A$. By Lemma 8.4.12, $x - y = x \circ (-y) = x \circ y'$ and hence

$$\lambda_{x-y}(a) = \lambda_{x \circ y'}(a) = \lambda_x \lambda_y^{-1}(a) = a.$$

Thus $x - y \in \mathrm{Soc}(A)$ and therefore the claim follows, as $\mathrm{Soc}(A)$ is central in $(A, +)$.

We now prove that $\lambda_a(x) \in \mathrm{Soc}(A)$ for all $a \in A$ and $x \in \mathrm{Soc}(A)$. First we note that $\lambda_a(x)$ is central, as

$$\begin{aligned} b + \lambda_a(x) &= b - a + a \circ x = a \circ (a' \circ b + x) \\ &= a \circ (x + a' \circ b) = a \circ x - a + b = -a + a \circ x + b = \lambda_a(x) + b \end{aligned}$$

for all $b \in A$, by Lemma 8.4.12. Moreover, by Lemma 8.4.12,

$$\lambda_{\lambda_a(x)}(b) = \lambda_{a \circ x \circ a'}(b) = \lambda_a \lambda_x \lambda_a^{-1}(b) = b$$

for all $b \in A$. Since $x * a = -x + x \circ a - a = -x + x + a - a = 0 \in \mathrm{Soc}(A)$, the claim follows from Proposition 8.4.9. □

As a corollary, we obtain that the socle of a skew brace A is a trivial skew brace of abelian type.

Proposition 8.4.14 *Let A be a skew brace. Then $Soc(A) = \ker \lambda \cap \ker \mu$.*

Proof Let $a \in \mathrm{Soc}(A)$ and $b \in A$. Then $\lambda_a = \mathrm{id}$ and $a \in Z(A, +)$. By Lemma 8.4.12,

$$\mu_a(b) = \lambda_b(a)' \circ b \circ a = (b \circ a \circ b')' \circ b \circ a = b.$$

Thus $a \in \ker \lambda \cap \ker \mu$.

Conversely, let $a \in \ker \lambda \cap \ker \mu$ and $b \in A$. Then $b' = \mu_a(b') = \lambda_{b'}(a)' \circ b' \circ a$, so $\lambda_{b'}(a) = b' \circ a \circ b$. By Lemma 8.4.12 and (8.3),

$$b + a = b \circ \lambda_b^{-1}(a) = b \circ \lambda_{b'}(a) = b \circ b' \circ a \circ b = a \circ b = a + \lambda_a(b) = a + b.$$

Hence $a \in \mathrm{Soc}(A)$. □

Definition 8.4.15 Let A be a skew brace. The *annihilator* of A is defined as the set $\mathrm{Ann}(A) = \mathrm{Soc}(A) \cap Z(A, \circ)$.

Note that $\mathrm{Ann}(A) \subseteq \mathrm{Fix}(A)$.

Proposition 8.4.16 *The annihilator of a skew brace A is an ideal of A.*

Proof Let $x, y \in \mathrm{Ann}(A)$. Note that $x - y = x \circ y' \in Z(A, \circ)$. Hence $\mathrm{Ann}(A)$ is a subbrace of A. Since $\mathrm{Ann}(A) \subseteq Z(A, +) \cap Z(A, \circ)$, we only need to note that $\lambda_a(x) = x \in \mathrm{Ann}(A)$ for all $a \in A$. □

If A is a skew brace and I is an ideal of A, then $a + I = a \circ I$ for all $a \in A$. Indeed, $a \circ x = a + \lambda_a(x) \in a + I$ and $a + x = a \circ \lambda_a^{-1}(x) = a \circ \lambda_{a'}(x) \in a \circ I$ for all $a \in A$ and $x \in I$. This allows us to prove that there exists a unique skew brace structure over A/I such that the map

$$\pi \colon A \to A/I, \quad a \mapsto a + I = a \circ I$$

is a homomorphism of skew braces. The skew brace A/I is the *quotient skew brace* of A modulo I. It is possible to prove the isomorphism theorems for skew braces, see Exercises 8.6.16, 8.6.17, 8.6.18, and 8.6.19.

8.5 Regular Subgroups of the Holomorph

For an additive group A, the *holomorph* of A is the semidirect product

$$\mathrm{Hol}(A) = A \rtimes \mathrm{Aut}(A).$$

This means that the operation is

$$(a, f)(b, g) = (a + f(b), fg), \quad a, b \in A, \quad f, g \in \mathrm{Aut}(A).$$

Every subgroup G of $\mathrm{Hol}(A)$ acts on A by

$$(x, f) \cdot a = \pi_1((x, f)(a, \mathrm{id})) = \pi_1(x + f(a), f) = x + f(a),$$

$a, x \in A, \quad f \in \mathrm{Aut}(A)$, where $\pi_1 \colon \mathrm{Hol}(A) \to A$, $(a, f) \mapsto a$.

A subgroup G of $\mathrm{Hol}(A)$ is said to be *regular* if it acts regularly on A, this means that given $a, b \in A$ there exists a unique $(x, f) \in G$ such that

$$b = (x, f) \cdot a = x + f(a).$$

Lemma 8.5.1 *If G is a regular subgroup of $\mathrm{Hol}(A)$, then $\pi_1 \colon G \to A$ is bijective.*

Proof We claim that restriction $\pi_1|_G$ of π_1 on G is injective. Let $(a, f), (b, g) \in G$ be such that $\pi_1(a, f) = \pi_1(b, g)$. Then $a = b$. Since G is a subgroup,

$$(-f^{-1}(a), f^{-1}) = (a, f)^{-1} \in G, \quad (-g^{-1}(a), g^{-1}) = (a, g)^{-1} \in G,$$

and hence $f = g$ since

$$(-f^{-1}(a), f^{-1}) \cdot a = 0 = (-g^{-1}(a), g^{-1}) \cdot a$$

and G is a regular subgroup. Now we prove that $\pi_1|_G$ is surjective. Let $a \in A$. Since G is regular, there exists $(x, f) \in G$ such that $x + f(a) = (x, f) \cdot a = 0$, so $(-f(a), f) \in G$ for some $f \in \mathrm{Aut}(A)$. Then $(a, f^{-1}) = (-f(a), f)^{-1} \in G$ and $\pi_1|_G(a, f^{-1}) = a$. □

Now we establish an important connection between skew braces and regular subgroups.

Theorem 8.5.2 *If A is a skew brace, then $\Delta = \{(a, \lambda_a) : a \in A\}$ is a regular subgroup of $Hol(A, +)$. Conversely, if A is an additive group and G is a regular subgroup of $Hol(A)$, then A is a skew brace with*

$$a \circ b = a + f(b),$$

where $(\pi_1|_G)^{-1}(a) = (a, f) \in G$.

Proof Assume first that A is a skew brace. Using (8.3) and that λ is a group homomorphism, it follows that $\Delta = \{(a, \lambda_a) : a \in A\}$ is a subgroup of $\mathrm{Hol}(A, +)$, as

$$\begin{aligned}(a, \lambda_a)^{-1} &= (\lambda_a^{-1}(-a), \lambda_a^{-1}) = (a', \lambda_{a'}) \in \Delta,\\ (a, \lambda_a)(b, \lambda_b) &= (a + \lambda_a(b), \lambda_a \circ \lambda_b) = (a \circ b, \lambda_{a \circ b}) \in \Delta.\end{aligned}$$

To see that Δ is a regular subgroup, note that $(c, \lambda_c) \cdot a = b$ implies that $c = b \circ a'$, as $(A, \circ)$ is a group.

Assume now that A is an additive group and that G is a regular subgroup of $\mathrm{Hol}(A)$. By Lemma 8.5.1, the restriction $\pi_1|_G$ is bijective. Use the bijection $\pi_1|_G$ to transport the operation of G into A:

$$a \circ b = \pi_1|_G\left((\pi_1|_G)^{-1}(a)(\pi_1|_G)^{-1}(b)\right) = a + f(b),$$

where $a, b \in A$ and $(\pi_1|_G)^{-1}(a) = (a, f) \in G$. Then $(A, \circ)$ is a group isomorphic to G and moreover A is a skew brace, as

$$\begin{aligned}a \circ (b + c) &= a + f(b + c) = a + f(b) + f(c)\\ &= a + f(b) - a + a + f(c) = a \circ b - a + a \circ c\end{aligned}$$

holds for all $a, b, c \in A$. □

The following lemma is from [17].

Lemma 8.5.3 *Let A be a group. If H and K are conjugate regular subgroups of $Hol(A)$, then H and K are conjugate by an automorphism of A.*

Proof Assume that H and K are conjugate in $\mathrm{Hol}(A)$. Let $(b, g) \in \mathrm{Hol}(A)$ be such that $(b, g)^{-1}H(b, g) = K$. Since $b \in A$, the regularity of H implies that there exists $(a, f) \in H$ such that $a + f(b) = 0$. Since $(a, f) \in H$,

$$\begin{aligned} K = (b, g)^{-1}H(b, g) &= (b, g)^{-1}(a, f)^{-1}H(a, f)(b, g) \\ &= (0, f \circ g)^{-1}H(0, f \circ g) = (f \circ g)^{-1}H(f \circ g). \end{aligned}$$ □

Proposition 8.5.4 *Let A be an additive group. There exists a bijective correspondence between isomorphism classes of skew brace structures with additive group A and conjugacy classes of regular subgroups of $Hol(A)$.*

Proof Assume that the additive group A has two isomorphic skew brace structures given by $(a, b) \mapsto a \circ b$ and $(a, b) \mapsto a \times b$. Let $f : A \to A$ be a bijective map such that $f(a + b) = f(a) + f(b)$ and $f(a \circ b) = f(a) \times f(b)$ for all $a, b \in A$. We claim that the regular subgroups $\{(a, \lambda_a) : a \in A\}$ and $\{(a, \lambda'_a) : a \in A\}$, where $\lambda_a(b) = -a + a \circ b$ and $\lambda'_a(b) = -a + a \times b$ are conjugate. Since f is an isomorphism of skew braces,

$$f \circ \lambda_a \circ f^{-1} = \lambda'_{f(a)}$$

for all $a \in A$. This implies that $(0, f)(a, \lambda_a)(0, f)^{-1} = (f(a), \lambda'_{f(a)})$ for all $a \in A$ and hence and the first claim follows.

Conversely, let H and K be conjugate regular subgroups of $\mathrm{Hol}(A)$. Since H and K are conjugate in $\mathrm{Hol}(A)$, by Lemma 8.5.3 there exists $\varphi \in \mathrm{Aut}(A)$ such that $\varphi H \varphi^{-1} = K$. The skew brace structure on A corresponding to the subgroup H is given by $a \circ b = a + f(b)$, where $(a, f) = (\pi_1|_H)^{-1}(a) \in H$, see Lemma 8.5.1. Since

$$\varphi(a, f)\varphi^{-1} = (\varphi(a), \varphi \circ f \circ \varphi^{-1}) \in K,$$

it follows that $(\pi_1|_K)^{-1}(\varphi(a)) = (\varphi(a), \varphi \circ f \circ \varphi^{-1})$. Since $\varphi \in \mathrm{Aut}(A)$,

$$\begin{aligned} \varphi(a) \times \varphi(b) &= \varphi(a) + (\varphi \circ f \circ \varphi^{-1})(\varphi(b)) \\ &= \varphi(a) + \varphi(f(b)) = \varphi(a + f(b)) = \varphi(a \circ b) \end{aligned}$$

and hence the skew braces corresponding to H and K are isomorphic. □

Now we present an algorithm used to enumerate skew braces. It is based on Theorem 8.5.2. The use of Lemma 8.5.3 in Proposition 8.5.4 significantly improves the performance.

Algorithm 8.5.5 Let A be a finite group. To construct all skew braces with additive group A we proceed as follows:

(1) Compute the holomorph $\mathrm{Hol}(A)$ of A.
(2) Compute the list of regular subgroups of $\mathrm{Hol}(A)$ of order $|A|$ up to conjugation.
(3) For each representative G of regular subgroups of $\mathrm{Hol}(A)$ construct the map $p\colon A \to G$ given by $a \mapsto (a, f) \in G$. Then the set A is a skew brace with additive group A and multiplication given by $a \circ b = p^{-1}(p(a)p(b))$ for all $a, b \in A$.

To enumerate all isomorphism classes of skew brace structures with a fixed additive group the third step of Algorithm 8.5.5 is not needed. Algorithm 8.5.5 can be used to compute the number $s(n)$ of non-isomorphic skew braces of size n. With small modifications it could be used to compute the number $a(n)$ of non-isomorphic skew braces of abelian type of size n, or the number of non-isomorphic radical rings, or the number of non-isomorphic skew braces of nilpotent type. Some values for $s(n)$ and $a(n)$ appear in Table 8.1.

8.6 Exercises

8.6.1 Prove that there exists a bijective correspondence between skew left braces and skew right braces.

8.6.2 Let p be a prime number. Prove that $\mathbb{Z}/(p^2)$ is a skew brace of abelian type with the operation $x \circ y = x + y + pxy$.

8.6.3 Let A be a skew brace. Prove that

$$\mu_b(a) = \lambda^{-1}_{\lambda_a(b)}(-a \circ b + a + a \circ b).$$

8.6.4 Let A be an additive (not necessarily abelian) group. Prove that a structure of skew brace over A is equivalent to an operation $A \times A \to A$, $(a, b) \mapsto a * b$, such that

$$a * (b + c) = a * b + b + a * c - b$$

holds for all $a, b, c \in A$, and the operation $a \circ b = a + a * b + b$ turns A into a group.

8.6.5 Let $(A, +, \circ)$ be a triple, where $(A, +)$ and $(A, \circ)$ are groups, and let $\lambda\colon A \to \mathbb{S}_A, a \mapsto \lambda_a, \lambda_a(b) = -a + a \circ b$. Prove that the following statements are equivalent:

(1) A is a skew brace.
(2) $\lambda_{a\circ b}(c) = \lambda_a\lambda_b(c)$ for all $a, b, c \in A$.
(3) $\lambda_a(b + c) = \lambda_a(b) + \lambda_a(c)$ for all $a, b, c \in A$.

8.6.6 Let A and B be skew braces. Prove that $f\colon A \to B$ is a homomorphism of skew braces if and only if $f(a + b) = f(a) + f(b)$ and $f(\lambda_a(b)) = \lambda_{f(a)}(f(b))$ for all $a, b \in A$.

Table 8.1 The number of non-isomorphic skew braces

n	1	2	3	4	5	6	7	8	9	10	11	12
$a(n)$	1	1	1	4	1	2	1	27	4	2	1	10
$s(n)$	1	1	1	4	1	6	1	47	4	6	1	38
n	13	14	15	16	17	18	19	20	21	22	23	24
$a(n)$	1	2	1	357	1	8	1	11	2	2	1	96
$s(n)$	1	6	1	1605	1	49	1	43	8	6	1	855
n	25	26	27	28	29	30	31	32	33	34	35	36
$a(n)$	4	2	37	9	1	4	1	25281	1	2	1	46
$s(n)$	4	6	101	29	1	36	1	1223061	1	6	1	400
n	37	38	39	40	41	42	43	44	45	46	47	48
$a(n)$	1	2	2	106	1	6	1	9	4	2	1	1708
$s(n)$	1	6	8	944	1	78	1	29	4	6	1	66209
n	49	50	51	52	53	54	55	56	57	58	59	60
$a(n)$	4	8	1	11	1	80	2	91	2	2	1	28
$s(n)$	4	51	1	43	1	1028	12	815	8	6	1	418
n	61	62	63	64	65	66	67	68	69	70	71	72
$a(n)$	1	2	11	15095601	1	4	1	11	1	4	1	489
$s(n)$	1	6	11	?	1	36	1	43	1	36	1	17790
n	73	74	75	76	77	78	79	80	81	82	83	84
$a(n)$	1	2	5	9	1	6	1	1985	804	2	1	34
$s(n)$	1	6	14	29	1	78	1	74120	8436	6	1	606
n	85	86	87	88	89	90	91	92	93	94	95	96
$a(n)$	1	2	1	90	1	16	1	9	2	2	1	195971
$s(n)$	1	6	1	800	1	294	1	29	8	6	1	?
n	97	98	99	100	101	102	103	104	105	106	107	108
$a(n)$	1	8	4	51	1	4	1	106	2	2	1	494
$s(n)$	1	53	4	711	1	36	1	944	8	6	1	11223
n	109	110	111	112	113	114	115	116	117	118	119	120
$a(n)$	1	6	2	1671	1	6	1	11	11	2	1	395
$s(n)$	1	94	8	65485	1	78	1	43	47	6	1	22711
n	121	122	123	124	125	126	127	128	129	130	131	132
$a(n)$	4	2	1	9	49	36	1	?	2	4	1	24
$s(n)$	4	6	1	29	213	990	1	?	8	36	1	324
n	133	134	135	136	137	138	139	140	141	142	143	144
$a(n)$	1	2	37	108	1	4	1	27	1	2	1	10215
$s(n)$	1	6	101	986	1	36	1	395	1	6	1	3013486
n	145	146	147	148	149	150	151	152	153	154	155	156
$a(n)$	1	2	9	11	1	19	1	90	4	4	2	40
$s(n)$	1	6	123	43	1	401	1	800	4	36	12	782
n	157	158	159	160	161	162	163	164	165	166	167	168
$a(n)$	1	2	1	209513	1	1374	1	11	2	2	1	443
$s(n)$	1	6	1	?	1	45472	1	43	12	6	1	28505

8.6.7 Let A be a skew brace such that $\lambda_a(a) = a$ for all $a \in A$. Prove that A is two sided. Hint: Prove first that $a' = -a$ for all $a \in A$.

8.6.8 Prove Equality (8.8).

8.6.9 Let k, n be positive integers, and A be a finite skew brace of size n and k orbits. Then $|k| \geq \frac{\log \log n}{\log 4}$. Hint: Use Newman's paper [89].

8.6.10 Recall that two-sided skew braces of abelian type are equivalent to radical rings. Prove that under this equivalence, (left) ideals of the radical ring correspond to (left) ideals of the associated skew brace.

8.6.11 Prove that the intersection of ideals of a skew brace A is an ideal of A.

8.6.12 Let A be a skew brace and I be a characteristic subgroup of the additive group of A. Prove that I is a left ideal of A.

8.6.13 Give an example of a left ideal that is not a strong left ideal of a skew brace. Give an example of a strong left ideal that is not an ideal of a skew brace.

8.6.14 Prove that the socle of a skew brace A is the kernel of the group homomorphism $(A, \circ) \to \operatorname{Aut}(A, +) \times \mathbb{S}_A$, $a \mapsto (\lambda_a, \mu_a^{-1})$.

8.6.15 Prove that the socle of a skew brace A is the kernel of the group homomorphism $(A, \circ) \to \operatorname{Aut}(A, +) \times \operatorname{Aut}(A, +)$, $a \mapsto (\lambda_a, \xi_a)$, where $\xi_a(b) = a + \lambda_a(b) - a$.

8.6.16 Let $f\colon A \to B$ be a skew brace homomorphism. Prove that $A/\ker f \cong f(A)$.

8.6.17 Let A be a skew brace and B be a subbrace of A. Prove that if I is an ideal of A, then $B \circ I$ is a subbrace of A, $B \cap I$ is an ideal of B and $(B \circ I)/I \cong B/(B \cap I)$.

8.6.18 Let A be a skew brace and I and J be ideals of A. Prove that if $I \subseteq J$, then $A/J \cong (A/I)/(J/I)$.

8.6.19 Let A be a skew brace and let I be an ideal of A. Prove that there is a bijective correspondence between (left) ideals of A containing I and (left) ideals of A/I.

8.7 Open Problems

Open Problem 8.7.1 Estimate $s(n)$ and $a(n)$ for $n \to \infty$.

Open Problem 8.7.2 Construct and enumerate skew braces of sizes 64, 96, 128, and 160.

8.8 Notes

Skew braces of abelian type (i.e., left braces) were introduced by Rump in [99] for studying involutive solutions to the YBE. Rump's definition was reformulated by Cedó, Jespers and Okniński in [39]. With this definition at hand, Guarnieri and Vendramin introduced arbitrary skew braces in [64].

Example 8.1.7 is motivated by the paper of Weinstein and Xu [114].

Theorem 8.1.20 was proved by Rump in [99].

Theorem 8.1.22 was proved by Lau [84] and independently by Kinyon (unpublished). It answers a question of Cedó, Gateva-Ivanova, and Smoktunowicz, see [37].

Theorem 8.2.1 for skew braces of abelian type appears implicit in the work [99] of Rump, see also [39]. The general case was proved by Guarnieri and Vendramin in [64].

Theorem 8.2.3 was proved by Smoktunowicz and Vendramin in [106].

In the case of skew braces of abelian type, Theorem 8.3.1 is implicit in the work of Rump, see [99, 100] or [39]. Similar results appear in the work of Etingof, Schedler, and Soloviev [56]; Lu, Yan, and Zhu [87]; and Soloviev [107]. In [55] Etingof and Gelaki give a method of constructing finite-dimensional complex semisimple triangular Hopf algebras. They show how any non-abelian group that admits a bijective 1-cocycle gives rise to a semisimple minimal triangular Hopf algebra which is not a group algebra.

The socle was defined by Rump in [99]. The annihilator first appeared in the work [34] of Catino, Colazzo, and Stefanelli.

The connection between skew braces and regular subgroups of the holomorph is a fundamental part of the theory of skew braces. It was first observed in [31] by Caranti, Dalla Volta, and Salla in the context of radical algebras. This was later generalized by Catino and Rizzo [35] and Bachiller [8] for skew braces of abelian type. The result for skew braces appeared in [64].

Algorithm 8.5.5 and most of the numbers of Table 8.1 appeared in [64]. It should be noted that the number of skew braces of size 57 of [64] is incorrect; the correct value is $s(57) = 8$, as Table 8.1 shows.

Lemma 8.5.3 appears in [17] and it is needed to compute the number $a(n)$ of isomorphism classes of skew braces of abelian type of size $n \in \{32, 81, 96, 144, 160, 162\}$ and the number $s(n)$ of isomorphism classes of skew braces of size

$$n \in \{32, 54, 80, 81, 108, 112, 120, 136, 144, 147, 150, 152, 162, 168\}.$$

Skew braces of abelian type of size 64 were computed by Ballester-Bolinches, Esteban-Romero, and Pérez-Calabuig (see [15]).

Exercise 8.6.5 combines results of Bachiller, Rump [99], and Gateva-Ivanova [61]. Exercise 8.6.7 for skew braces of abelian type comes from [39]. Exercises 8.6.14 and 8.6.15 appear in [10].

Exercise 8.6.9 is a direct translation into the context of skew braces (or, more precisely, the context of groups acting by automorphisms on groups) of Neumann's [89] result based on Landau's method [83].

Chapter 9
The Structure Skew Brace of a Solution

In this chapter, we study the structure monoid $M(X, r)$ and the structure group $G(X, r)$ of a solution (X, r) to the YBE. We prove that if (X, r) is involutive, then $M(X, r)$ is cancellative. We show that $G(X, r)$ has a natural structure of skew brace. Then we use the bijective correspondence between braided groups and skew braces, which we study in the first section of this chapter, to prove that the solution (X, r) induces a natural solution $(G(X, r), r_G)$ to the YBE such that r_G also is a braiding operator on $G(X, r)$ with a nice universal property.

We also prove that the kernel of the natural group homomorphism from the structure group $G(X, r)$ to the permutation group $\mathcal{G}(X, r)$ of a solution (X, r) to the YBE is an ideal of the skew brace $G(X, r)$ and hence $\mathcal{G}(X, r)$ has a natural structure of skew brace induced by the skew brace $G(X, r)$.

In the last section, we study the structure group of a finite solution to the YBE.

9.1 Braided Groups and Skew Braces

Definition 9.1.1 A *braided group* is a pair (G, r), where G is a group with operation $m\colon G \times G \to G$, $m(x, y) = xy$, and $r\colon G \times G \to G \times G$ is a bijective map such that

(1) $r(xy, z) = (\mathrm{id} \times m)r_1r_2(x, y, z)$ for all $x, y, z \in G$,
(2) $r(x, yz) = (m \times \mathrm{id})r_2r_1(x, y, z)$ for all $x, y, z \in G$,
(3) $r(1, x) = (x, 1)$ and $r(x, 1) = (1, x)$ for all $x \in G$, and
(4) $m \circ r = m$,

where $r_1 = r \times \mathrm{id}_G$ and $r_2 = \mathrm{id}_G \times r$. The map r is called a *braiding operator* on G.

Lemma 9.1.2 *Let (G, r) be a braided group. We write $r(x, y) = (x \triangleright_r y, x \triangleleft_r y)$, for all $x, y \in G$. Let $\xi\colon G \to \mathbb{S}_G$ be the map defined by $\xi(x)(y) = x \triangleright_r y$ for all*

F. Cedó and L. Vendramin, *Groups, Radical Rings, and the Yang-Baxter Equation*, Progress in Mathematics 361, https://doi.org/10.1007/978-3-032-01524-2_9

$x, y \in G$. Let $\eta\colon G \to \mathbb{S}_G$ be the map defined by $\eta(x)(y) = y \lhd_r x$ for all $x, y \in G$. Then ξ is a left action of G on itself, η is a right action of G on itself and (G, r) is a solution to the YBE.

Proof By the definition of a braiding operator, we have that

$$\begin{aligned} r(xy, z) &= (\mathrm{id} \times m)r_1r_2(x, y, z) \\ &= (\mathrm{id} \times m)r_1(x, y \rhd_r z, y \lhd_r z) \\ &= (\mathrm{id} \times m)(x \rhd_r (y \rhd_r z), x \lhd_r (y \rhd_r z), y \lhd_r z) \\ &= (x \rhd_r (y \rhd_r z), (x \lhd_r (y \rhd_r z))(y \lhd_r z)), \end{aligned}$$

and thus $(xy) \rhd_r z = x \rhd_r (y \rhd_r z)$. Since $r(1, x) = (x, 1)$, we have that the map ξ is a left action of G on itself. By the definition of a braiding operator, we have that

$$\begin{aligned} r(x, yz) &= (m \times \mathrm{id})r_2r_1(x, y, z) \\ &= (m \times \mathrm{id})r_2(x \rhd_r y, x \lhd_r y, z) \\ &= (m \times \mathrm{id})(x \rhd_r y, (x \lhd_r y) \rhd_r z, (x \lhd_r y) \lhd_r z) \\ &= ((x \rhd_r y)((x \lhd_r y) \rhd_r z), (x \lhd_r y) \lhd_r z), \end{aligned} \tag{9.1}$$

and thus $x \lhd_r (yz) = (x \lhd_r y) \lhd_r z$. Since $r(x, 1) = (1, x)$, we have that the map η is a right action of G on itself. Furthermore, $xy = (x \rhd_r y)(x \lhd_r y)$ for all $x, y \in G$. Hence, by Theorem 4.1.15, (G, r) is a solution to the YBE. □

Theorem 9.1.3 *Let G be a group. Then there is a bijection between the set of braiding operators on G and the set of skew brace structures on the set G with multiplicative group G.*

Proof Let $\mathcal{BO}(G)$ be the set of all the braiding operators on G. Let $\mathcal{B}(G)$ be the set of all the skew brace structures $(G, +, \circ)$ on the group G, where $x \circ y = xy$ for all $x, y \in G$. For $r \in \mathcal{BO}(G)$, we write $r(x, y) = (x \rhd_r y, x \lhd_r y)$ for all $x, y \in G$. We define

$$f\colon \mathcal{BO}(G) \to \mathcal{B}(G)$$

by $f(r) = (G, +_r, \circ)$, where $x +_r y = x(x^{-1} \rhd_r y)$ for all $x, y \in G$ and $r \in \mathcal{BO}(G)$. Now, by (9.1) and Lemma 9.1.2, we have

$$\begin{aligned} x \rhd_r (y +_r z) &= x \rhd_r (y(y^{-1} \rhd_r z)) \\ &= (x \rhd_r y)((x \lhd_r y) \rhd_r (y^{-1} \rhd_r z)) \\ &= (x \rhd_r y)(((x \lhd_r y)y^{-1}) \rhd_r z) \\ &= (x \rhd_r y)(((x \rhd_r y)^{-1}x) \rhd_r z) \\ &= (x \rhd_r y)((x \rhd_r y)^{-1} \rhd_r (x \rhd_r z)) \\ &= (x \rhd_r y) +_r (x \rhd_r z). \end{aligned} \tag{9.2}$$

Note that $1 +_r x = 1(1^{-1} \triangleright_r x) = x$ and $x +_r 1 = x(x^{-1} \triangleright_r 1) = x1 = x$ for all $x \in G$. By Lemma 9.1.2, for every $x \in G$, we have

$$x +_r (x \triangleright_r x^{-1}) = x(x^{-1} \triangleright_r (x \triangleright_r x^{-1})) = xx^{-1} = 1$$

and by (9.1),

$$\begin{aligned}(x \triangleright_r x^{-1}) +_r x &= (x \triangleright_r x^{-1})((x \triangleright_r x^{-1})^{-1} \triangleright_r x)\\ &= (x \triangleright_r x^{-1})((x \triangleleft_r x^{-1}) \triangleright_r x)\\ &= x \triangleright_r (x^{-1}x) = 1.\end{aligned}$$

By Lemma 9.1.2 and (9.2), we have that

$$\begin{aligned}(x +_r y) +_r z &= x(x^{-1} \triangleright_r y)(((x^{-1} \triangleright_r y)^{-1}x^{-1}) \triangleright_r z)\\ &= x(x^{-1} \triangleright_r y)((x^{-1} \triangleright_r y)^{-1} \triangleright_r (x^{-1} \triangleright_r z))\\ &= x((x^{-1} \triangleright_r y) +_r (x^{-1} \triangleright_r z))\\ &= x(x^{-1} \triangleright_r (y +_r z)) = x +_r (y +_r z)\end{aligned}$$

for all $x, y, z \in G$. Hence $(G, +_r)$ is a group. By Lemma 9.1.2,

$$x +_r (x \triangleright_r y) = x(x^{-1} \triangleright_r (x \triangleright_r y)) = xy$$

for all $x, y \in G$. By (9.2), we have

$$\begin{aligned}x(y +_r z) &= x +_r (x \triangleright_r (y +_r z))\\ &= x +_r (x \triangleright_r y) +_r (x \triangleright_r z)\\ &= xy -_r x +_r xz\end{aligned}$$

for all $x, y, z \in G$. Hence $(G, +_r, \circ)$ is a skew brace and f is a well-defined map. Note that the lambda map of this skew brace is defined by $\lambda_x(y) = -_r x +_r xy = x \triangleright_r y$ for all $x, y \in G$.

Now we define $g\colon \mathcal{B}(G) \to \mathcal{BO}(G)$, by $g(G, +, \circ) = r_+$ and

$$r_+(x, y) = (-x + xy, (-x + xy)^{-1}xy) = (\lambda_x(y), \mu_y(x))$$

for all $(G, +, \circ) \in \mathcal{B}(G)$ and $x, y \in G$. We shall prove that r_+ is a braiding operator on G. Note that, by Proposition 8.1.12,

$$\begin{aligned}r_+(xy, z) &= (\lambda_{xy}(z), \mu_z(xy))\\ &= (\lambda_x\lambda_y(z), \lambda_x(\lambda_y(z))^{-1}xyz)\\ &= (\lambda_x\lambda_y(z), \lambda_x(\lambda_y(z))^{-1}x\lambda_y(z)\lambda_y(z)^{-1}yz)\end{aligned}$$

$$\begin{aligned}&= (\lambda_x\lambda_y(z), \mu_{\lambda_y(z)}(x)\mu_z(y))\\&= (\mathrm{id} \times m)(r_+)_1(r_+)_2(x, y, z).\end{aligned}$$

By Propositions 8.1.12 and 8.1.13,

$$\begin{aligned}r_+(x, yz) &= (\lambda_x(yz), \mu_{yz}(x))\\&= (\lambda_x(y + \lambda_y(z)), \mu_z\mu_y(x))\\&= (\lambda_x(y) + \lambda_x\lambda_y(z), \mu_z\mu_y(x))\\&= (\lambda_x(y)\lambda_{\lambda_x(y)^{-1}xy}(z), \mu_z\mu_y(x))\\&= (\lambda_x(y)\lambda_{\mu_y(x)}(z), \mu_z\mu_y(x))\\&= (m \times \mathrm{id})(r_+)_2(r_+)_1(x, y, z).\end{aligned}$$

We also have that

$$r_+(1, x) = (\lambda_1(x), \mu_x(1)) = (x, 1),$$

$$r_+(x, 1) = (\lambda_x(1), \mu_1(x)) = (1, x),$$

and

$$\lambda_x(y)\mu_y(x) = \lambda_x(y)\lambda_x(y)^{-1}xy = xy$$

for all $x, y \in G$. Hence r_+ is a braiding operator on G, thus g is a well-defined map. Now it is easy to check that $fg = \mathrm{id}_{\mathcal{B}(G)}$ and $gf = \mathrm{id}_{\mathcal{BO}(G)}$. □

9.2 The Structure Monoid of a Solution

Let (X, r) be a solution to the YBE. Write as usual $r(x, y) = (\sigma_x(y), \tau_y(x))$. The *structure monoid* of (X, r) is the monoid $M(X, r)$ presented as a monoid with

$$M(X, r) = \langle X : xy = \sigma_x(y)\tau_y(x) \text{ for all } x, y \in X\rangle.$$

The *derived structure monoid* of (X, r) is the monoid $A(X, r)$ presented as a monoid with

$$A(X, r) = \langle x\sigma_x(y) = \sigma_x(y)\sigma_{\sigma_x(y)}\tau_y(x) \text{ for all } x, y \in X\rangle.$$

Definition 9.2.1 Let M and A be monoids. Let $\theta\colon M \to \mathrm{Aut}(A)$ be a monoid homomorphism. A map $\varphi\colon M \to A$ is a 1-*cocycle* with respect to θ if $\varphi(1) = 1$ and

$$\varphi(xy) = \varphi(x)\theta(x)(\varphi(y))$$

for all $x, y \in M$.

Lemma 9.2.2 *Let M and A be monoids. Let $\theta\colon M \to \operatorname{Aut}(A)$ be a monoid homomorphism and let $\varphi\colon M \to A$ be a bijective 1-cocycle with respect to θ. For a congruence η on M define*

$$\varphi(\eta) = \{(\varphi(x), \varphi(y)) : (x, y) \in \eta\} \subseteq A \times A.$$

If the congruence η satisfies the following properties:

(1) $\eta \subseteq \ker\theta = \{(x, y) \in M \times M : \theta(x) = \theta(y)\}$ *and*
(2) $\varphi(\eta) = \{(\theta(z)(\varphi(x)), \theta(z)(\varphi(y))) : (x, y) \in \eta\}$ *for all $z \in M$,*

then $\varphi(\eta)$ is a congruence on A. Moreover, θ induces a monoid homomorphism $\overline{\theta}\colon M/\eta \to \operatorname{Aut}(A/\varphi(\eta))$ and φ induces a bijective 1-cocycle $\overline{\varphi}\colon M/\eta \to A/\varphi(\eta)$ with respect to $\overline{\theta}$.

Proof Since η is an equivalence relation on M and φ is bijective, we have that $\varphi(\eta)$ is an equivalence relation on A.

Let $(a, b) \in \varphi(\eta)$ and $c, d \in A$. Since φ is bijective, there exists $z \in M$ such that $c = \varphi(z)$. By assumption (2), there exists $(x, y) \in \eta$ such that

$$(a, b) = (\theta(z)(\varphi(x)), \theta(z)(\varphi(y))).$$

Now

$$ca = \varphi(z)\theta(z)(\varphi(x)) = \varphi(zx) \text{ and } cb = \varphi(z)\theta(z)(\varphi(y)) = \varphi(zy).$$

Since η is a congruence and $(x, y) \in \eta$, we have that $(zx, zy) \in \eta$. By assumption (1), $\theta(zx) = \theta(zy)$. By the bijectivity of φ, there exists $t \in M$ such that

$$d = \theta(zx)(\varphi(t)) = \theta(zy)(\varphi(t)).$$

Hence

$$cad = \varphi(zx)\theta(zx)(\varphi(t)) = \varphi(zxt)$$

and

$$cbd = \varphi(zy)\theta(zy)(\varphi(t)) = \varphi(zyt).$$

Since η is a congruence and $(x, y) \in \eta$ we have that $(zxt, zyt) \in \eta$ and thus $(cad, cbd) \in \varphi(\eta)$. Therefore $\varphi(\eta)$ is a congruence on A. We define $\overline{\theta}\colon M/\eta \to \operatorname{Aut}(A/\varphi(\eta))$ by

$$\overline{\theta}(\bar{x})(\bar{a}) = \overline{\theta(x)(a)}$$

for all $x \in M$ and $a \in A$, where $\bar{x}$ denotes the η-class of x and $\bar{a}$ denotes the $\varphi(\eta)$-class of a. Let $(x, y) \in \eta$ and $(a, b) \in \varphi(\eta)$. By assumption (1), $\theta(x) = \theta(y)$. By the second assumption,

$$(\theta(x)(a), \theta(y)(b)) = (\theta(x)(a), \theta(x)(b)) \in \varphi(\eta).$$

Hence $\overline{\theta}(\bar{x}) \colon A/\varphi(\eta) \to A/\varphi(\eta)$ is well-defined. Similarly, one can see that the map $\overline{\theta}(\bar{x})' \colon A/\varphi(\eta) \to A/\varphi(\eta)$ defined by $\overline{\theta}(\bar{x})'(\bar{a}) = \overline{\theta(x)^{-1}(a)}$ is well-defined and it is the inverse of $\overline{\theta}(\bar{x})$. It is easy to see that $\overline{\theta}(\bar{x})$ is a monoid homomorphism. Hence $\overline{\theta}$ is well-defined. It is easy to check that $\overline{\theta}$ is a monoid homomorphism.

We define $\overline{\varphi} \colon M/\eta \to A/\varphi(\eta)$ by $\overline{\varphi}(\bar{x}) = \overline{\varphi(x)}$ for all $x \in M$. By the definition of $\varphi(\eta)$ it is clear that $\overline{\varphi}$ is a well-defined bijection. Let $x, y \in M$. We have

$$\overline{\varphi}(\bar{x}\bar{y}) = \overline{\varphi}(\overline{xy}) = \overline{\varphi(xy)} = \overline{\varphi(x)\theta(x)(\varphi(y))} = \overline{\varphi}(\bar{x})\overline{\theta}(\bar{x})(\overline{\varphi}(\bar{y})).$$

Hence $\overline{\varphi}$ is a bijective 1-cocycle with respect to $\overline{\theta}$. Therefore the result follows. □

Proposition 9.2.3 *Let (X, r) be a solution to the YBE. Then there exist a unique monoid homomorphism $\theta \colon M(X, r) \to \operatorname{Aut}(A(X, r))$ and a unique bijective 1-cocycle $\varphi \colon M(X, r) \to A(X, r)$ with respect to θ such that $\theta(x)(y) = \sigma_x(y)$ and $\varphi(x) = x$ for all $x, y \in X$.*

Proof Let $FM(X)$ be the free monoid on X. Then the map

$$\theta' \colon FM(X) \to \operatorname{Aut}(FM(X))$$

defined by

$$\theta'(x_1 \cdots x_n)(y_1 \cdots y_m) = \sigma_{x_1} \cdots \sigma_{x_n}(y_1) \cdots \sigma_{x_1} \cdots \sigma_{x_n}(y_m),$$

for all $x_1, \dots, x_n, y_1, \dots, y_m \in X$, is the unique monoid homomorphism from $FM(X)$ to $\operatorname{Aut}(FM(X))$ such that $\theta'(x)(y) = \sigma_x(y)$ for all $x, y \in X$. Let

$$\varphi' \colon FM(X) \to FM(X)$$

be the map defined by

$$\varphi'(x_1 \cdots x_n) = x_1\sigma_{x_1}(x_2) \cdots \sigma_{x_1} \cdots \sigma_{x_{n-1}}(x_n)$$

for all $x_1, \dots, x_n \in X$. It is easy to check that φ is the unique 1-cocycle with respect to θ' such that $\varphi(x) = x$ for all $x \in X$, and clearly it is bijective.

Let η be the congruence on $FM(X)$ generated by the set

$$\{(xy, \sigma_x(y)\tau_y(x)) : x, y \in X\}.$$

Note that $FM(X)/\eta = M(X,r)$. Since $\sigma_x\sigma_y = \sigma_{\sigma_x(y)}\sigma_{\tau_y(x)}$ for all $x, y \in X$, we have that $\eta \subseteq \ker\theta$. Let

$$\varphi'(\eta) = \{(\varphi'(a), \varphi'(b)) : (a,b) \in \eta\}.$$

Let $x, y, z \in X$. By Lemma 4.1.5, we have that

$$\begin{aligned}\theta'(z)(\varphi'(xy)) &= \theta'(z)(x\sigma_x(y)) = \sigma_z(x)\sigma_z\sigma_x(y)\\ &= \sigma_z(x)\sigma_{\sigma_z(x)}\sigma_{\tau_x(z)}(y)\\ &= \varphi'(\sigma_z(x)\sigma_{\tau_x(z)}(y))\end{aligned}$$

and

$$\begin{aligned}\theta'(z)(\varphi'(\sigma_x(y)\tau_y(x))) &= \theta'(z)(\sigma_x(y)\sigma_{\sigma_x(y)}\tau_y(x))\\ &= \sigma_z\sigma_x(y)\sigma_z\sigma_{\sigma_x(y)}\tau_y(x)\\ &= \sigma_{\sigma_z(x)}\sigma_{\tau_x(z)}(y)\sigma_{\sigma_z\sigma_x(y)}\sigma_{\tau_{\sigma_x(y)}(z)}\tau_y(x)\\ &= \sigma_{\sigma_z(x)}\sigma_{\tau_x(z)}(y)\sigma_{\sigma_{\sigma_z(x)}\sigma_{\tau_x(z)}(y)}\tau_{\sigma_{\tau_x(z)}(y)}\sigma_z(x)\\ &= \varphi'(\sigma_{\sigma_z(x)}(\sigma_{\tau_x(z)}(y))\tau_{\sigma_{\tau_x(z)}(y)}(\sigma_z(x))).\end{aligned}$$

Now it is easy to see that

$$\varphi'(\eta) = \{(\theta'(c)(\varphi'(a)), \theta'(c)(\varphi'(b))) : (a,b) \in \eta\}$$

for all $c \in FM(X)$. Hence η satisfies properties (1) and (2) of Lemma 9.2.2. Thus $\varphi'(\eta)$ is a congruence, and it is generated by the pairs

$$(\varphi'(xy), \varphi'(\sigma_x(y)\tau_y(x)) = (x\sigma_x(y), \sigma_x(y)\sigma_{\sigma_x(y)}\tau_y(x))$$

for all $x, y \in X$. Hence $FM(X)/\varphi'(\eta) = A(X,r)$ and the result follows by Lemma 9.2.2. □

A monoid M is said to be *cancellative* if

$$xy = xz \implies y = z$$

and

$$xy = zy \implies x = z$$

for all $x, y, z \in M$.

Proposition 9.2.4 *Let (X,r) be an involutive solution to the YBE. Then $M(X,r)$ is a cancellative monoid.*

Proof By Proposition 9.2.3, there exist a monoid homomorphism

$$\theta\colon M(X,r) \to \mathrm{Aut}(A(X,r))$$

and a bijective 1-cocycle

$$\varphi\colon M(X,r) \to A(X,r)$$

with respect to θ such that $\theta(x)(y) = \sigma_x(y)$ and $\varphi(x) = x$ for all $x, y \in X$. Note that

$$A(X,r) = \langle X : x\sigma_x(y) = \sigma_x(y)\sigma_{\sigma_x(y)}\tau_y(x) \text{ for all } x, y \in X\rangle$$

and $\sigma_{\sigma_x(y)}\tau_y(x) = x$ because (X,r) is involutive. Hence

$$A(X,r) = \langle X : xy = yx \text{ for all } x, y \in X\rangle$$

is the free abelian monoid on X. In particular, $A(X,r)$ is cancellative.

Let $a, b, c, d \in M(X,r)$ be elements such that

$$abc = adc.$$

Hence $\theta(a)\theta(b)\theta(c) = \theta(abc) = \theta(adc) = \theta(a)\theta(d)\theta(c)$, and thus $\theta(b) = \theta(d)$. Now we have that

$$\begin{aligned}
\varphi(a)\theta(a)(\varphi(b))\theta(a)\theta(b)(\varphi(c)) &= \varphi(abc)\\
&= \varphi(adb)\\
&= \varphi(a)\theta(a)(\varphi(d))\theta(a)\theta(d)(\varphi(c)).
\end{aligned}$$

Since $A(X,r)$ is cancellative and $\theta(b) = \theta(d)$, we have that

$$\theta(a)(\varphi(b)) = \theta(a)\varphi(d).$$

Since $\theta(a)$ and φ are bijective, we get that $b = d$. Therefore the result follows. □

9.3 The Structure Group of a Solution

Let (X,r) be a solution to the YBE. Write as usual $r(x,y) = (\sigma_x(y), \tau_y(x))$. The *structure group* of (X,r) is the group

$$G(X,r) = \operatorname{gr}(X : xy = \sigma_x(y)\tau_y(x) \text{ for all } x, y \in X).$$

The *derived structure group* of (X,r) is the group

$$G'(X,r) = \operatorname{gr}(X : x\sigma_x(y) = \sigma_x(y)\sigma_{\sigma_x(y)}\tau_y(x) \text{ for all } x, y \in X).$$

In this section, we shall see that there is a natural braiding operator on $G(X,r)$ induced by r that satisfies a nice universal property.

Theorem 9.3.1 *Let (X, r) be a solution to the YBE. Then there exist a unique group homomorphism $\theta\colon G(X, r) \to \operatorname{Aut}(G'(X, r))$ and a unique bijective 1-cocycle $\varphi\colon G(X, r) \to G'(X, r)$ with respect to θ such that*

$$\theta(i(x))(i'(y)) = i'(\sigma_x(y)) \text{ and } \varphi(i(x)) = i'(x)$$

for all $x, y \in X$, where $i\colon X \to G(X, r)$ and $i'\colon X \to G'(X, r)$ are the natural maps.

Proof Let $X' = \{x' : x \in X\}$ be a copy of X and let Y be the disjoint union of X and X'. Note that the group $G(X, r)$ can be presented as a monoid as

$$G(X, r) = \langle Y : xy = \sigma_x(y)\tau_y(x),\ xx' = x'x = 1,\ \text{for all } x, y \in X\rangle,$$

and the group $G'(X, r)$ can be presented as a monoid as

$$G'(X, r) = \langle Y : x\sigma_x(y) = \sigma_x(y)\sigma_{\sigma_x(y)}\tau_y(x),\ xx' = x'x = 1 \text{ for all } x, y \in X\rangle.$$

Let $FM(Y)$ be the free monoid on Y. Let $\theta'\colon FM(Y) \to \operatorname{Aut}(FM(Y))$ be the unique monoid homomorphism such that

$$\theta'(x)(y) = \sigma_x(y),\ \theta'(x)(y') = \sigma_x(y)',\ \theta'_{x'}(y) = \sigma_x^{-1}(y) \text{ and } \theta'_{x'}(y') = \sigma_x^{-1}(y)'$$

for all $x, y \in X$. Let $\varphi'\colon FM(Y) \to FM(Y)$ be the unique 1-cocycle with respect to θ' such that $\varphi'(1) = 1$,

$$\varphi'(x) = x \text{ and } \varphi'(x') = \sigma_x^{-1}(x)'$$

for all $x \in X$. By Lemma 4.3.12, the map $T\colon X \to X$, defined by $T(x) = \sigma_x^{-1}(x)$ for all $x \in X$, is bijective. Hence the restriction of φ' to Y is bijective. Note that a natural degree exists on $FM(Y)$. For $y_1, \ldots, y_n \in Y$, we say that the element

$$y_1 \cdots y_n \in FM(Y)$$

has degree n. It is clear that $\theta'(a)(b)$ and b have the same degree for all $a, b \in FM(Y)$. For every non-negative integer n, let

$$Y_n = \{a \in FM(Y) : a \text{ has degree } n\}.$$

By induction on n, we shall prove that the restriction of φ' to Y_n is bijective. By definition, $Y_0 = \{1\}$ and $\varphi'(1) = 1$. We have seen that the restriction of φ' to $Y_1 = Y$ is bijective. Suppose that $n > 1$ and that the restriction of φ' to Y_{n-1} is bijective. Let $a \in Y_n$. There exist $y \in Y$ and $b \in Y_{n-1}$ such that $a = yb$. Let $z \in Y$ such that $\varphi(z) = y$. By the inductive hypothesis, there exists $c \in Y_{n-1}$ such that $b = \theta'(z)(\varphi'(c))$. Hence

$$\varphi'(zc) = \varphi'(z)\theta'(z)(\varphi'(c)) = yb = a.$$

Thus the restriction of φ' to Y_n is surjective. Let $y_1, y_2 \in Y$ and $b_1, b_2 \in Y_{n-1}$ such that $\varphi'(y_1 b_1) = \varphi'(y_2 b_2)$. We have that

$$\varphi'(y_1)\theta'(y_1)(\varphi'(b_1) = \varphi'(y_1 b_1) = \varphi'(y_2 b_2) = \varphi'(y_2)\theta'(y_2)(\varphi'(b_2)).$$

Hence $\varphi'(y_1) = \varphi'(y_2)$ and $\theta'(y_1)(\varphi'(b_1)) = \theta'(y_2)(\varphi'(b_2))$. Since the restriction of φ' to Y_1 is bijective, we get that $y_1 = y_2$ and thus $\varphi'(b_1) = \varphi'(b_2)$. By the inductive hypothesis, $b_1 = b_2$. Hence the restriction of φ' to Y_n is injective. Therefore, the restriction of φ' to Y_n is bijective. By induction, φ' is a bijective 1-cocycle with respect to θ'.

Let η be the monoid congruence generated by

$$(xy, \sigma_x(y)\tau_y(x)) : x, y \in X\} \cup \{(xx', 1) : x \in X\} \cup \{(x'x, 1) : x \in X\}.$$

Note that

$$\theta'(xy)(z) = \sigma_x\sigma_y(z) = \sigma_{\sigma_x(y)}\sigma_{\tau_y(x)}(z) = \theta'(\sigma_x(y)\tau_y(x))(z)$$

and

$$\theta'(xy)(z') = \sigma_x\sigma_y(z)' = \sigma_{\sigma_x(y)}\sigma_{\tau_y(x)}(z)' = \theta'(\sigma_x(y)\tau_y(x))(z')$$

for all $x, y, z \in X$. Hence $\eta \subseteq \ker \theta'$. As in the proof of Proposition 9.2.3, for $x, y, z \in X$, we have that

$$\theta'(z)(\varphi'(xy)) = \varphi'(\sigma_z(x)\sigma_{\tau_x(z)}(y))$$

and

$$\theta'(z)(\varphi'(\sigma_x(y)\tau_y(x)) = \varphi'(\sigma_{\sigma_z(x)}(\sigma_{\tau_x(z)}(y))\tau_{\sigma_{\tau_x(z)}(y)}(\sigma_z(x))).$$

For $x, z \in X$, we have

$$\begin{aligned}\theta'(z)(\varphi'(xx')) &= \theta'(z)(\varphi'(x)\theta'(x)(\varphi'(x')))\\ &= \theta'(z)(x)\theta'(z)(\theta'(x)(\sigma_x^{-1}(x)'))\\ &= \sigma_z(x)\sigma_z(x)'\end{aligned}$$

and $\theta'(z)(\varphi'(1)) = 1$. Now it is easy to see that

$$\varphi'(\eta) = \{(\theta'(c)(\varphi'(a)), \theta'(c)(\varphi'(b))) : (a, b) \in \eta\}$$

for all $c \in FM(Y)$. Hence η satisfies properties (1) and (2) of Lemma 9.2.2. Thus $\varphi'(\eta)$ is a congruence, and it is generated by the pairs

$$(\varphi'(xy), \varphi'(\sigma_x(y), \tau_y(x))) = (x\sigma_x(y), \sigma_{\sigma_x(y)}\tau_y(x)),$$
$$(\varphi(xx'), \varphi'(1)) = (xx', 1),$$

and

$$(\varphi'(x'x), \varphi'(1)) = (\sigma_x^{-1}(x)'\sigma_x^{-1}(x), 1)$$

for all $x, y \in X$. Hence $FM(Y)/\varphi'(\eta) = G'(X, r)$ and the result follows by Lemma 9.2.2. □

Theorem 9.3.2 *Let (X, r) be a solution to the YBE. Let $i\colon X \to G(X, r)$ be the natural map. Then there exists a unique structure of skew brace with multiplicative group the structure group $G(X, r)$ such that*

$$\lambda_{i(x)}(i(y)) = i(\sigma_x(y))$$

for all $x, y \in X$. Furthermore, if $(B, +, \circ)$ is a skew brace and $j\colon X \to B$ is a map such that $\lambda_{j(x)}(j(y)) = j(\sigma_x(y))$ and $j(x) \circ j(y) = j(\sigma_x(y)) \circ j(\tau_y(x))$ for all $x, y \in X$, then there exists a unique homomorphism of skew braces $f\colon G(X, r) \to B$ such that $fi = j$.

Proof By Theorem 9.3.1, there exist a unique group homomorphism

$$\theta\colon G(X, r) \to \operatorname{Aut}(G'(X, r))$$

and a unique bijective 1-cocycle

$$\varphi\colon G(X, r) \to G'(X, r)$$

with respect to θ such that $\theta(i(x))(i'(y)) = i'(\sigma_x(y))$ and $\varphi(i(x)) = i'(x)$ for all $x, y \in X$, where $i\colon X \to G(X, r)$ and $i'\colon X \to G'(X, r)$ are the natural maps.

We define on $G(X, r)$ a second group structure given by

$$a + b = \varphi^{-1}(\varphi(a)\varphi(b))$$

for all $a, b \in G(X, r)$. Note that

$$\varphi(a(b + c)) = \varphi(a)\theta(a)(\varphi(b)\varphi(c)) = \varphi(a)\theta(a)(\varphi(b))\theta(a)(\varphi(c))$$

and

$$\begin{aligned}\varphi(ab - a + ac)) &= \varphi(ab)\varphi(a)^{-1}\varphi(ac)\\ &= \varphi(a)\theta(a)(\varphi(b))\varphi(a)^{-1}\varphi(a)\theta(a)(\varphi(c))\\ &= \varphi(a)\theta(a)(\varphi(b))\theta(a)(\varphi(c))\end{aligned}$$

for all $a, b, c \in G(X, r)$. Since φ is bijective, we have that

$$a(b + c) = ab - a + ac$$

for all $a, b, c \in G(X, r)$. Hence $(G(X, r), +, \cdot)$ is a skew brace.
Let $x, y \in X$. We have that

$$\begin{aligned}\lambda_{i(x)}(i(y)) &= -i(x) + i(x)i(y)\\ &= \varphi^{-1}(\varphi(i(x))^{-1}\varphi(i(x))\theta(i(x))(\varphi(i(y))))\\ &= \varphi^{-1}(\theta(i(x))(\varphi(i(y))))\\ &= \varphi^{-1}(\theta(i(x))(i'(y)))\\ &= \varphi^{-1}(i'(\sigma_x(y)))\\ &= i(\sigma_x(y)).\end{aligned}$$

Let $(B, +, \circ)$ be a skew brace and $j : X \to B$ be a map such that

$$\lambda_{j(x)}(j(y)) = j(\sigma_x(y))$$

and

$$j(x) \circ j(y) = j(\sigma_x(y)) \circ j(\tau_y(x))$$

for all $x, y \in X$. By the presentation of $G(X, r)$ (as a group), there exists a unique group homomorphism $f : G(X, r) \to (B, \circ)$ such that $f(i(x)) = j(x)$ for all $x \in X$. We shall prove that $f(a + b) = f(a) + f(b)$ for all $a, b \in G(X, r)$. Note that

$$f(a + b) = f(a\lambda_a^{-1}(b)) = f(a) \circ f(\lambda_a^{-1}(b)) = f(a) \circ f(\lambda_{a^{-1}}(b))$$

and

$$f(a) + f(b) = f(a) \circ \lambda_{f(a)}^{-1}(f(b)) = f(a) \circ \lambda_{f(a^{-1})}(f(b))$$

Hence it is enough to prove that

$$f(\lambda_a(b)) = \lambda_{f(a)}(f(b))$$

for all $a, b \in G(X, r)$. Note that

$$f(\lambda_{i(x)}(i(y))) = f(i(\sigma_x(y)) = j(\sigma_x(y)) = \lambda_{j(x)}(j(y)) = \lambda_{f(i(x))}(f(i(y)))$$

for all $x, y \in X$. We also have that

$$f(\lambda_{i(x)^{-1}}(i(y))) = f(\lambda_{i(x)}^{-1}(i(y))) = f(i(\sigma_x^{-1}(y)))$$

$$= j(\sigma_x^{-1}(y)) = \lambda_{j(x)}^{-1}(j(y)) = \lambda_{f(i(x)^{-1})}(f(i(y)))$$

for all $x, y \in X$. By Lemma 4.3.12, the map $T\colon X \to X$, defined by $T(x) = \sigma_x^{-1}(x)$ for all $x \in X$, is bijective. Hence for all $y \in X$ there exists $z \in X$ such that $y = \sigma_z^{-1}(z)$, and thus

$$-i(y) = -i(\sigma_z^{-1}(z)) = -\lambda_{i(z)}^{-1}(i(z)) = i(z)^{-1}$$

and

$$\begin{aligned} f(-i(y)) &= f(i(z)^{-1}) = f(i(z))^{-1} = j(z)^{-1} \\ &= -\lambda_{j(z)^{-1}}(j(z)) = -\lambda_{f(i(z)^{-1})}(f(i(z))) = -f(\lambda_{i(z)^{-1}}(i(z))) \\ &= -f(i(\sigma_z^{-1}(z))) = -f(i(y)). \end{aligned}$$

Now we have that

$$\begin{aligned} f(\lambda_{i(x)}(i(y)^{-1})) &= f(-\lambda_{i(x)}\lambda_{i(y)^{-1}}(i(y)) = f(-\lambda_{i(x)}(i(\sigma_y^{-1}(y)))) \\ &= f(-i(\sigma_x\sigma_y^{-1}(y))) = -j(\sigma_x\sigma_y^{-1}(y)) \\ &= -\lambda_{j(x)}(j(\sigma_y^{-1}(y))) = \lambda_{f(i(x))}(-f(i(\sigma_y^{-1}(y)))) \\ &= \lambda_{f(i(x))}(f(-i(\sigma_y^{-1}(y)))) = \lambda_{f(i(x))}(f(i(y)^{-1})) \end{aligned}$$

for all $x, y \in X$. We also have that

$$\begin{aligned} f(\lambda_{i(x)^{-1}}(i(y)^{-1})) &= f(-\lambda_{i(x)^{-1}}\lambda_{i(y)^{-1}}(i(y))) = f(-\lambda_{i(x)^{-1}}(i(\sigma_y^{-1}(y)))) \\ &= f(-i(\sigma_x^{-1}\sigma_y^{-1}(y))) = -j(\sigma_x^{-1}\sigma_y^{-1}(y)) \\ &= -\lambda_{j(x)}^{-1}(j(\sigma_y^{-1}(y))) = \lambda_{f(i(x))}^{-1}(-f(i(\sigma_y^{-1}(y)))) \\ &= \lambda_{f(i(x)^{-1})}(f(-i(\sigma_y^{-1}(y)))) = \lambda_{f(i(x)^{-1})}(f(i(y)^{-1})) \end{aligned}$$

for all $x, y \in X$. For $a, b \in G(X, r)$ there exist $x_1, \dots, x_n, y_1, \dots, y_m \in X$ and $\varepsilon_1, \dots, \varepsilon_n, \varepsilon'_1, \dots, \varepsilon'_m \in \{-1, 1\}$ such that

$$a = i(x_1)^{\varepsilon_1} \cdots i(x_n)^{\varepsilon_n} \text{ and } b = i(y_1)^{\varepsilon'_1} \cdots i(y_m)^{\varepsilon'_m}.$$

We shall prove that $f(\lambda_a(b)) = \lambda_{f(a)}(f(b))$ by induction on m. For $m = 1$, we have that

$$\begin{aligned} \lambda_{f(a)}(f(b)) &= \lambda_{f(i(x_1)^{\varepsilon_1})} \cdots \lambda_{f(i(x_n)^{\varepsilon_n})}(f(i(y_1)^{\varepsilon'_1})) \\ &= f(\lambda_{i(x_1)^{\varepsilon_1}} \cdots \lambda_{i(x_n)^{\varepsilon_n}}(i(y_1)^{\varepsilon'_1})) \\ &= f(\lambda_a(b). \end{aligned}$$

Suppose that $m > 1$ and that $f(\lambda_c(d)) = \lambda_{f(c)}(f(d))$ for all $c \in G(X, r)$ and all $d = i(z_1)^{\nu_1} \cdots i(z_{m-1})^{\nu_{m-1}}$, where $z_1, \dots, z_{m-1} \in X$ and $\nu_1, \dots, \nu_{m-1} \in \{-1, 1\}$. By the inductive hypothesis, we have that

$$\begin{aligned}
\lambda_{f(a)}(f(b)) &= \lambda_{f(a)}(f(i(y_1)^{\varepsilon'_1}) \circ f(i(y_2)^{\varepsilon'_2} \cdots i(y_m)^{\varepsilon'_m})) \\
&= \lambda_{f(a)}(f(i(y_1)^{\varepsilon'_1}) + \lambda_{f(i(y_1)^{\varepsilon'_1})}(f(i(y_2)^{\varepsilon'_2} \cdots i(y_m)^{\varepsilon'_m})) \\
&= \lambda_{f(a)}(f(i(y_1)^{\varepsilon'_1})) + \lambda_{f(a)}\lambda_{f(i(y_1)^{\varepsilon'_1})}(f(i(y_2)^{\varepsilon'_2} \cdots i(y_m)^{\varepsilon'_m})) \\
&= f(\lambda_a(i(y_1)^{\varepsilon'_1})) + \lambda_{f(ai(y_1)^{\varepsilon'_1})}(f(i(y_2)^{\varepsilon'_2} \cdots i(y_m)^{\varepsilon'_m})) \\
&= f(\lambda_a(i(y_1)^{\varepsilon'_1})) \circ \lambda_{f(\lambda_a(i(y_1)^{\varepsilon'_1})^{-1}ai(y_1)^{\varepsilon'_1})}(f(i(y_2)^{\varepsilon'_2} \cdots i(y_m)^{\varepsilon'_m}))) \\
&= f(\lambda_a(i(y_1)^{\varepsilon'_1})) \circ f(\lambda_{\lambda_a(i(y_1)^{\varepsilon'_1})^{-1}ai(y_1)^{\varepsilon'_1}}(i(y_2)^{\varepsilon'_2} \cdots i(y_m)^{\varepsilon'_m})) \\
&= f(\lambda_a(i(y_1)^{\varepsilon'_1})\lambda_{\lambda_a(i(y_1)^{\varepsilon'_1})^{-1}ai(y_1)^{\varepsilon'_1}}(i(y_2)^{\varepsilon'_2} \cdots i(y_m)^{\varepsilon'_m})) \\
&= f(\lambda_a(i(y_1)^{\varepsilon'_1}) + \lambda_{ai(y_1)^{\varepsilon'_1}}(i(y_2)^{\varepsilon'_2} \cdots i(y_m)^{\varepsilon'_m})) \\
&= f(\lambda_a(i(y_1)^{\varepsilon'_1} + \lambda_{i(y_1)^{\varepsilon'_1}}(i(y_2)^{\varepsilon'_2} \cdots i(y_m)^{\varepsilon'_m})) \\
&= f(\lambda_a(i(y_1)^{\varepsilon'_1} i(y_2)^{\varepsilon'_2} \cdots i(y_m)^{\varepsilon'_m})) = f(\lambda_a(b)).
\end{aligned}$$

Hence by induction $f(\lambda_a(b)) = \lambda_{f(a)}(f(b))$ for all $a, b \in G(X, r)$. Therefore the result follows. □

Definition 9.3.3 Given a solution (X, r) to the YBE, its *structure skew brace* is the skew brace structure on the group $G(X, r)$ given in Theorem 9.3.2.

Theorem 9.3.4 *Let (X, r) be an involutive solution to the YBE. Then the additive group of the structure skew brace $G(X, r)$ is the additive free abelian group on X.*

Proof As we have seen at the beginning of the proof of Theorem 9.3.2, the additive group of the structure skew brace $G(X, r)$ is isomorphic to the derived structure group

$$G'(X, r) = \mathrm{gr}(X : x\sigma_x(y) = \sigma_x(y)\sigma_{\sigma_x(y)}\tau_y(x) \text{ for all } x, y \in X)$$

of (X, r). Since (X, r) is involutive,

$$\sigma_{\sigma_x(y)}\tau_y(x) = x$$

for all $x, y \in X$. Hence

$$G'(X, r) = \mathrm{gr}(X : x\sigma_x(y) = \sigma_x(y)x, \text{ for all } x, y \in X)$$

is the free abelian group on X, and the result follows. □

Theorem 9.3.5 *Let (X, r) be a solution to the YBE. Let $i \colon X \to G(X, r)$ be the natural map. Then there exists a unique braiding operator r_G on $G(X, r)$ such that*

$r_G(i \times i) = (i \times i)r$. Furthermore, if (H, s) is a braided group and $j: X \to H$ is a map such that $s(j \times j) = (j \times j)r$, then there exists a unique group homomorphism $f: G(X,r) \to H$ such that $s(f \times f) = (f \times f)r_G$ and $j = fi$. Furthermore, if $r^2 = \mathrm{id}_{X^2}$, then $r_G^2 = \mathrm{id}_{G(X,r)^2}$.

Proof By Theorem 9.3.2, there exists a unique structure of skew brace on $G(X,r)$ with multiplicative group the structure group $G(X,r)$ such that

$$\lambda_{i(x)}(i(y)) = i(\sigma_x(y))$$

for all $x, y \in X$. Let $G = G(X,r)$. Let $r_G: G \times G \to G \times G$ be the map defined by

$$r_G(a,b) = (\lambda_a(b), \lambda_a(b)^{-1}ab)$$

for all $a, b \in G$. By the proof of Theorem 9.1.3, r_G is a braiding operator on the structure group G. Note that by the defining relations of the structure group G,

$$\lambda_{i(x)}(i(y))^{-1}i(x)i(y) = i(\sigma_x(y))^{-1}i(x)i(y) = i(\tau_y(x))$$

for all $x, y \in X$. Hence $r_G(i \times i) = (i \times i)r$.

Let (H, s) be a braided group and $j: X \to H$ is a map such that $s(j \times j) = (j \times j)r$. By the proof of Theorem 9.1.3, $(H, +_s, \cdot)$, where

$$a +_s b = a(a^{-1} \triangleright_s b) \text{ and } s(a,b) = (a \triangleright_s b, a \triangleleft_s b)$$

for all $a, b \in H$, is a skew brace. The lambda map of this skew brace satisfies $\lambda_a(b) = a \triangleright_s b$ for all $a, b \in H$. Since $s(j \times j) = (j \times j)r$, we have that

$$j(\sigma_x(y)) = j(x) \triangleright_s j(y) = \lambda_{j(x)}(j(y))$$

for all $x, y \in X$. Since s is a braiding operator on H, we have that

$$ab = (a \triangleright_s b)(a \triangleleft_s b)$$

for all $a, b \in H$. In particular,

$$j(x)j(y) = (j(x) \triangleright_s j(y))(j(x) \triangleleft_s j(y)) = j(\sigma_x(y))j(\tau_y(x))$$

for all $x, y \in X$. By Theorem 9.3.2, there exists a unique homomorphism of skew braces $f: G \to H$ such that $fi = j$. Note that

$$\begin{aligned} s(f(a), f(b)) &= (\lambda_{f(a)}(f(b)), \lambda_{f(a)}(f(b))^{-1}f(a)f(b)) \\ &= (f(\lambda_a(b), f(\lambda_a(b)^{-1}ab)) \end{aligned}$$

for all $a, b \in G$. Hence $s(f \times f) = (f \times f)r_G$.

Suppose that $r^2 = \mathrm{id}_{X^2}$. By Theorem 9.3.4, the additive group of G is the free abelian group on X. Hence

$$\begin{aligned}\lambda_a(b)^{-1}ab &= \lambda_a(b)^{-1}(a + \lambda_a(b)) = \lambda_a(b)^{-1}a - \lambda_a(b)^{-1} \\ &= -\lambda_a(b)^{-1} + \lambda_a(b)^{-1}a = \lambda_{\lambda_a(b)^{-1}}(a) = \lambda^{-1}_{\lambda_a(b)}(a)\end{aligned}$$

for all $a, b \in G$. Thus

$$r_G(a, b) = (\lambda_a(b), \lambda^{-1}_{\lambda_a(b)}(a))$$

for all $a, b \in G$ and it is easy to check that $r_G^2 = \mathrm{id}_{G^2}$. Therefore the result follows. □

9.4 The Permutation Group of a Solution

Let (X, r) be a solution to the YBE. We write

$$r(x, y) = (\sigma_x(y), \tau_y(x))$$

for all $x, y \in X$. Consider the structure group $G(X, r)$ of the solution (X, r). Let $i \colon X \to G(X, r)$ be the natural map. The *permutation group* of (X, r) is the subgroup

$$\mathcal{G}(X, r) = \langle(\sigma_x, \tau_x^{-1}) : x \in X\rangle \subseteq \mathbb{S}_X \times \mathbb{S}_X.$$

Since

$$\sigma_x\sigma_y = \sigma_{\sigma_x(y)}\sigma_{\tau_y(x)} \quad \text{and} \quad \tau_x^{-1}\tau_y^{-1} = \tau^{-1}_{\sigma_x(y)}\tau^{-1}_{\tau_y(x)}$$

for all $x, y \in X$, there exists a unique group homomorphism $h \colon G(X, r) \to \mathcal{G}(x, r)$ such that $hi(x) = (\sigma_x, \tau_x^{-1})$ for all $x \in X$. We write

$$h(a) = (\sigma_a, \tau_a^{-1})$$

for all $a \in G(X, r)$. By Theorem 9.3.2, $G(X, r)$ has a unique structure of skew brace with multiplicative group the structure group $G(X, r)$ and $\lambda_{i(x)}(i(y)) = i(\sigma_x(y))$ for all $x, y \in X$. We shall see that $\ker h$ is an ideal of the skew brace $G(X, r)$. Note that

$$\mu_{i(y)}(i(x)) = \lambda_{i(x)}(i(y))^{-1}i(x)i(y) = i(\sigma_x(y))^{-1}i(x)i(y) = i(\tau_y(x))$$

for all $x, y \in X$, by the defining relations of $G(X, r)$.

Lemma 9.4.1 *With the above notation,*

$$\ker h \subseteq \ker\lambda \cap \ker\mu = \mathrm{Soc}(G(X, r)).$$

Proof Let $a \in \ker h$. There exists $x_1, \dots, x_n \in X$ and $\epsilon_1, \dots, \epsilon_n \in \{-1, 1\}$ such that $a = i(x_1)^{\epsilon_1} \cdots i(x_n)^{\epsilon_n}$. Hence

$$(\mathrm{id}_X, \mathrm{id}_X) = h(a) = (\sigma_{x_1}^{\epsilon_1} \cdots \sigma_{x_n}^{\epsilon_n}, \tau_{x_1}^{-\epsilon_1} \cdots \tau_{x_n}^{-\epsilon_n}).$$

Therefore

$$\lambda_a(i(z)) = i(\sigma_{x_1}^{\epsilon_1} \cdots \sigma_{x_n}^{\epsilon_n}(z)) = i(z)$$

and

$$\mu_a^{-1}(i(z)) = i(\tau_{x_1}^{-\epsilon_1} \cdots \tau_{x_n}^{-\epsilon_n}(z)) = i(z)$$

for all $z \in X$. Since λ_a is an automorphism of the additive group of the skew brace $G(X, r)$ and this additive group is generated by $i(X)$, we have that $\lambda_a = \mathrm{id}$. On the other hand, we know that $r_G : G(X, r) \times G(X, r) \to G(X, r) \times G(X, r)$, defined by $r_G(b, c) = (\lambda_b(c), \mu_c(b))$ for all $b, c \in G(X, r)$, is a braiding operator on the structure group $G(X, r)$. Hence $\mu_a(1) = 1$ and

$$\begin{aligned} 1 = \mu_a(i(x)^{-1}i(x)) &= \mu_{\lambda_{i(x)}(a)}(i(x)^{-1})\mu_a(i(x)) \\ &= \mu_{i(x)a\mu_a(i(x))^{-1}}(i(x)^{-1})i(x) \\ &= \mu_{i(x)ai(x)^{-1}}(i(x)^{-1})i(x) \end{aligned}$$

for all $x \in X$. Hence $\mu_a(i(x)^{-1}) = i(x)^{-1}$ for all $x \in X$ and $a \in \ker h$. Let $b \in G(X, r)$. There exist $y_1, \dots, y_m \in X$ and $\eta_1, \dots, \eta_m \in \{-1, 1\}$ such that $b = i(y_1)^{\eta_1} \cdots i(y_m)^{\eta_m}$. We shall prove that $\mu_a(b) = b$ for all $a \in \ker h$, by induction on m. We have proved that this holds for $m = 1$. Suppose that $m > 1$ and that

$$\mu_a(i(y_1)^{\eta_1} \cdots i(y_{m-1})^{\eta_{m-1}}) = i(y_1)^{\eta_1} \cdots i(y_{m-1})^{\eta_{m-1}}$$

for all $a \in \ker h$, $y_1, \dots, y_{m-1} \in X$ and $\eta_1, \dots, \eta_{m-1} \in \{-1, 1\}$. By the inductive hypothesis, we have that

$$\begin{aligned} \mu_a(b) &= \mu_{\lambda_{i(y_m)^{\eta_m}}(a)}((y_1)^{\eta_1} \cdots i(y_{m-1})^{\eta_{m-1}})\mu_a(i(y_m)^{\eta_m}) \\ &= \mu_{i(y_m)^{\eta_m}a(\mu_a(i(y_m)^{\eta_m})^{-1}}(i(y_1)^{\eta_1} \cdots i(y_{m-1})^{\eta_{m-1}})i(y_m)^{\eta_m} \\ &= \mu_{i(y_m)^{\eta_m}ai(y_m)^{-\eta_m}}(i(y_1)^{\eta_1} \cdots i(y_{m-1})^{\eta_{m-1}})i(y_m)^{\eta_m} \\ &= i(y_1)^{\eta_1} \cdots i(y_{m-1})^{\eta_{m-1}}i(y_m)^{\eta_m} = b. \end{aligned}$$

Hence, by induction, $\mu_a = \mathrm{id}$ for all $a \in \ker h$. Thus

$$\ker h \subseteq \ker \lambda \cap \ker \mu = \mathrm{Soc}(G(X, r)),$$

by Proposition 8.4.14. □

Lemma 9.4.2 *With the above notation,* $\ker h$ *is an ideal of the skew brace* $G(X, r)$.

Proof By Lemma 9.4.1, $\ker h \subseteq \operatorname{Soc}(G(X,r))$. Hence $\ker h \subseteq Z(G(X,r),+)$ and thus $\ker h$ is a normal subgroup of the additive group of the skew brace $G(X,r)$. It is clear that $\ker h$ also is a normal subgroup of the multiplicative group of $G(X,r)$. Let $a \in \ker h$ and $b, c \in G(X,r)$. By Lemma 9.4.1, $\ker h \subseteq \ker \mu$. Hence we have that

$$\lambda_b(a) = ba(\mu_a(b))^{-1} = bab^{-1} \in \ker h.$$

Therefore $\ker h$ is an ideal of the skew brace $G(X,r)$. □

Let $\tilde{h}\colon G(X,r)/\ker h \to \mathcal{G}(X,r)$ be the isomorphism of groups defined by

$$\tilde{h}(a \ker h) = h(a)$$

for all $a \in G(X,r)$. We define an addition on $\mathcal{G}(X,r)$ by

$$g_1 + g_2 = \tilde{h}(\tilde{h}^{-1}(g_1) + \tilde{h}^{-1}(g_2))$$

for all $g_1, g_2 \in \mathcal{G}(X,r)$, i.e.,

$$h(a) + h(b) = h(a+b)$$

for all $a, b \in G(X,r)$. Hence $\mathcal{G}(X,r)$ with this addition is a skew brace, and then h and $\tilde{h}$ are skew brace homomorphisms.

In particular, the lambda map of $\mathcal{G}(X,r)$ satisfies that

$$\lambda_{(\sigma_a,\tau_a^{-1})}(\sigma_b, \tau_b^{-1}) = \lambda_{h(a)}h(b) = h(\lambda_a(b)) = (\sigma_{\lambda_a(b)}, \tau_{\lambda_a(b)}^{-1})$$

for all $a, b \in G(X,r)$.

Proposition 9.4.3 *Let (X,r) be an involutive solution to the YBE and*

$$h\colon G(X,r) \to \mathcal{G}(X,r)$$

be the unique homomorphism of groups such that $hi(x) = (\sigma_x, \tau_x^{-1})$ for all $x \in X$. Then $\ker(h) = \operatorname{Soc}(G(X,r))$ *and*

$$\mathcal{G}(X,r) \cong G(X,r)/\operatorname{Soc}(G(X,r)) \cong \langle \sigma_x : x \in X \rangle.$$

Proof By Theorem 9.3.4, the natural map $i\colon X \to G(X,r)$ is injective and $G(X,r)$ is a skew brace of abelian type. Hence $\operatorname{Soc}(G(X,r)) = \ker \lambda$. By Lemma 9.4.1, $\ker h \subseteq \ker \lambda$. Let $a \in \ker \lambda$. We have that $h(a) = (\sigma_a, \tau_a^{-1})$. There exist $x_1, \dots, x_k \in X$ and $\epsilon_1, \dots, \epsilon_k \in \{1, -1\}$ such that

$$a = i(x_1)^{\epsilon_1} \cdots i(x_k)^{\epsilon_k}.$$

Thus

$$i(\sigma_a(y)) = i(\sigma_{x_1}^{\epsilon_1} \cdots \sigma_{x_k}^{\epsilon_k}(y)) = \lambda_{i(x_1)}^{\epsilon_1} \cdots \lambda_{i(x_k)}^{\epsilon_k}(i(y)) = \lambda_a(i(y)) = i(y)$$

for all $y \in X$. Since i is injective, we have that

$$\sigma_a(y) = y$$

for all $y \in X$. Hence $\sigma_a = \mathrm{id}_X$. By Proposition 4.2.6,

$$\tau_x^{-1} = T^{-1}\sigma_x T$$

for all $x \in X$, where $T(y) = \sigma_y^{-1}(y)$ for all $y \in X$. Hence

$$\tau_a^{-1} = \tau_{x_1}^{-\epsilon_1} \cdots \tau_{x_k}^{-\epsilon_k} = T^{-1}\sigma_{x_1}^{\epsilon_1} \cdots \sigma_{x_k}^{\epsilon_k} T = T^{-1}\sigma_a T = \mathrm{id}_X .$$

Therefore $h(a) = (\mathrm{id}_X, \mathrm{id}_X)$ and consequently $\ker h = \ker \lambda = \mathrm{Soc}(G(X, r))$. Note also that the homomorphism of groups

$$\mathcal{G}(X, r) \to \langle \sigma_x : x \in X \rangle \colon (\sigma_a, \tau_a^{-1}) \mapsto \sigma_a$$

is an isomorphism. □

Thus, for involutive solutions (X, r), we identify $\mathcal{G}(X, r)$ with $\langle \sigma_x : x \in X \rangle$. In particular the lambda map of $\mathcal{G}(X, r)$ satisfies

$$\lambda_{\sigma_a}(\sigma_b) = \sigma_{\lambda_a(b)}$$

for all $a, b \in G(X, r)$.

9.5 Structure Skew Brace of a Finite Solution

In this section, we study the structure of the skew brace of a finite solution to the YBE.

Theorem 9.5.1 *Let (X, r) be a finite solution to the YBE. Then $\mathrm{Soc}(G(X, r))$ is a normal subgroup of finite index of the structure group $G(X, r)$. In particular, the structure group $G(X, r)$ is abelian-by-finite.*

Proof We write $r(x, y) = (\sigma_x(y), \tau_y(x))$ for all $x, y \in X$. Let $i \colon X \to G(X, r)$ be the natural map. We have seen that there is a unique group homomorphism

$$h \colon G(X, r) \to \mathcal{G}(X, r) \subseteq \mathbb{S}_X \times \mathbb{S}_X$$

such that $hi(x) = (\sigma_x, \tau_x^{-1})$ for all $x \in X$. By Lemma 9.4.1, $\ker h \subseteq \mathrm{Soc}(G(X, r))$. The result follows since $\mathcal{G}(X, r)$ is finite. □

We now use Hall's theorem to obtain information related to the structure of finite skew braces of nilpotent type.

Theorem 9.5.2 *Let A be a finite skew brace of nilpotent type. Then the multiplicative group of A is solvable.*

Proof Let K be the additive group of A and G be the multiplicative group of A. Assume that $|A| = p_1^{\alpha_1} \cdots p_n^{\alpha_n}$ for distinct primes $p_1, \dots, p_n$ and positive integers $\alpha_1, \dots, \alpha_n$. Since K is nilpotent and finite, each $K_i \in \mathrm{Syl}_{p_i}(K)$ is normal in K, so each K_i is a left ideal of A. It follows that for each $i \in \{1, \dots, n\}$ both K_i and $\prod_{j \neq i} K_j$ are subbraces of coprime order. In particular, for each $i \in \{1, \dots, n\}$ there exists a p_i-complement of G. Then G is solvable by Hall's theorem. □

Theorem 9.5.3 *Let (X, r) be a finite involutive solution to the YBE. Then the structure group $G(X, r)$ is solvable.*

Proof By Theorem 9.3.4, $G(X, r)$ is a skew brace of abelian type. By Theorem 9.5.1, $\mathrm{Soc}(G(X, r))$ is a normal abelian subgroup of the structure group $G(X, r)$ of finite index. Hence $G(X, r)/\mathrm{Soc}(G(X, r))$ is a finite skew brace of abelian type. Therefore the result follows by Theorem 9.5.2. □

9.6 Exercises

9.6.1 Check the missing steps of the proofs of this chapter's results.

9.7 Notes

The structure group and the permutation group of a solution to the YBE were introduced by Etingof, Schedler, and Soloviev in [56] to study involutive solutions to the YBE. The results in [56] were generalized to arbitrary solutions to the YBE by Lu, Yan, and Zhu in [87] and by Soloviev in [107]. We use the notation $G(X, r)$, for the structure group of a solution (X, r) to the YBE, due to Lu, Yan, and Zhu, and the notation $\mathcal{G}(X, r)$, for the permutation group of a solution (X, r) to the YBE, due to Gateva-Ivanova.

Lemma 9.1.2 and Theorem 9.1.3 are due to Lu, Yan and Zhu [87], taking into account Theorem 8.3.1. Lemma 9.2.2 and Proposition 9.2.3 are due to Lebed and Vendramin [85]. Proposition 9.2.4 is due to Gateva-Ivanova and Van den Bergh for finite involutive solutions. Theorem 9.3.5 is due to Lu, Yan and Zhu [87]. The original proof uses graphical figures. The proof here uses ideas of Lebed and Vendramin in [85], based on their Lemma 9.2.2, and the equivalence between braided groups, skew braces, and bijective 1-cocycles with respect to group actions. The involutive case of Theorem 9.3.2 is due to Rump [98, 99]. The general case of Theorem 9.3.2 is due to Guarnieri and Vendramin [64]. Theorem 9.3.4 is due to Etingof, Schedler and Soloviev [56], taking into account Theorem 8.3.1. Lemmas 9.4.1 and 9.4.2 are due essentially to Soloviev [107]. Theorem 9.5.1 is due to Etingof, Schedler, and Soloviev [56] and Gateva–Ivanova and Van den Bergh [62] in the involutive case, and to Lu, Yan, and Zhu [87] and Soloviev [107] in the general case. Theorem 9.5.2 was proved by Byott in the context of Hopf–Galois structures in [29]. Theorem 9.5.3 was proved by Etingof, Schedler, and Soloviev in [56].

Chapter 10
Bieberbach Groups

In this chapter, we study some classes of torsion-free groups and their relations.

10.1 Left Ordered Groups

A group G is *left ordered* if there is a total ordering $\leq$ on G such that $x \leq y$ implies $zx \leq zy$ for all $x, y, z \in G$. In this case, we say that G is a left ordered group with respect to the total order $\leq$. Similarly, one defines *right ordered* groups.

A group G is *ordered* if there is a total ordering $\leq$ on G such that $x \leq y$ implies $zx \leq zy$ and $xz \leq yz$ for all $x, y, z \in G$.

Example 10.1.1 The group $\mathbb{Z}$ is an ordered group with respect to the natural order.

Example 10.1.2 If G is a left ordered group and H is a subgroup of G, then H is left ordered.

Remark 10.1.3 Let G be a left ordered group with respect to the total order $\leq$. We define another total order $\leq'$ on G by

$$x \leq' y \text{ if and only if } x^{-1} \leq y^{-1}$$

for all $x, y \in G$. Note that if $x \leq' y$ then

$$(xz)^{-1} = z^{-1}x^{-1} \leq z^{-1}y^{-1} = (yz)^{-1},$$

and thus $xz \leq' yz$ for all $z \in G$. Hence G is a right ordered group with respect to the total order $\leq'$.

F. Cedó and L. Vendramin, *Groups, Radical Rings, and the Yang-Baxter Equation*, Progress in Mathematics 361, https://doi.org/10.1007/978-3-032-01524-2_10

Proposition 10.1.4 *Let G be a group and let $N \trianglelefteq G$. If N and G/N are left ordered groups, then G is left ordered.*

Proof Suppose that N is a left ordered group with respect to the total order $\leq_N$ and $\bar{G} = G/N$ is a left ordered group with respect to the total order $\leq_{\bar{G}}$. We define

$$x \leq y \iff \begin{cases} 1 \leq_N x^{-1}y & \text{if } xN = yN, \\ xN \leq_{\bar{G}} yN & \text{otherwise,} \end{cases}$$

for all $x, y \in G$. A straightforward computation shows that then G is a left ordered group with respect to the total order $\leq$. □

Example 10.1.5 Let us show that $G = \mathrm{gr}(x, y : xyx^{-1} = y^{-1})$ is left ordered. Let $f : G \to \mathbb{Z}$ be given by $x \mapsto 1$ and $y \mapsto 0$. Then $\ker f = \langle y \rangle$. The map

$$\{x, y\} \to \mathbf{GL}_2(\mathbb{C}), \quad x \mapsto \begin{pmatrix} -1 & 0 \\ 0 & 1 \end{pmatrix}, \quad y \mapsto \begin{pmatrix} 1 & 1 \\ 0 & 1 \end{pmatrix},$$

induces a group homomorphism $G \to \mathbf{GL}_2(\mathbb{C})$. In particular, y has infinite order and thus $\langle y \rangle \cong \mathbb{Z}$. Hence $\ker f$ and $G/\ker f$ are left ordered groups and, by Proposition 10.1.4, G is a left ordered group. We shall see that G is not an ordered group. Suppose that G is an ordered group with respect to a total order $\leq$. Suppose that $1 \leq y$. Then we have

$$1 = xx^{-1} \leq xyx^{-1} = y^{-1},$$

and thus $y \leq yy^{-1} = 1$, a contradiction because $y \neq 1$. Hence $y \leq 1$. But then we have that

$$y^{-1} = xyx^{-1} \leq xx^{-1} = 1,$$

and thus $1 = yy^{-1} \leq y$, a contradiction. Therefore G is not an ordered group.

The previous example is the Baumslag–Solitar group $B(1, -1)$. Recall that for $n, m \in \mathbb{Z}$ the Baumslag–Solitar group is defined as the group

$$B(m, n) = \mathrm{gr}(a, b : ba^m b^{-1} = a^n).$$

The map

$$\{a, b\} \to \mathbf{GL}_2(\mathbb{C}), \quad a \mapsto \begin{pmatrix} 1 & 1 \\ 0 & 1 \end{pmatrix} \quad \text{and} \quad b \mapsto \begin{pmatrix} \frac{1}{m} & 0 \\ 0 & \frac{1}{n} \end{pmatrix}$$

induces a group homomorphism $B(m, n) \to \mathbf{GL}_2(\mathbb{C})$.

A group G is said to be *poly-* $\mathbb{Z}$ if it has a finite subnormal series

$$\{1\} = N_0 \trianglelefteq N_1 \trianglelefteq \cdots \trianglelefteq N_n = G$$

such that $N_{i+1}/N_i \cong \mathbb{Z}$ for all $0 \leq i < n$. By Proposition 10.1.4 and induction on n, it is easy to see that every poly-$\mathbb{Z}$ group is left ordered.

A group G is said to be *indicable* if there exists a non-trivial group homomorphism $G \to \mathbb{Z}$, and G is said to be *locally indicable* if every non-trivial finitely generated subgroup of G is indicable.

Proposition 10.1.6 *Every poly-$\mathbb{Z}$ group is locally indicable.*

Proof Let G be a non-trivial poly-$\mathbb{Z}$ group. Let

$$\{1\} = N_0 \trianglelefteq N_1 \trianglelefteq \cdots \trianglelefteq N_n = G$$

be a subnormal series of G such that $N_{k+1}/N_k \cong \mathbb{Z}$ for all $0 \leq k < n$. Let m be a positive integer and $a_1, \ldots, a_m \in G \setminus \{1\}$. Let k be the smallest positive integer such that $a_1, \ldots, a_m \in N_k$. Let $H = \langle a_1, \ldots, a_m \rangle$ and let $f\colon H \to N_k/N_{k-1}$ be the map defined by $f(a) = aN_{k-1}$ for all $a \in H$. By the definition of k, f is a non-trivial group homomorphism. Since $N_k/N_{k-1} \cong \mathbb{Z}$, the result follows. □

Example 10.1.7 The additive group of $\mathbb{Q}$ is not poly-$\mathbb{Z}$ because it is not finitely generated. Every non-trivial finitely generated subgroup of $(\mathbb{Q}, +)$ is isomorphic to $\mathbb{Z}$. Hence $(\mathbb{Q}, +)$ is locally indicable.

Theorem 10.1.8 (Burns–Hale) *Let G be a group. Then G is left ordered if and only if for each finitely generated non-trivial subgroup H of G there exists a left ordered group L and a non-trivial group homomorphism $H \to L$.*

Proof If G is a left ordered group, take $L = G$.

Conversely, suppose that for each finitely generated non-trivial subgroup H of G there exists a left ordered group L and a non-trivial group homomorphism $H \to L$. We claim that for all $\{x_1, \ldots, x_n\} \subseteq G \setminus \{1\}$ there exist $\epsilon_1, \ldots, \epsilon_n \in \{-1, 1\}$ such that

$$1 \notin S(x_1^{\epsilon_1}, \ldots, x_n^{\epsilon_n}),$$

where $S(x_1^{\epsilon_1}, \ldots, x_n^{\epsilon_n})$ denotes the semigroup generated by the set $\{x_1^{\epsilon_1}, \ldots, x_n^{\epsilon_n}\}$. We proceed by induction on n. If $n = 1$, then $x_1 \in G \setminus \{1\}$. Let $\epsilon_1 = 1$. If $1 \in S(x_1)$, then x_1 is an element of finite order and hence $\langle x_1 \rangle \to L$ is the trivial homomorphism for every left ordered group L. Hence $1 \notin S(x_1)$. Now assume that the claim holds for some $n \geq 1$. Let $\{x_1, \ldots, x_{n+1}\} \subseteq G \setminus \{1\}$. By assumption, there exists a non-trivial group homomorphism $h\colon \langle x_1, \ldots, x_{n+1} \rangle \to L$ for some left ordered group L. In particular, $h(x_i) \neq 1$ for some $i \in \{1, \ldots, n+1\}$. Without loss of generality, we may assume that there exists an integer $1 \leq k \leq n+1$ such that $h(x_j) \neq 1$ for all $j \in \{1, \ldots, k\}$ and $h(x_j) = 1$ for all $j > k$. Suppose that L is left ordered with respect to a total order $\leq$. Since $h(x_j) \neq 1$ for all $j \leq k$, there are elements $\epsilon_j \in \{-1, 1\}$ such that $1 \leq h(x_j^{\epsilon_j})$ for all $j \leq k$. By the inductive hypothesis, there are elements $\epsilon_{k+1}, \ldots, \epsilon_{n+1} \in \{-1, 1\}$ such that $1 \notin S(x_{k+1}^{\epsilon_{k+1}}, \ldots, x_{n+1}^{\epsilon_{n+1}})$. Note that for every $x \in S(x_1^{\epsilon_1}, \ldots, x_{n+1}^{\epsilon_{n+1}}) \setminus S(x_{k+1}^{\epsilon_{k+1}}, \ldots, x_{n+1}^{\epsilon_{n+1}})$, $1 \leq h(x) \neq 1$. Hence $1 \notin S(x_1^{\epsilon_1}, \ldots, x_{n+1}^{\epsilon_{n+1}})$, and the claim follows by induction.

Consider the set

$\mathcal{F} = \{(F, f) : F$ is a finite subset of $G \setminus \{1\}$ and $f : F \to \{-1, 1\}$ such that for every finite subset B of $G \setminus \{1\}$ containing F, there exists a map $g : B \to \{-1, 1\}$ such that $1 \notin S(a^{g(a)} : a \in B)$ and $g(x) = f(x)$ for all $x \in F\}$.

Let $\mathcal{C} = \{(A, f) : A \subseteq G \setminus \{1\}$ and $f : A \to \{-1, 1\}$ such that $(F, f|_F) \in \mathcal{F}$ for all finite subset F of $A\}$. We define and order on $\mathcal{C}$ by $(A, f) \leq (B, g)$ if and only if $A \subseteq B$ and $g(a) = f(a)$ for all $a \in A$, i. e. $f = g|_A$. Note that there is a unique map $f_\emptyset : \emptyset \to \{-1, 1\}$. We have shown that $(\emptyset, f_\emptyset) \in \mathcal{C}$. Hence $\mathcal{C} \neq \emptyset$. Furthermore, it is easy to see that every chain of elements in $\mathcal{C}$ has an upper bound in $\mathcal{C}$. Thus, by Zorn's lemma, there exists a maximal element $(A, f) \in \mathcal{C}$. Suppose that $A \neq G \setminus \{1\}$. Let $x \in G \setminus (A \cup \{1\})$. Let $g_1 : A \cup \{x\} \to \{-1, 1\}$ and $g_{-1} : A \cup \{x\} \to \{-1, 1\}$ be the maps defined $g_i(a) = f(a)$ for all $a \in A$ and $g_i(x) = i$ for $i \in \{-1, 1\}$. By the maximality of (A, f), we have that $(A \cup \{x\}, g_i) \notin \mathcal{C}$. Hence there exist finite subsets F_1 and F_{-1} of $A \cup \{x\}$ and finite subsets B_1 and B_{-1} of $G \setminus \{1\}$ such that $F_1 \subseteq B_1$, $F_{-1} \subseteq B_{-1}$, $1 \in S(a^{h_1(a)} : a \in B_1)$ and $1 \in S(a^{h_{-1}(a)} : a \in B_2)$ for all $h_1 : B_1 \to \{-1, 1\}$ and all $h_{-1} : B_{-1} \to \{-1, 1\}$ such that $g_i(a) = h_i(a)$ for all $a \in F_i$. Let $C = \bigcup_{i \in \{-1,1\}} (A \cap F_i)$. Note that $C \cup \{x\} = F_1 \cup F_{-1} \subseteq B_1 \cup B_{-1}$. Since $(C, f|_C) \in \mathcal{F}$, there exists $h : B_1 \cup B_{-1} \to \{-1, 1\}$ such that $1 \notin S(a^{h(a)} : a \in B_1 \cup B_{-1})$ and $h(a) = f(a)$ for all $a \in C$. Let $i = h(x) \in \{-1, 1\}$. We have that $h(x) = g_i(x)$, and thus $h(a) = g_i(a)$ for all $a \in F_i$, a contradiction because $S(a^{h(a)} : a \in B_i) \subseteq S(a^{h(a)} : a \in B_1 \cup B_{-1})$. Therefore $A = G \setminus \{1\}$.

Let $P = \{a \in G \setminus \{1\} : f(a) = 1\}$. Note that if $b \in G \setminus \{1\}$ then

$$1 \notin S(b^{f(b)}, (b^{-1})^{f(b^{-1})}),$$

and thus $f(b)f(b^{-1}) = -1$. Hence G is the disjoint union of P, $P^{-1} = \{a^{-1} : a \in P\}$ and $\{1\}$. Note that for all $a, b \in P$, $1 \notin S(a, b, (ab)^{f(ab)})$. Hence $f(ab) = 1$ and thus $ab \in P$. This proves that P is a subsemigroup of G. We define a binary relation $\leq$ on G by, for all $a, b \in G$,

$$a \leq b \text{ if and only if } a^{-1}b \in P \cup \{1\}.$$

It is straightforward to check that $\leq$ is a total order on G and that G is a left ordered group with respect to $\leq$. □

An immediate corollary:

Corollary 10.1.9 *Locally indicable groups are left ordered groups.* □

10.2 The Unique Product Property and Diffuse Groups

A group G satisfies the *unique product property* if for all finite non-empty subsets A and B there exists $g \in G$ such that $g = ab$ for unique elements $a \in A$ and $b \in B$.

Proposition 10.2.1 *A group with the unique product property is torsion-free.*

Proof Assume that G has torsion and let $g \in G$ be an element of order $n \geq 2$. Let $A = B = \langle g \rangle$. Then $g^i g^j = g^{i+1} g^{j-1}$ for all i, j, and $g^i \neq g^{i+1}$, so G cannot have the unique product property. □

Proposition 10.2.2 *A left ordered group satisfies the unique product property.*

Proof Let G be a left ordered group with respect to $\leq$ and $A = \{a_1, \dots, a_n\}$ and $B = \{b_1, \dots, b_m\}$ be non-empty subsets of G. We may assume that $a_1 < a_2 < \cdots < a_n$ and $b_1 < b_2 < \cdots < b_m$. Let a_i be the unique element in A such that $a_i b_m$ is the largest element in AB. Note that $a_j b_k \leq a_j b_m \leq a_i b_m$. Furthermore, if $a_j b_k = a_i b_m$, then $a_j b_k = a_j b_m$ and thus $b_k = b_m$ and $a_j = a_i$. Hence G satisfies the unique product property. □

A group G satisfies the *double unique product property* if for any two given finite non-empty subsets A and B of G such that $|A| + |B| > 2$ there exist at least two unique products in AB, i. e. there exist two distinct elements $g_1, g_2 \in G$ such that $g_1 = a_1 b_1$ for unique elements $a_1 \in A$ and $b_1 \in B$, and $g_2 = a_2 b_2$ for unique elements $a_2 \in A$ and $b_2 \in B$.

Theorem 10.2.3 (Strojnowski) *Let G be a group. The following conditions are equivalent:*

(1) G satisfies the double unique product property.
(2) For all finite non-empty subset $A \subseteq G$ there exists at least a unique product in $AA = \{a_1 a_2 : a_1, a_2 \in A\}$.
(3) G satisfies the unique product property.

Proof The implication (1) $\implies$ (2) is trivial.

We shall prove (2) $\implies$ (3). Suppose that G satisfies (2). Let A, B be finite non-empty subsets of G. Let $C = BA$. By (2), there exist a unique $g \in G$ such that $g = (b_1 a_1)(b_2 a_2)$ for unique $b_1 a_1, b_2 a_2 \in C$, where $a_1, a_2 \in A$ and $b_1, b_2 \in B$. Note that this implies that $a_1 b_2 = ab$ for $a \in A$ and $b \in B$ if and only if $a = a_1$ and $b = b_2$. Hence G satisfies the unique product property.

We shall prove (3) $\implies$ (1). Suppose that G satisfies the unique product property but not the double unique product property. Thus there exist finite non-empty subsets $A, B \subseteq G$ with $|A| + |B| > 2$ and there is a unique $g \in G$ such that $g = ab$ for unique elements $a \in A$ and $b \in B$. Let $C = a^{-1}A$ and $D = Bb^{-1}$. Then $1 \in C \cap D$. Note that if $c \in C$, $d \in D$ and $cd \neq 1$, then there exist $a_1 \in A$ and $b_1 \in B$ such that $c = a^{-1}a_1, d = b_1 b^{-1}$ and $ab \neq a_1 b_1$. Hence there exist $a_2 \in A \setminus \{a_1\}$ and $b_2 \in B \setminus \{b_1\}$ such that $a_1 b_1 = a_2 b_2$. Let $c_1 = a^{-1}a_2$ and $d_1 = b_2 b^{-1}$. We have that $c \neq c_1$, $d \neq d_1$ and

$$cd = a^{-1}a_1 b_1 b^{-1} = a^{-1}a_2 b_2 b^{-1} = c_1 d_1.$$

Let $E = D^{-1}C$ and $F = DC^{-1}$. Every element of EF is of the form $(d_1^{-1}c_1)(d_2 c_2^{-1})$, where $c_1, c_2 \in C$ and $d_1, d_2 \in D$. Suppose that $c_1 d_2 \neq 1$. We have seen that then there exist $c_3 \in C \setminus \{c_1\}$ and $d_3 \in D \setminus \{d_2\}$ such that $c_1 d_2 = c_3 d_3$. Hence $d_1^{-1} c_3 \in E \setminus \{d_1^{-1} c_1\}$, $d_3 c_2^{-1} \in F \setminus \{d_2 c_2^{-1}\}$ and

$$(d_1^{-1} c_1)(d_2 c_2^{-1}) = (d_1^{-1} c_3)(d_3 c_2^{-1}).$$

Suppose that $c_2 d_1 \neq 1$. Then there exist $c_4 \in C \setminus \{c_2\}$ and $d_4 \in D \setminus \{d_1\}$ such that $c_2 d_1 = c_4 d_4$. Hence $d_4^{-1} \cdot 1 \in E \setminus \{d_1^{-1} \cdot 1\}$, $1 \cdot c_4^{-1} \in F \setminus \{1 \cdot c_2^{-1}\}$ and

$$(d_1^{-1} \cdot 1)(1 \cdot c_2^{-1}) = (d_4^{-1} \cdot 1)(1 \cdot c_4^{-1}).$$

Since $|C| + |D| = |A| + |B| > 2$, either there exists $c \in C \setminus \{1\}$ or there exists $d \in D \setminus \{1\}$. In the first case, we have

$$(1 \cdot 1)(1 \cdot 1) = (1 \cdot c)(1 \cdot c^{-1}),$$

and in the second case, we have

$$(1 \cdot 1)(1 \cdot 1) = (d^{-1} \cdot 1)(d \cdot 1).$$

Thus, we have found two finite non-empty subsets $E, F \subseteq G$ such that for every $e \in E$ and $f \in F$, there exist $e_1 \in E \setminus \{e\}$ and $f_1 \in F \setminus \{f\}$ such that $ef = e_1 f_1$, a contradiction, because G satisfies the unique product property. Therefore G also satisfies the double unique product property. □

In general, it is difficult to check whether a group satisfies the unique product property.

Definition 10.2.4 A group G is said to be *diffuse* if, given any finite non-empty subset $A \subseteq G$, there exists $a \in A$ such that for every $g \in G \setminus \{1\}$ either $ga \notin A$ or $g^{-1} a \notin A$. Such an element $a \in A$ satisfying this property is called an *extreme point* of A.

Proposition 10.2.5 (Linnell–Morris) *Let G be a diffuse group. Then, given any finite subset $A \subseteq G$ such that $|A| \geq 2$, there exist two distinct extreme points of A.*

Proof Suppose that there exists a finite subset $A \subseteq G$ with a unique extreme point $a \in A$ and $|A| \geq 2$. Let $B = a^{-1} A$. We shall see that 1 is the unique extreme point of B. Let $b \in B$ be an extreme point of B. Thus, for every $g \in G \setminus \{1\}$, either $gb \notin B$ or $g^{-1} b \notin B$. We have that $b = a^{-1} a_1$ for some $a_1 \in A$. Hence either $ga^{-1} a_1 \notin a^{-1} A$ or $g^{-1} a^{-1} a_1 \notin a^{-1} A$, and thus, either $aga^{-1} a_1 \notin A$ or $ag^{-1} a^{-1} a_1 \notin A$. Hence a_1 is an extreme point of A, and thus $a_1 = a$. This proves that $b = a^{-1} a_1 = 1$ is the only extreme point of B. Let $C = B \cup B^{-1}$, where $B^{-1} = \{b^{-1} : b \in B\}$. Let $b \in B \setminus \{1\}$. Since b is not an extreme point of B, there exists $g \in G \setminus \{1\}$ such that $gb, g^{-1} b \in B \subseteq C$. Hence b is not an extreme point of C. Note that

$$(b^{-1} g^{-1} b) b^{-1} = (gb)^{-1} \in B^{-1} \subseteq C \text{ and } (b^{-1} g b) b^{-1} = (g^{-1} b)^{-1} \in B^{-1} \subseteq C.$$

Hence b^{-1} is not an extreme point of C. Since $|B| = |A| \geq 2$, there exists $b \in B \setminus \{1\}$. Since $b \cdot 1, b^{-1} \cdot 1 \in C$, we have that 1 is not an extreme point of C. Hence C has no extreme points, a contradiction. Therefore the result follows. □

Lemma 10.2.6 *Let G be a left ordered group. Then G is diffuse.*

Proof Suppose that G is left ordered with respect to a total order $\leq$. Let $A \subseteq G$ be a finite non-empty subset. Let $a_1 \in A$ such that $a_1 \leq a$ for all $a \in A$. We shall prove that a_1 is an extreme point of A. Suppose that there exists $g \in G \setminus \{1\}$ such that $ga_1, g^{-1}a_1 \in A$. Hence $a_1 \leq ga_1$ and $a_1 \leq g^{-1}a_1$. Hence

$$g^{-1}a_1 \leq g^{-1}ga_1 = a_1 \leq g^{-1}a_1,$$

and thus $g^{-1}a_1 = a_1$, a contradiction because $g \neq 1$. Thus a_1 is an extreme point of A. Therefore G is diffuse. □

Lemma 10.2.7 *Let G be a diffuse group. Then G satisfies the unique product property.*

Proof Suppose that G does not satisfy the unique product property. There exist finite non-empty subsets $A, B \subseteq G$ such that for all $(a, b) \in A \times B$, there is an element $(a_1, b_1) \in A \times B \setminus \{(a, b)\}$ such that $ab = a_1b_1$. Let $C = AB = \{ab : a \in A,\ b \in B\}$ and $c \in C$. There are two distinct elements $(a, b), (a_1, b_1) \in A \times B$ such that $c = ab = a_1b_1$. Note that $aa_1^{-1} \neq 1$ and

$$aa_1^{-1}c = aa_1^{-1}a_1b_1 = ab_1 \in C \text{ and } (aa_1^{-1})^{-1}c = a_1a^{-1}ab = a_1b \in C.$$

Hence c is not an extreme point of C. Therefore the result follows. □

Note that the above result shows that every diffuse group G is torsion-free.

10.3 The Transfer Map

Let G be a group and H be a finite-index subgroup. We will define a group homomorphism $G \to H/[H, H]$, known as the *transfer map* of G on H.

Lemma 10.3.1 *Let G be a group and H be a subgroup of finite-index $n = (G : H)$. Let $S = \{s_1, \dots, s_n\}$ and $T = \{t_1, \dots, t_n\}$ be left transversals of H in G. If $g \in G$, there exist unique $h_{1,g}, \dots, h_{n,g} \in H$ and a permutation $\sigma_g \in \mathbb{S}_n$ such that*

$$gt_i = s_{\sigma_g(i)}h_{i,g}, \quad i \in \{1, \dots, n\}.$$

Furthermore, if $s_i = t_i$ for all i, then $\sigma_{g_1g_2} = \sigma_{g_1}\sigma_{g_2}$ and $h_{i,g_1g_2} = h_{\sigma_{g_2}(i),g_1}h_{i,g_2}$ for all $g_1, g_2 \in G$ and all i.

Proof If $i \in \{1, \dots, n\}$, then there exists a unique $j \in \{1, \dots, n\}$ such that $gt_i \in s_jH$. Thus there exists a unique $h_{i,g} \in H$ such that $gt_i = s_jh_{i,g}$. Take $\sigma_g(i) = j$ and thus there is a well-defined map $\sigma_g \colon \{1, \dots, n\} \to \{1, \dots, n\}$. To prove that $\sigma_g \in \mathbb{S}_n$ it is enough to check that σ_g is injective. If $\sigma_g(i) = \sigma_g(k) = j$, since $gt_i = s_jh_{i,g}$ and $gt_k = s_jh_{k,g}$, it follows that $t_i^{-1}t_k = h_{i,g}^{-1}h_{k,g} \in H$. Hence $i = k$, as $t_iH = t_kH$.

Suppose that $s_i = t_i$ for all i. Let $g_1, g_2 \in G$. We have that

$$t_{\sigma_{g_1g_2}(i)}h_{i,g_1g_2} = g_1g_2t_i = g_1t_{\sigma_{g_2}(i)}h_{i,g_2} = t_{\sigma_{g_1}\sigma_{g_2}(i)}h_{\sigma_{g_2}(i),g_1}h_{i,g_2}.$$

Hence $\sigma_{g_1g_2}(i) = \sigma_{g_1}\sigma_{g_2}(i)$ and $h_{i,g_1g_2} = h_{\sigma_{g_2}(i),g_1}h_{i,g_2}$, thus the result follows. □

Let G be a group and H be a subgroup of G of finite-index n. If $T = \{t_1, \dots, t_n\}$ is a left transversal of H in G, we define the map

$$\nu_T \colon G \to H/[H, H], \quad \nu_T(g) = \prod_{i=1}^{n} \bar{h}_i$$

where $gt_i = t_jh_i$ and $\bar{h} = h[H, H] \in H/[H, H]$. Note that the product is well-defined since $H/[H, H]$ is an abelian group. We now prove that the map does not depend on the transversal.

Lemma 10.3.2 *Let G be a group and H be a subgroup of G of finite index. if T and S are left transversals of H in G, then $\nu_T = \nu_S$.*

Proof Assume that $n = (G : H)$, $T = \{t_1, \dots, t_n\}$ and $S = \{s_1, \dots, s_n\}$, where $s_i = t_ik_i$, for some $k_i \in H$. Let $g \in G$. There exist $\sigma_g \in \mathbb{S}_n$ and $h_{i,g} \in H$ such that $gs_i = s_{\sigma_g(i)}h_{i,g}$ for all i. Let $l_{i,g} = k_{\sigma_g(i)}h_{i,g}k_i^{-1} \in H$. Then

$$gt_i = gs_ik_i^{-1} = s_{\sigma_g(i)}h_{i,g}k_i^{-1} = t_{\sigma_g(i)}k_{\sigma_g(i)}h_{i,g}k_i^{-1} = t_{\sigma_g(i)}l_{i,g}$$

for all $i \in \{1, \dots, n\}$. Moreover,

$$s_{\sigma_g(i)}^{-1}gs_i = k_{\sigma_g(i)}^{-1}t_{\sigma_g(i)}^{-1}gt_ik_i.$$

Since $H/[H, H]$ is abelian,

$$\begin{aligned}\nu_S(g) &= \prod_{i=1}^{n} \overline{s_{\sigma_g(i)}^{-1}gs_i} = \prod_{i=1}^{n} \bar{k}_{\sigma_g(i)}^{-1}\overline{t_{\sigma_g(i)}^{-1}gt_i}\bar{k}_i \\ &= \prod_{i=1}^{n} \bar{k}_{\sigma_g(i)}^{-1} \prod_{i=1}^{n} \bar{k}_i \prod_{i=1}^{n} \overline{t_{\sigma_g(i)}^{-1}gt_i} = \prod_{i=1}^{n} \overline{t_{\sigma_g(i)}^{-1}gt_i} = \nu_T(g).\end{aligned}$$

□

By Lemma 10.3.2, if H is a finite-index subgroup of G, the map

$$\nu\colon G \to H/[H,H], \quad \nu(g) = \nu_T(g),$$

where T is some left transversal of H in G, is well-defined.

Theorem 10.3.3 *Let G be a group and H be a finite-index subgroup of G. Then $\nu(xy) = \nu(v)\nu(y)$ for all $x, y \in G$.*

Proof Let $T = \{t_1, \dots, t_n\}$ be a left transversal of H in G, where $n = (G : H)$. By Lemma 10.3.1, for every $g \in G$ there exist unique elements $h_{1,g}, \dots, h_{n,g} \in H$ and a permutation $\sigma_g \in \mathbb{S}_n$ such that $gt_i = t_{\sigma_g(i)}h_{i,g}$. Furthermore $\sigma_{xy} = \sigma_x\sigma_y$ and $h_{i,xy} = h_{\sigma_y(i),x}h_{i,y}$ for all $x, y \in G$ and all i. Since $H/[H,H]$ is abelian,

$$\nu(xy) = \prod_{i=1}^{n} h_{i,xy} = \prod_{i=1}^{n} h_{\sigma_y(i),x}h_{i,y} = \prod_{i=1}^{n} h_{\sigma_y(i),x} \prod_{i=1}^{n} h_{i,y} = \nu(x)\nu(y).$$

□

If G is a group and H is a finite-index subgroup of G, the *transfer homomorphism* is the group homomorphism

$$\nu\colon G \to H/[H,H], \quad \nu(g) = \nu_T(g),$$

for some left transversal T of H in G.

Lemma 10.3.4 *Let G be a group and H be a subgroup of G with $(G : H) = n$. Let $T = \{t_1, \dots, t_n\}$ be a left transversal of H in G. For each $g \in G$ there exist a positive integer m, m distinct elements $s_1, \dots, s_m \in T$ and positive integers $n_1, \dots, n_m$ such that*

$$s_i^{-1}g^{n_i}s_i \in H, \quad n_1 + \cdots + n_m = n \quad \textit{and} \quad \nu(g) = \prod_{i=1}^{m} s_i^{-1}g^{n_i}s_i.$$

Proof Let $g \in G$. For each i there exist $h_1, \dots, h_n \in H$ and a permutation $\sigma \in \mathbb{S}_n$ such that $gt_i = t_{\sigma(i)}h_i$. Write σ as a product

$$\sigma = \alpha_{k+1} \cdots \alpha_m$$

of disjoint cycles and $|\{j : \sigma(j) = j\}| = k$. Note that if $\sigma(j) = j$, then $gt_j = t_jh_j$, and thus $t_j^{-1}gt_j \in H$. Thus we take $\{s_1, \dots, s_k\} = \{t_j : \sigma(j) = j\}$ and $n_1 = \cdots = n_k = 1$.

Fix $i \in \{k+1, \dots, m\}$ and write $\alpha_i = (j_1 \cdots j_{n_i})$. Since

$$gt_{j_k} = t_{\sigma(j_k)}h_{j_k} = \begin{cases} t_{j_1}h_{n_k} & \text{if } k = n_i, \\ t_{j_{k+1}}h_k & \text{otherwise,} \end{cases}$$

it follows that

$$\begin{aligned}
t_{j_1}^{-1}g^{n_i}t_{j_1} &= t_{j_1}^{-1}g^{n_i-1}gt_{j_1}\\
&= t_{j_1}^{-1}g^{n_i-1}t_{j_2}h_{j_1}\\
&= t_{j_1}^{-1}g^{n_i-2}gt_{j_2}h_{j_1}\\
&= t_{j_1}^{-1}g^{n_i-2}t_{j_3}h_{j_2}h_{j_1}\\
&\vdots\\
&= t_{j_1}^{-1}gt_{j_{n_i}}h_{j_{n_i-1}}\cdots h_{j_2}h_{j_1}\\
&= t_{j_1}^{-1}t_{j_1}h_{j_{n_i}}\cdots h_{j_2}h_{j_1} \in H.
\end{aligned}$$

So we let $s_i = t_{j_1} \in T$. Now the claim follows, since $\nu(g) = \bar{h}_1 \cdots \bar{h}_n$. □

Proposition 10.3.5 *Let G be a group and H be a central subgroup of index n. Then $\nu(g) = g^n$ for all $g \in G$.*

Proof Let $g \in G$. Let T be a left transversal of H in G. By Lemma 10.3.4, there exist $s_1, \ldots, s_m \in T$ and positive integers $n_1, \ldots, n_m$ such that $\sum_{i=1}^m n_i = n$, $s_i^{-1}g^{n_i}s_i \in H$ and $\nu(g) = \prod_{i=1}^m s_i^{-1}g^{n_i}s_i$. Since H is central in G, then it is normal in G. Thus

$$g^{n_i} = s_i(s_i^{-1}g^{n_i}s_i)s_i^{-1} \in H \subseteq Z(G)$$

and hence

$$\nu(g) = \prod_{i=1}^m s_i^{-1}g^{n_i}s_i = \prod_{i=1}^m g^{n_i} = g^{\sum_{i=1}^m n_i} = g^n.$$

□

The next result is an application to finite groups.

Proposition 10.3.6 *Let G be a finite group and H a central subgroup of index n, where n is coprime with $|H|$. Then $G \cong N \times H$.*

Proof Let $\nu\colon G \to H$ be the transfer homomorphism (see Theorem 10.3.3). By Proposition 10.3.5, $\nu(g) = g^n$ for all $g \in G$. In particular, since the order of $h \in H$ and n are coprime, there exists a positive integer k such that $h^{kn} = h$. Hence ν is surjective. Let $N = \ker \nu$. Then $|G| = |N||H| = n|H|$. Hence $|N| = n$. Since n and $|H|$ are coprime, we have that $N \cap H = \{1\}$ and $G = NH \cong N \times H$, because N and H are normal subgroups of G. □

An application to infinite groups.

Theorem 10.3.7 *Let G be a torsion-free group that contains a finite-index subgroup isomorphic to $\mathbb{Z}$. Then $G \cong \mathbb{Z}$.*

Proof We may assume that G contains a finite-index normal subgroup isomorphic to $\mathbb{Z}$. Indeed, if H is a finite-index subgroup of G such that $H \cong \mathbb{Z}$, and T is a left transversal of H in G, then $K = \bigcap_{x \in T} xHx^{-1}$ is a finite-index normal subgroup of G (see Exercise 10.5.1). Since $K \subseteq H \subseteq G$ and $(G : K) = (G : H)(H : K)$ is finite, we have that $K \cong \mathbb{Z}$. The action of G on K by conjugation induces a group homomorphism $\epsilon \colon G \to \mathrm{Aut}(K)$. Since $\mathrm{Aut}(K) \cong \mathrm{Aut}(\mathbb{Z}) = \{-1, 1\}$, there are two cases to consider.

Assume first that $\epsilon(g) = \mathrm{id}$ for all $g \in G$. In this case, $K \subseteq Z(G)$, let $\nu \colon G \to K$ be the transfer homomorphism. By Proposition 10.3.5, $\nu(g) = g^n$, where $n = (G : K)$. Since G has no torsion, ν is injective. Thus $G \cong \mathbb{Z}$ because it is isomorphic to a non-trivial subgroup of K.

Assume now that there exists $g \in G$ such that $\epsilon(g) \neq \mathrm{id}$. Let $N = \ker \epsilon \neq G$. Since $K \cong \mathbb{Z}$ is abelian, $K \subseteq N$. The result proved in the previous paragraph applied to $\epsilon|_N$ implies that $N \cong \mathbb{Z}$, as N contains a finite-index subgroup isomorphic to $\mathbb{Z}$. Let $g \in G \setminus N$. Since N is normal in G, G acts by conjugation on N and hence there exists a group homomorphism $c_g \in \mathrm{Aut}(N) \cong \{-1, 1\}$ defined by $c_g(n) = gng^{-1}$ for all $n \in N$. Since $K \subseteq N$ and g acts non-trivially on K,

$$c_g(n) = gng^{-1} = n^{-1}$$

for all $n \in N$. Since $g^2 \in N$,

$$g^2 = gg^2g^{-1} = g^{-2}.$$

Therefore $g^4 = 1$, a contradiction since $g \neq 1$ and G has no torsion. Therefore $\epsilon(g) = \mathrm{id}$ for all $g \in G$ and the result follows. □

10.4 Bieberbach Groups

We begin this section with an elementary property of the subgroups of finite index.

Lemma 10.4.1 *Let G be a group. Let H be a subgroup of finite index of G. Then H is finitely generated if and only if G is finitely generated.*

Proof Suppose that H is finitely generated. Let A be a finite subset of H such that $H = \langle A \rangle$. Let T be a left transversal of H in G. It is clear that $G = \langle T \cup A \rangle$, and thus G is finitely generated. The converse follows by Lemma 5.1.27. □

The following result is due to Schur.

Lemma 10.4.2 *Let G be a group such that $(G : Z(G)) = n < \infty$. Then $[G, G]$ is finite and $|[G, G]| \leq n^{n(n-1)}$.*

Proof Let $\nu\colon G \to Z(G)$ be the transfer homomorphism. By Proposition 10.3.5, $\nu(g) = g^n$ for all $g \in G$. Since $[G, G] \in \ker \nu$, we have that $h^n = 1$ for all $h \in [G, G]$. Let $T = \{t_1, \ldots, t_n\}$ be a left transversal of $Z(G)$ in G. Let

$$A = \{[g, h] : g, h \in G\}.$$

Note that

$$A = \{[t_i, t_j] : 1 \leq i, j \leq n\}.$$

Since $[t_i, t_j]^{-1} = [t_j, t_i]$, every element of $[G, G]$ is a product of elements of A. Note that $|A \setminus \{1\}| = k \leq n(n-1)$. Let $A \setminus \{1\} = \{a_1, \ldots, a_k\}$. We shall prove that

$$[G, G] = \{a_1^{n_1} \cdots a_k^{n_k} : 0 \leq n_i < n \text{ for all } i = 1, \ldots, k\}.$$

Let $B = \{a_1^{n_1} \cdots a_k^{n_k} : 0 \leq n_i < n \text{ for all } i = 1, \ldots, k\}$ and $h \in [G, G]$. There exist $a_{i_1}, \ldots, a_{i_m} \in A \setminus \{1\}$ such that $h = a_{i_1} \cdots a_{i_m}$. We shall prove that $h = a_1^{n_1} \cdots a_k^{n_k} \in B$, for some non-negative integers $n_1, \ldots, n_k$ such that $n_1 + \cdots + n_k \leq m$, by induction on m. For $m = 1$, it is clear that $h = a_{i_1} \in B$. Suppose that $m > 1$ and that

$$a_{j_1} \cdots a_{j_l} = a_1^{n_1} \cdots a_k^{n_k} \in B,$$

for some non-negative integers $n_1, \ldots, n_k < n$ such that $n_1 + \cdots + n_k \leq l$, for all $l < m$. Since

$$\begin{aligned} [u, v] &= [u, v][u, v]^{-1}[x, y][u, v] \\ &= [u, v][[u, v]^{-1}x[u, v], [u, v]^{-1}y[u, v]] \end{aligned} \tag{10.1}$$

for all $x, y, u, v \in G$, we may assume that i_1 is the smallest positive integer in all the representations of h as a product of m elements in $A \setminus \{1\}$. By the inductive hypothesis

$$a_{i_2} \cdots a_{i_m} = a_1^{n_1} \cdots a_k^{n_k} \in B,$$

for some non-negative integers $n_1, \ldots, n_k < n$ such that $n_1 + \cdots + n_k \leq m - 1$. Hence, by the choice of i_1 and (10.1), we have that

$$h = a_{i_1} a_1^{n_1} \cdots a_k^{n_k} = a_{i_1}^{n_{i_1}+1} a_{i_1+1}^{n_{i_1+1}} \cdots a_k^{n_k}.$$

Note that $n_{i_1} + 1 \leq n$ and $n_{i_1} + 1 + n_{i_1+1} + \cdots + n_k \leq m$. If $n_{i_1} + 1 < n$, then clearly $h \in B$. If $n_{i_1} + 1 = n$, then

$$h = a_{i_1+1}^{n_{i_1+1}} \cdots a_k^{n_k} \in B.$$

Hence, by induction, $[G, G] = B$. Clearly $|[G, G]| \leq n^k \leq n^{n(n-1)}$, and the result follows. □

The *FC-center* of a group G is the set

$$\Delta(G) = \{g \in G : (G : C_G(g)) < \infty\}.$$

Note that $\Delta(G)$ is the set of all elements in G that have finitely many conjugates, i.e.,

$$\Delta(G) = \{g \in G : |\{xgx^{-1} : x \in G\}| < \infty\}.$$

Lemma 10.4.3 *Let G be a group. Then $\Delta(G)$ and $T(\Delta(G)) = \{g \in \Delta(G) : o(g) < \infty\}$ are characteristic subgroups of G and $\Delta(G)/T(\Delta(G))$ is an abelian torsion-free group.*

Proof Note that $1 \in T(\Delta(G))$. Let $g, h \in \Delta(G)$. Then

$$xgh^{-1}x^{-1} = (xgx^{-1})(xhx^{-1})^{-1}$$

for all $x \in G$. Since g and h have finitely many conjugates, we get that gh^{-1} also has finitely many conjugates. Hence $\Delta(G)$ is a subgroup of G. Let $\alpha \in \mathrm{Aut}(G)$. Then

$$x\alpha(g)x^{-1} = \alpha(\alpha^{-1}(x)g\alpha^{-1}(x)^{-1})$$

for all $x \in G$. Hence $\alpha(g) \in \Delta(G)$ and thus $\Delta(G)$ is a characteristic subgroup of G. Let $H = \langle g, h\rangle$. Note that $\Delta(H) = H$ and $Z(H) = C_H(g) \cap C_H(h)$. Thus $(H : Z(H)) < \infty$. By Lemma 10.4.2, $[H, H]$ is finite. Hence $T(H)$ is a finite subgroup. In particular, if $g, h \in T(\Delta(G))$, then $\langle g, h\rangle \subseteq T(\Delta(G))$, and this shows that $T(\Delta(G))$ is a subgroup of G. Clearly $T(\Delta(G))$ is characteristic in $\Delta(G)$. Since $\Delta(G)$ is characteristic in G, it follows that $T(\Delta(G))$ is characteristic in G. Since $[g, h] \in T(\Delta(G))$ for all $g, h \in \Delta(G)$, it is clear that $\Delta(G)/T(\Delta(G))$ is torsion-free abelian. □

A group G is said to be a *Bieberbach* group if it is a finitely generated torsion-free group with an abelian subgroup of finite index.

Proposition 10.4.4 *Let G be a Bieberbach group. Then $\Delta(G)$ is a torsion-free finitely generated abelian normal subgroup of G of finite index. Furthermore, $A \subseteq \Delta(G)$ for every abelian subgroup A of G of finite index.*

Proof Let A be an abelian subgroup of G of finite index. Note that for every $a \in A$, $A \subseteq C_G(a)$. Hence $A \subseteq \Delta(G)$. By Lemmas 10.4.3 and 10.4.1, the result follows. □

The *dimension* of a Bieberbach group G is the rank of $\Delta(G)$.

Lemma 10.4.5 *Let G be a Bieberbach group. Then every subgroup H of G is a Bieberbach group and* $\dim(H) \leq \dim(G)$.

Proof Let H be a subgroup of G. Then $(H : H \cap \Delta(G)) < \infty$. Let $n = \dim(G)$. Thus $\Delta(G) \cong \mathbb{Z}^n$. Hence $H \cap \Delta(G)$ is a finitely generated subgroup of $\Delta(G)$, and thus $H \cap \Delta(G) \cong \mathbb{Z}^m$ for some $m \leq n$. Since $(H : H \cap \Delta(G)) < \infty$ and $H \cap \Delta(G)$ is finitely generated, we have that H also is finitely generated by Lemma 10.4.1. Hence H is a Bieberbach group. Note that $H \cap \Delta(G) \subseteq \Delta(H)$. Since $(\Delta(H) : H \cap \Delta(G)) < \infty$, the rank of $\Delta(H)$ is m. Thus $\dim(H) = m \leq n = \dim(G)$. □

Lemma 10.4.6 *Let G be a group with a torsion-free abelian normal subgroup A of finite index. Then G is isomorphic to a subgroup H of the semidirect product $A \rtimes_\alpha G/A$, where the action α is naturally induced by conjugation. Furthermore,*

$$H \cap (A \times \{A\}) = \{a^n : a \in A\} \times \{A\},$$

where $n = (G : A)$.

Proof Let $n = (G : A)$ and let T be a transversal of A in G such that $1 \in T$. For each $g \in G$ there exist unique $a_g \in A$ and $t_g \in T$ such that $g = a_g t_g$. We denote $a_{t_g t_h}$ by $c(t_g, t_h)$ for all $g, h \in G$. Hence

$$t_g t_h = c(t_g, t_h) t_{gh}$$

for all $g, h \in G$. Note that $c(t_1, t) = c(1, t) = 1 = c(t, 1) = c(t, t_1)$ for all $t \in T$. The associativity of G implies that

$$c(t_{g_1}, t_{g_2}) c(t_{g_1 g_2}, t_{g_3}) = t_{g_1} c(t_{g_2}, t_{g_3}) t_{g_1}^{-1} c(t_{g_1}, t_{g_2 g_3})$$

for all $g_1, g_2, g_3 \in G$. Since A is abelian and $n = (G : A)$, multiplying this equality for all $t_{g_3} \in T$, we have that

$$c(t_{g_1}, t_{g_2})^n \prod_{t \in T} c(t_{g_1 g_2}, t) = t_{g_1} \left(\prod_{t \in T} c(t_{g_2}, t) \right) t_{g_1}^{-1} \prod_{t \in T} c(t_{g_1}, t). \tag{10.2}$$

for all $g_1, g_2 \in G$. Let $f : G \to A \rtimes_\alpha G/A$ be the map defined by

$$f(g) = f(a_g t_g) = \left(a_g^n \prod_{t \in T} c(t_g, t), gA \right)$$

for all $g \in G$. It is clear that $f(g) \in A \times \{A\}$ if and only if $t_g = 1$, and in this case $f(g) = (g^n, A)$. Note that for $g_1, g_2 \in G$ we have

$$\begin{aligned} f(g_1 g_2) &= f(a_{g_1} t_{g_1} a_{g_2} t_{g_2}) \\ &= f(a_{g_1} t_{g_1} a_{g_2} t_{g_1}^{-1} c(t_{g_1}, t_{g_2}) t_{g_1 g_2}) \\ &= \left(a_{g_1}^n t_{g_1} a_{g_2}^n t_{g_1}^{-1} c(t_{g_1}, t_{g_2})^n \prod_{t \in T} c(t_{g_1 g_2}, t), g_1 g_2 A \right) \end{aligned}$$

and

$$f(g_1) f(g_2) = f(a_{g_1} t_{g_1}) f(a_{g_2} t_{g_2})$$

$$= \left(a_{g_1}^n \prod_{t\in T} c(t_{g_1}, t), g_1 A\right)\left(a_{g_2}^n \prod_{t\in T} c(t_{g_2}, t), g_2 A\right)$$

$$= \left(a_{g_1}^n \prod_{t\in T} c(t_{g_1}, t) t_{g_1} a_{g_2}^n t_{g_1}^{-1} t_{g_1} \left(\prod_{t\in T} c(t_{g_2}, t)\right) t_{g_1}^{-1}, g_1 g_2 A\right).$$

By (10.2), we get that $f(g_1 g_2) = f(g_1) f(g_2)$. Hence f is a homomorphism of groups. Let $g \in \ker f$. Then $gA = A$, and thus $t_g = 1$. Hence

$$(1, A) = f(g) = \left(a_g^n \prod_{t\in T} c(1, t), A\right) = (a_g^n, A),$$

and, since A is torsion-free, we get that $a_g = 1$. Hence f is injective and the result follows. □

We denote by $\mathrm{Isom}(\mathbb{R}^n)$ the group of isometries of the Euclidean space $\mathbb{R}^n$. Let $T(\mathbb{R}^n)$ be the subgroup of translations of $\mathrm{Isom}(\mathbb{R}^n)$ and let $O(\mathbb{R}^n)$ be the subgroup of orthogonal automorphism of $\mathbb{R}^n$. It is known that $T(\mathbb{R}^n)$ is normal in $\mathrm{Isom}(\mathbb{R}^n)$, $\mathrm{Isom}(\mathbb{R}^n) = T(\mathbb{R}^n)O(\mathbb{R}^n)$ and $T(\mathbb{R}^n) \cap O(\mathbb{R}^n) = \{\mathrm{id}\}$. Hence every $g \in \mathrm{Isom}(\mathbb{R}^n)$ can be written uniquely as $g = tf$ with $t \in T(\mathbb{R}^n)$ and $f \in O(\mathbb{R}^n)$.

Theorem 10.4.7 *Let G be an n-dimensional Bieberbach group. Then G is isomorphic to a subgroup H of the group* $\mathrm{Isom}(\mathbb{R}^n)$ *such that $\Delta(G)$ corresponds to the translations of H, that is $\Delta(H) = H \cap T(\mathbb{R}^n)$.*

Proof By Lemma 10.4.6, it is enough to show that $\Delta(G) \rtimes G/\Delta(G)$ is isomorphic to a subgroup H of $\mathrm{Isom}(\mathbb{R}^n)$ and $\Delta(H) = H \cap T(\mathbb{R}^n)$. Since G is an n-dimensional Bieberbach group, there is an isomorphism $f\colon \Delta(G) \to \mathbb{Z}^n$. Let $\sigma : G \to \mathrm{Aut}(\mathbb{Z}^n)$ be the map defined by $\sigma(g) = \sigma_g$ and $\sigma_g(a_1, \dots, a_n) = f(gf^{-1}(a_1, \dots, a_n)g^{-1})$ for all $g \in G$ and $(a_1, \dots, a_n) \in \mathbb{Z}^n$. It is straightforward to check that $\sigma_g \in \mathrm{Aut}(\mathbb{Z}^n)$ and that σ is a homomorphism of groups. Note that $\ker \sigma = C_G(\Delta(G))$. By Proposition 10.4.4, $\Delta(G) = C_G(\Delta(G))$. Hence $\sigma(G) \cong G/\Delta(G)$. Note that every automorphism of $\mathbb{Z}^n$ can be naturally extended to an automorphism of $\mathbb{R}^n$. Hence, we can consider $\sigma(G)$ as a finite subgroup of $\mathrm{Aut}(\mathbb{R}^n)$.

Let $\langle u \mid v\rangle$ be the standard inner product of $u, v \in \mathbb{R}^n$. We define a new inner product on $\mathbb{R}^n$ by

$$\langle u \mid v\rangle' = \sum_{g\in\sigma(G)} \langle g(u) \mid g(v)\rangle$$

for all $u, v \in \mathbb{R}^n$. Let $e'_1, \dots, e'_n$ be an orthonormal basis for $\langle . \mid . \rangle'$. Let $\alpha \in \mathrm{Aut}(\mathbb{R}^n)$ be the map defined by $\alpha(a_1, \dots, a_n) = \sum_{i=1}^n a_i e'_i$. Note that

$$\langle u \mid v\rangle = \langle \alpha(u) \mid \alpha(v)\rangle'$$

for all $u, v \in \mathbb{R}^n$. If $g \in \sigma(G)$, then

$$\begin{aligned}\langle \alpha^{-1} g\alpha(u), \alpha^{-1} g\alpha(v)\rangle &= \langle g\alpha(u), g\alpha(v)\rangle' \\ &= \langle \alpha(u), \alpha(v)\rangle' = \langle u, v\rangle\end{aligned}$$

for all $u, v \in \mathbb{R}^n$. Hence $\alpha^{-1} g\alpha$ is in the orthogonal group of $\mathbb{R}^n$ for all $g \in \sigma(G)$.
Let $\Psi\colon \Delta(G) \rtimes G/\Delta(G) \to \mathrm{Isom}(\mathbb{R}^n)$ be the map defined by

$$\Psi(h, g\Delta(G))(a_1, \ldots, a_n) = \alpha^{-1}(f(h)) + \alpha^{-1}\sigma(g)\alpha(a_1, \ldots, a_n)$$

for all $h \in \Delta(g)$, $g \in G$ and $(a_1, \ldots, a_n) \in \mathbb{R}^n$. For $h_1, h_2 \in \Delta(G)$, $g_1, g_2 \in G$ and $(a_1, \ldots, a_n) \in \mathbb{R}^n$, we have that

$$\begin{aligned}
&\Psi(h_1, g_1\Delta(G))\Psi(h_2, g_2\Delta(G))(a_1, \ldots, a_n)\\
&\qquad = \Psi(h_1, g_1\Delta(G))(\alpha^{-1}(f(h_2)) + \alpha^{-1}\sigma(g_2)\alpha(a_1, \ldots, a_n))\\
&\qquad = \alpha^{-1}(f(h_1)) + \alpha^{-1}\sigma(g_1)\alpha(\alpha^{-1}(f(h_2)) + \alpha^{-1}\sigma(g_2)\alpha(a_1, \ldots, a_n))\\
&\qquad = \alpha^{-1}(f(h_1)) + \alpha^{-1}\sigma(g_1)(f(h_2)) + \alpha^{-1}\sigma(g_1)\sigma(g_2)\alpha(a_1, \ldots, a_n))\\
&\qquad = \alpha^{-1}(f(h_1)) + \alpha^{-1}(f(g_1h_2g_1^{-1})) + \alpha^{-1}\sigma(g_1g_2)\alpha(a_1, \ldots, a_n))\\
&\qquad = \alpha^{-1}(f(h_1g_1h_2g_1^{-1})) + \alpha^{-1}\sigma(g_1g_2)\alpha(a_1, \ldots, a_n))\\
&\qquad = \Psi(h_1g_1h_2g_1^{-1}, g_1g_2\Delta(G))(a_1, \ldots, a_n)\\
&\qquad = \Psi((h_1, g_1\Delta(G))(h_2, g_2\Delta(G))(a_1, \ldots, a_n).
\end{aligned}$$

Hence Ψ is a homomorphism of groups. Let $h \in \Delta(G)$ and $g \in G$ such that $\Psi(h, g\Delta(G)) = \mathrm{id}$. In particular,

$$(0, \ldots, 0) = \Psi(h, g\Delta(G))(0, \ldots 0) = \alpha^{-1}(f(h)),$$

and thus $h = 1$. Now we have that

$$\alpha^{-1}(f(x)) = \Psi(1, g\Delta(G))(\alpha^{-1}(f(x))) = \alpha^{-1}\sigma_g(f(x)) = \alpha^{-1}(f(gxg^{-1}))$$

for all $x \in \Delta(G)$. Hence $g \in C_G(\Delta(G)) = \Delta(G)$. Therefore $g\Delta(G) = \Delta(G)$. This shows that Ψ is injective, and the result follows. □

Theorem 10.4.8 *Let G be a Bieberbach group. Then the following conditions are equivalent:*

(1) Every non-trivial subgroup of G has non-trivial center.
(2) G is locally indicable.
(3) G is a poly-$\mathbb{Z}$ group.
(4) G is left ordered.
(5) G is diffuse.

Proof (1) $\implies$ (2). Suppose that every non-trivial subgroup of G has a non-trivial center. Let H be a non-trivial finitely generated subgroup of G. Let $z \in Z(H) \setminus \{1\}$.

By Lemma 10.4.5, H is a Bieberbach group. Let $\nu\colon H \to \Delta(H)$ be the transfer homomorphism. Let $n = (H : \Delta(H))$. Then $\nu(z) = z^n \neq 1$. Hence $\nu(H)$ is an infinite finitely generated abelian group. Therefore there exists a surjective homomorphism of groups $f\colon \nu(H) \to \mathbb{Z}$ and thus $f\nu\colon H \to \mathbb{Z}$ is a surjective homomorphism of groups. This shows that G is locally indicable.

(2) $\Longrightarrow$ (3). Suppose that G is locally indicable. We shall show that G is a poly-$\mathbb{Z}$ group by induction on the dimension of G. If G has dimension 1, then by Theorem 10.3.7, $G \cong \mathbb{Z}$. Suppose that $\dim(G) = n > 1$ and that every locally indicable Bieberbach group of dimension $n-1$ is a poly-$\mathbb{Z}$ group. Since G is locally indicable and finitely generated, a surjective homomorphism $f\colon G \to \mathbb{Z}$ exists. Let $H = \ker f$. Let $g \in G$ be an element such that $f(g) = 1$. Note that $G = H\langle g\rangle$ and $H \cap \langle g\rangle = \{1\}$. By Lemma 10.4.5, H is a Bieberbach group. Let $A_1 = H \cap \Delta(G)$ and $A_2 = \langle g\rangle \cap \Delta(G)$. Note that A_1A_2 is an abelian subgroup of finite index in G. Since A_1 is a subgroup of finite index of H, we have that the rank of A_1 is the dimension of H. Since A_2 is a non-trivial subgroup of $\langle g\rangle$, we have that $A_2 \cong \mathbb{Z}$. Hence $\dim(G) = \dim(H) + 1$. Therefore H has dimension $n-1$ and it is locally indicable. By the inductive hypothesis, H is a poly-$\mathbb{Z}$ group. Therefore, G also is a poly-$\mathbb{Z}$ group. Thus the result follows by induction.

(3) $\Longrightarrow$ (4). Since $\mathbb{Z}$ is an ordered group, the result is an easy consequence of Proposition 10.1.4.

(4) $\Longrightarrow$ (5). This follows by Lemma 10.2.6

(5) $\Longrightarrow$ (1). Suppose that there exists a non-trivial subgroup H of G such that $Z(H) = \{1\}$. We shall prove that H is not diffuse, and therefore G is not diffuse, and the result follows.

By Lemma 10.4.5, H is a Bieberbach group. Let $n = \dim(H)$. By Theorem 10.4.7, H is isomorphic to a subgroup H_1 of $\mathrm{Isom}(\mathbb{R}^n)$ such that $\Delta(H)$ corresponds to the translations of H_1.

Note that if $u \in \mathbb{R}^n$ and $r \in \mathbb{R}$ is positive, then

$$B(u,r) = \{t \in \Delta(H_1) : \|t(0) - u\| < r\}$$

is finite. Since $(H_1 : \Delta(H_1)) = n$,

$$B'(0,r) = \{h \in H_1 : \|h(0)\| < r\}$$

is finite. We shall prove that $B'(0,r)$ has no extreme points for all sufficiently large $r > 0$.

Let $r_0 > 0$ such that $B(u, r_0) \neq \emptyset$ for all $u \in \mathbb{R}^n$. Note that every element $h \in H_1$ can be uniquely written as $h = t_h f_h$ with $t_h \in T(\mathbb{R}^n)$ and $f_h \in O(\mathbb{R}^n)$. Furthermore $H_1/\Delta(H_1) \cong (T(\mathbb{R}^n)H_1)/T(\mathbb{R}^n) \cong \langle f_h : h \in H_1\rangle = \{f_h : h \in H_1\}$.

Let $F = \{v \in \mathbb{R}^n : f_g(v) = v \text{ for all } g \in H_1\}$. Let $v \in F$. Let

$$w = \sum_{t \in B(v,r_0)} t(0) = \left(\prod_{t \in B(v,r_0)} t\right)(0).$$

Suppose that $v = (v_1, \ldots, v_n) \neq (0, \ldots, 0)$. Since $sv \in F$ for all $s \in \mathbb{R}$, we may assume that $v_i > r_0$. If $t \in B(v, r_0)$ and $t(0) = (a_1, \ldots, a_n)$, then $0 < v_i - r_0 < a_i < v_i + r_0$. Hence $w \neq 0$. Note that if $t \in B(v, r_0)$, then

$$\|f_g t f_g^{-1}(0) - v\| = \|f_g t(0) - f_g(v)\| = \|t(0) - v\| < r_0$$

for all $g \in H_1$. Now we have,

$$\begin{aligned} f_g(w) &= \sum_{t \in B(v,r_0)} f_g(t(0)) \\ &= \sum_{t \in B(v,r_0)} f_g t f_g^{-1}(0) \\ &= \sum_{t \in B(v,r_0)} t(0) = w \neq 0. \end{aligned}$$

Furthermore,

$$\begin{aligned} g\left(\prod_{t \in B(v,r_0)} t\right) g^{-1}(0) &= f_g\left(\prod_{t \in B(v,r_0)} t\right) f_g^{-1}(0) \\ &= \sum_{t \in B(v,r_0)} f_g t f_g^{-1}(0) \\ &= \sum_{t \in B(v,r_0)} t(0) \\ &= \left(\prod_{t \in B(v,r_0)} t\right)(0) \neq 0, \end{aligned}$$

for all $g \in H_1$. Hence $\prod_{t \in B(v,r_0)} t \in Z(H_1) \setminus \{\mathrm{id}\}$. This contradicts the fact that $Z(H_1) = \{\mathrm{id}\}$. Hence $F = \{(0, \ldots, 0)\}$.

Hence, for every nonzero vector $u \in \mathbb{R}^n$, there exists $h_u \in H_1$ such that $f_{h_u}(u) \neq u$, and thus

$$\|f_{h_u}(u) + u\| < 2\|u\|.$$

Hence, there exists $0 < \delta_u < 1$ such that

$$\|f_{h_u}(u) + u\| \leq 2\delta_u\|u\|.$$

Let $S = \{u \in \mathbb{R}^n : \|u\| = 1\}$. Let $V_u = \{v \in \mathbb{R}^n : \|u - v\| < \frac{1-\delta_u}{2}\}$. Note that

$$\|f_{h_u}(v) + v\| \leq \|f_{h_u}(v) - f_{h_u}(u)\| + \|f_{h_u}(u) + u\| + \|v - u\|$$

$$<1 - \delta_u + 2\delta_u = 1 + \delta_u.$$

Since S is compact and $S \subseteq \bigcup_{u\in S} V_u$, there exists a finite subset $K \subseteq S$ such that

$$S \subseteq \bigcup_{u\in K} V_u.$$

Let $\delta = \max\{\frac{1+\delta_u}{2} : u \in K\}$. Note that $0 < \delta < 1$. Then it is easy to check that for all $u \in \mathbb{R}^n$ there exists $h_u \in H_1 \setminus \Delta(H_1)$ such that

$$\| f_{h_u}(u) + u\| \leq 2\delta \|u\|.$$

Let

$$r > \frac{r_0}{1-\delta}.$$

Let $h \in B'(0, r)$. Hence $\|h(0)\| < r$. Let $u = h(0)$. We have seen that there exists $h_u \in H_1 \setminus \Delta(H_1)$ such that

$$\| f_{h_u}(u) + u\| \leq 2\delta \|u\|.$$

Let $v = \frac{1}{2}(-f_{h_u}(u) + u)$. Let $t \in B(v - h_u(0), r_0)$. Since $\|u\| = \|f_{h_u}(u)\| < r$, we have that

$$\|u - t(0) - h_u(0)\| \leq \|u - v\| + \|v - t(0) - h_u(0))\| \leq \delta r + r_0 < r$$

and

$$\|f_{h_u}(u) + t(0) + h_u(0)\| \leq \|f_{h_u}(u) + v\| + \| - v + t(0) + h_u(0)\| \leq \delta r + r_0 < r.$$

Let $g = th_u$. Note that $g \in H_1$. Since $f_{h_u} \neq \mathrm{id}$, we have that $g \neq 1$. Note that

$$\|gh(0)\| = \|g(u)\| = \|t(0) + h_u(0) + f_{h_u}(u)\| < r$$

and

$$\begin{aligned}\|g^{-1}h(0)\| &= \|h_u^{-1}(0) - f_{h_u}^{-1}t(0) + f_{h_u}^{-1}(u)\| \\ &= \| - f_{h_u}h_u(0) - f_{h_u}^{-1}t(0) + f_{h_u}^{-1}(u)\| = \| - h_u(0) - t(0) + u\| < r.\end{aligned}$$

Hence $gh, g^{-1}h \in B'(0, r)$, and the result follows. □

Theorem 10.4.9 *Let (X, r) be a finite involutive solution to the YBE. Then the structure group $G(X, r)$ is a Bieberbach group.*

Proof By Theorem 9.5.1, $G(X,r)$ has an abelian subgroup of finite index. Hence, to prove that $G(X,r)$ is a Bieberbach group, it is enough to show that $G(X,r)$ is torsion-free. Suppose that $G(X,r)$ is not torsion-free. Hence there exists $g \in G(X,r)$ of prime order p. Since $1 = g^p = g + \lambda_g(g) + \cdots + \lambda_{g^{p-1}}(g)$ and the additive group of the skew brace $G(X,r)$ is torsion-free, we have that $\lambda_g(g) \neq g$. Hence λ_g has order p. Let $X = \{x_1, \ldots, x_n\}$ with $|X| = n$. The restriction of λ_g to X is a product of disjoint cycles of length p in $\mathbb{S}_X$. Hence we may assume that there exists a non-negative integer m such that

$$\lambda_g(x_{kp+i}) = \begin{cases} x_{kp+i+1} & \text{if } 1 \leq i < p \\ x_{kp+1} & \text{if } i = p \end{cases}$$

for all $0 \leq k \leq m$ and $\lambda_g(x_t) = x_t$ for all $mp + p < t \leq n$. Since $g^p = 1$, in the additive group of the skew brace $G(X,r)$ we have that $g = z_1x_1 + \cdots + z_nx_n$ for some $z_1, \ldots, z_n \in \mathbb{Z}$ and

$$g + \lambda_g(g) + \cdots + \lambda_{g^{n-1}}(g) = 0.$$

Hence

$$\sum_{i=1}^{p} z_{kp+i} = 0 \quad \text{and} \quad z_t = 0$$

for all $0 \leq k \leq m$ and $mp + p < t \leq n$. Thus $g = \sum_{i=1}^{mp+p} z_i x_i$. Let

$$h_k = \sum_{j=1}^{p-1} \left(\sum_{i=1}^{j} z_{kp+i} \right) x_{kp+j}$$

for all $0 \leq k \leq m$. Let $h = \sum_{k=0}^{m} h_k$. We have that

$$\begin{aligned}
gh = g + \lambda_g(h) &= \sum_{k=0}^{m} \sum_{j=1}^{p} z_{kp+j} x_{kp+j} + \sum_{k=0}^{m} \sum_{j=1}^{p-1} \left(\sum_{i=1}^{j} z_{kp+i} \right) \lambda_g(x_{kp+j}) \\
&= \sum_{k=0}^{m} \sum_{j=1}^{p} z_{kp+j} x_{kp+j} + \sum_{k=0}^{m} \sum_{j=1}^{p-1} \left(\sum_{i=1}^{j} z_{kp+i} \right) x_{kp+j+1} \\
&= \sum_{k=0}^{m} z_{kp+1} x_{kp+1} + \sum_{k=0}^{m} \sum_{j=1}^{p-1} \left(\sum_{i=1}^{j+1} z_{kp+i} \right) x_{kp+j+1} \\
&= \sum_{k=0}^{m} \sum_{j=0}^{p-1} \left(\sum_{i=1}^{j+1} z_{kp+i} \right) x_{kp+j+1} \\
&= h + \sum_{k=0}^{m} \left(\sum_{i=1}^{p} z_{kp+i} \right) x_{kp+p} \\
&= h.
\end{aligned}$$

Hence $g = 1$, a contradiction. Therefore the structure group $G(X, r)$ is torsion-free and the result follows. □

10.5 Exercises

10.5.1 Let G be a group. Let H_1 and H_2 be two finite-index subgroups of G. Prove that $H_1 \cap H_2$ has finite index in G.

10.5.2 Let G be a group and $\{N_\alpha : \alpha\}$ be a collection of normal subgroups of G such that $\cap_\alpha N_\alpha = \{1\}$. Prove that if G/N_α is a left ordered group for all α, then G is a left ordered group.

10.5.3 Let K be a field. Prove that if G is a group satisfying the unique product property, then every invertible element of $K[G]$ is of the form ag for $a \in K \setminus \{0\}$ and $g \in G$ (i.e., $K[G]$ has only trivial units).

10.5.4 Let G be a group such that $(G : Z(G)) = n$. Prove that $[x, y]^n = 1$ for all $x, y \in G$.

10.5.5 Let H be a central subgroup of a finite group G. Prove that if $|H|$ and $|G/H|$ are coprime, then $G \cong H \times G/H$.

10.6 Open Problems

Open problem 10.6.1 Let G be a group satisfying the unique product property. Is G diffuse?

Open problem 10.6.2 (*Kaplansky's units conjecture*) Let G be a torsion-free group. Does $\mathbb{C}[G]$ admit non-trivial units?

10.7 Notes

Burns and Hale proved Theorem 10.1.8 and Corollary 10.1.9 in [27].

Bergman gave an example of a left ordered group that is not locally indicable in [22]. He proved that this example contains finitely generated perfect subgroups.

Strojnowski proved Theorem 10.2.3 in [109]. As a consequence one gets that for every field K and every group G satisfying the unique product property, the invertible elements of $K[G]$ are of the form kg, with $k \in K \setminus \{0\}$ and $g \in G$ (trivial units) (see [92, page 591]).

Bowditch introduced diffuse groups in [24]. He proved Lemma 10.2.7. The converse is unknown (see Open problem 10.6.1). Dunfield gave an example of a diffuse group that is not a left ordered group in an appendix in [79]. Promislow gave an example of a torsion-free group that does not satisfy the unique product property in [94].

Linnell and Morris proved Proposition 10.2.5 in [86]. They also proved that an amenable group is diffuse if and only if it is locally indicable. This gives an alternative proof of the equivalence of conditions (2) and (5) in Theorem 10.4.8. For the proof of (5) implies (1) in Theorem 10.4.8, we follow the proof of [79, Lemma 3.4]. Theorem 10.4.9 is due to Gateva-Ivanova and Van den Bergh [62]. For its proof, we follow [76, Corollary 8.2.6].

Gardam in [60] found a counterexample to the unit conjecture over fields of characteristic two. The conjecture, commonly attributed to Kaplansky, states that for every field K and every torsion-free group G, the group algebra $K[G]$ has only trivial units; see Exercise 10.5.3. It is not known whether the conjecture is true for fields of zero characteristic; see Open problem 10.6.2.

Chapter 11
Garside Groups

In this chapter, we prove that the structure group of a finite involutive solution to the YBE is a Garside group.

11.1 Ore Monoids

A cancellative monoid M is said to be a *right Ore monoid* if $xM \cap yM \neq \emptyset$ for all $x, y \in M$. It is said that M is a *left Ore monoid* if $Mx \cap My \neq \emptyset$ for all $x, y \in M$.

Let G be a group and M be a submonoid of G. It is said that G is a *right group of fractions* of M if for every $g \in G$ there exist $a, b \in M$ such that $g = ab^{-1}$. It is said that G is a *left group of fractions* of M if for every $g \in G$ there exist $a, b \in M$ such that $g = a^{-1}b$.

A monoid M has a right (left) group of fractions if there exists an injective homomorphism of monoids $f\colon M \to G$, and G is a right (left) group of fractions of $f(M)$. In this case, we say that G is a right (left) group of fractions of M.

Proposition 11.1.1 *A monoid M has a right group of fractions if and only if it is a cancellative right Ore monoid.*

Proof Suppose that M has a right group of fractions. We may assume that M is a submonoid of a group G and for every $g \in G$ there exist $a, b \in M$ such that $g = ab^{-1}$. Clearly, M is cancellative. Let $x, y \in M$. Then there exist $a, b \in M$ such that $x^{-1}y = ab^{-1}$. Hence

$$xa = yb \in xM \cap yM,$$

and thus M is a right Ore monoid.

Conversely, suppose that M is a cancellative right Ore monoid. We define a binary relation $\sim$ on $M \times M$ by $(a, b) \sim (c, d)$ if and only if there exist $x, y \in M$ such that

F. Cedó and L. Vendramin, *Groups, Radical Rings, and the Yang-Baxter Equation*, Progress in Mathematics 361, https://doi.org/10.1007/978-3-032-01524-2_11

$$ax = cy \quad \text{and} \quad bx = dy.$$

We shall prove that $\sim$ is an equivalence relation. Since $1 \in M$, it is clear that $(a, b) \sim (a, b)$. Clearly, $\sim$ is a symmetric relation. Let $a, b, c, d, e, f \in M$ be elements such that

$$(a, b) \sim (c, d) \quad \text{and} \quad (c, d) \sim (e, f).$$

Hence there exist $x, y, z, t \in M$ such that

$$ax = cy, \quad bx = dy, \quad cz = et \quad \text{and} \quad dz = ft.$$

Since $yM \cap zM \neq \emptyset$, there exist $u, v \in M$ such that $yu = zv$. Hence

$$axu = cyu = czv = etv \quad \text{and} \quad bxu = dyu = dzv = ftv,$$

and thus $(a, b) \sim (e, f)$. Therefore $\sim$ is an equivalence relation. We denote by a/b the $\sim$-equivalence class of (a, b). Let

$$G = (M \times M)/\!\sim\, = \{a/b : a, b \in M\}.$$

We define a multiplication on G by

$$(a/b)(c/d) = (ax)/(dy),$$

where $bx = cy$ for some $x, y \in M$ (these elements exist because $bM \cap cM \neq \emptyset$). We shall prove that this is a well-defined operation. Let $a_1, b_1, c_1, d_1, x_1, y_1 \in M$ be elements such that

$$(a, b) \sim (a_1, b_1), \quad (c, d) \sim (c_1, d_1) \quad \text{and} \quad b_1x_1 = c_1y_1.$$

We should prove that

$$(ax, dy) \sim (a_1x_1, d_1y_1).$$

Since $bxM \cap b_1x_1M \neq \emptyset$, there exist $u, v \in M$ such that

$$cyu = bxu = b_1x_1v = c_1y_1v.$$

Note that

$$(axu, bxu) \sim (a, b) \sim (a_1, b_1) \sim (a_1x_1v, b_1x_1v).$$

Hence there exist $t, z \in M$ such that

$$axut = a_1x_1vz \quad \text{and} \quad bxut = b_1x_1vz.$$

Since $bxu = b_1x_1v$ and M is cancellative, we have that $t = z$ and thus $axu = a_1x_1v$. Note also that

$$(cyu, dyu) \sim (c, d) \sim (c_1, d_1) \sim (c_1y_1v, d_1y_1v).$$

Hence there exist $t', z' \in M$ such that

$$cyut' = c_1y_1vz' \quad \text{and} \quad dyut' = d_1y_1vz'.$$

Since $cyu = c_1y_1v$ and M is cancellative, we have that $t' = z'$ and thus $dyu = d_1y_1v$. Hence

$$(ax, dy) \sim (a_1x_1, d_1y_1),$$

and thus the multiplication on G is well-defined.

We shall prove that G with this multiplication is a group. Let $a, b, c, d, e, f \in M$. We have that

$$((a/b)(c/d))(e/f) = ((ax)/(dy))(e/f) = (axz)/(ft),$$

where $x, y, z, t \in M$ satisfy

$$bx = cy \quad \text{and} \quad dyz = et.$$

On the other hand, we have

$$(a/b)((c/d)(e/f)) = (a/b)((cyz)/(ft)) = (axz)/(ft),$$

because $d(yz) = et$ and $b(xz) = (cyz)1$. Hence the multiplication is associative. Note that

$$(a/b)(1/1) = a/b = (1/1)(a/b)$$

and

$$(a/b)(b/a) = a/a = 1/1$$

for all $a, b \in M$. Hence G is a group. Let $f \colon M \to G$ be the map defined by $f(a) = a/1$ for all $a \in M$. Since

$$f(1) = 1/1 \quad \text{and} \quad f(ab) = (ab)/1 = (a/1)(b/1) = f(a)f(b)$$

for all $a, b \in M$, we have that f is a homomorphism of monoids. Let $a, b \in M$ be elements such that $a/1 = b/1$. Then there exist $x, y \in M$ such that $ax = by$ and $1x = 1y$. Hence $x = y$ and, since M is cancellative, $a = b$. Therefore f is injective. Note that for every $a, b \in M$,

$$a/b = (a/1)(1/b) = (a/1)(b/1)^{-1} = f(a)f(b)^{-1}.$$

Hence G is a right group of fractions of M. □

Proposition 11.1.2 *Let M be a left and right Ore monoid. Let G be a left group of fractions of M. Then G is a right group of fractions of M.*

Proof We may assume that M is a submonoid of G. Let $g \in G$. There exist $a, b \in M$ such that $g = a^{-1}b$. Since M is a right Ore monoid, there exist $x, y \in M$ such that $ax = by$. Hence $g = xy^{-1}$ and therefore the result follows. □

Proposition 11.1.3 *Let G be a right group of fractions of a submonoid M. Let H be a group and let $f\colon M \to H$ be a homomorphism of monoids. Then there exists a unique group homomorphism $g\colon G \to H$ such that $g(a) = f(a)$ for all $a \in M$.*

Proof We define $g(ab^{-1}) = f(a)f(b)^{-1}$ for all $a, b \in M$. It is an easy exercise to prove that $g\colon G \to H$ is a well-defined group homomorphism and that it is the unique homomorphism such that $g(a) = f(a)$ for all $a \in M$. □

An easy consequence of Proposition 11.1.3 is that if G and H are right groups of fractions of a monoid M, then $G \cong H$.

11.2 Garside Monoids and Garside Groups

Definition 11.2.1 A *Garside monoid* is a pair (M, Δ), where M is a cancellative monoid such that

(1) there exists a map $d\colon M \to \mathbb{N}$ such that $d(xy) \geq d(x) + d(y)$ and $d(x) \neq 0$ if $x \neq 1$,
(2) any two elements of M have a left- and a right-lcm and a left- and a right-gcd,
(3) Δ is a *Garside element* of M, this meaning that the left- and right-divisors of Δ coincide and generate M, and
(4) the family of all divisors of Δ is finite.

A non-invertible element of a monoid is an *atom* if it is not a product of two non-invertible elements. A monoid S is *atomic* if every non-invertible element of S is a product of atoms.

Note that if (M, Δ) is a Garside monoid, then condition (1) in the definition implies that $d(1) = 0$ and M is *conical*, that is $ab = 1$ implies $a = b = 1$. Furthermore, M is atomic and the set of atoms of M is finite by (4). Moreover, by condition (1), for every $a \in M$, the supremum of the number of atoms in the factorization of a as a product of atoms is finite. Condition (2) in the definition implies that M is a right and left Ore monoid. Thus, by Proposition 11.1.1, M has a left and a right group of fractions.

Definition 11.2.2 A group G is said to be a *Garside group* if it is the (left) group of fractions of a submonoid M and there exists $\Delta \in M$ such that (M, Δ) is a Garside monoid. In this case, we say that (M, Δ) is a Garside submonoid of G.

Lemma 11.2.3 *Let (M, Δ) be a Garside monoid. Let n be a positive integer and let $s_1, \dots, s_n, s \in M$ be divisors of Δ such that*

(1) $s_1 \notin \{\Delta, 1\}$, $s \neq 1$,
(2) *s is a left divisor of $s_1 \cdots s_n$, and*
(3) *for every $1 \leq i < n$, $\forall a \in M \setminus \{1\}$, a is a left divisor of s_{i+1} implies that $s_i a$ is not a divisor of Δ.*

Then s is a left divisor of s_1.

Proof We shall prove the result by induction on n. For $n = 1$, the result is obvious. Suppose that $n > 1$ and that the result holds for $n - 1$. Suppose that s is not a left divisor of s_1. Let $b \in M$ be the right-lcm of s and s_1. Since s and s_1 are divisors of Δ, we have that b is a divisor of Δ. There exists $c \in M$ such that $b = s_1 c$. Since s is a left divisor of $s_1 \cdots s_n$, we have that b is a left divisor of $s_1 \cdots s_n$. Since M is cancellative, we have that c is a left divisor of $s_2 \cdots s_n$. Since s is not a left divisor of s_1, $c \neq 1$. Hence by assumption (3), $s_2 \neq 1$. If $s_2 = \Delta$, then there exist $b', b'' \in M \setminus \{1\}$ such that $s_1 b' = \Delta$ and $b'b'' = \Delta = s_2$, in contradiction with assumption (3). Hence $s_2 \notin \{\Delta, 1\}$. By the inductive hypothesis, c is a left divisor of s_2. Since $s_1 c = b$ is a divisor of Δ, we get a contradiction. Therefore s is a left divisor of s_1 and the result follows by induction. □

Theorem 11.2.4 *Let G be a Garside group and let (M, Δ) be a Garside submonoid of G such that G is a left group of fractions of M. Then for every element $g \in G$ there exist unique $z \in \mathbb{Z}$, non-negative integer n and $s_1, \dots, s_n \in M$ such that*

$$g = \Delta^z s_1 \cdots s_n,$$

$s_1, \dots, s_n$ are divisors of Δ in M, $s_1 \neq \Delta$, $s_n \neq 1$ and for all $1 \leq i < n$, $a \in M \setminus \{1\}$ is a left divisor of s_{i+1} implies that $s_i a$ is not a divisor of Δ.

Proof Note that if $a \in M$ is a divisor of Δ, then there exist $a', a'', c, c' \in M$ such that

$$c'c = ca = aa' = a'a'' = \Delta,$$

$a\Delta = \Delta a''$ and $\Delta a = c'\Delta$. Furthermore, $a^{-1} = a'\Delta^{-1} = \Delta^{-1}c$. Hence, for every element $g \in G$, there exist $z \in \mathbb{Z}$ and $b \in M$ such that $g = \Delta^z b$. By condition (1) in the definition of Garside monoid, we may assume that z is the greatest integer satisfying that $g = \Delta^z b$ for some $b \in M$.

We shall prove by induction on $d(b)$, where $d\colon M \to \mathbb{N}$ is the map of the condition (1) in the definition of Garside monoid, that there exist a non-negative integer n and $s_1, \dots, s_n$ divisors of Δ in M, such that $b = s_1 \cdots s_n$, $s_1 \neq \Delta$, $s_n \neq 1$ and for all $1 \leq i < n$, $a \in M \setminus \{1\}$ is a left divisor of s_{i+1} implies that $s_i a$ is not a divisor of Δ. If $b = 1$, then $n = 0$. Suppose that $b \neq 1$. Let s_1 be the left-gcd of b and Δ. We have that $s_1 \neq 1$. There exists $b_1 \in M$ such that $b = s_1 b_1$. Now $d(b) = d(s_1 b_1) \geq d(s_1) + d(b_1)$. Since $d(s_1) \neq 0$, we have that $d(b_1) < d(b)$. By the inductive hypothesis, there exist a positive integer n, and $s_2, \dots, s_n$ divisors of Δ such that $b_1 = s_2 \cdots s_n$,

$s_2 \neq \Delta$, $s_n \neq 1$ and for all $2 \leq i < n$, $a \in M \setminus \{1\}$ is a left divisor of s_{i+1} implies that $s_i a$ is not a divisor of Δ. Hence $b = s_1 \cdots s_n$. By the maximality of z, $s_1 \neq \Delta$. Since s_1 is the left-gcd of b and Δ, it is clear that if $a \in M \setminus \{1\}$ is a left divisor of s_2, then $s_1 a$ is not a divisor of Δ.

Now we shall prove the uniqueness of z, n, and $s_1, \dots, s_n$. Let $z, z' \in \mathbb{Z}$, n_1, n_2 non-negative integers and $s_1, \dots, s_{n_1}, s'_1, \dots, s'_{n_2} \in M$ divisors of Δ such that

$$g = \Delta^z s_1 \cdots s_{n_1} = \Delta^{z'} s'_1 \cdots s'_{n_2},$$

$s_1 \neq \Delta$, $s_{n_1} \neq 1$ and for all $1 \leq i < n_1$, $a \in M \setminus \{1\}$ is a left divisor of s_{i+1} implies that $s_i a$ is not a divisor of Δ, and $s'_1 \neq \Delta$, $s'_{n_2} \neq 1$ and for all $1 \leq j < n_2$, $a' \in M \setminus \{1\}$ is a left divisor of s'_{j+1} implies that $s_j a'$ is not a divisor of Δ. We may assume that $z \geq z'$. Hence

$$\Delta^{z-z'} s_1 \cdots s_{n_1} = s'_1 \cdots s'_{n_2}.$$

Suppose that $z > z'$. Hence $n_2 > 0$ and $s'_1 \notin \{\Delta, 1\}$. By Lemma 11.2.3, Δ is a left divisor of s'_1, but s'_1 is a divisor of Δ and thus $s'_1 = \Delta$, a contradiction. Hence $z = z'$ and

$$s_1 \cdots s_{n_1} = s'_1 \cdots s'_{n_2}.$$

Note that $n_1 = 0$ if and only if $n_2 = 0$. If $n_1 \neq 0$, then $s_1 \notin \{\Delta, 1\}$ and by Lemma 11.2.3, s'_1 is a left divisor of s_1, and moreover $n_2 \neq 0$, and again by Lemma 11.2.3, s_1 is a left divisor of s'_1. Hence $s_1 = s'_1$ and

$$s_2 \cdots s_{n_1} = s'_2 \cdots s'_{n_2}.$$

Now by induction on $\min\{n_1, n_2\}$ it is easy to see that $n_1 = n_2$ and $s_i = s'_i$ for all $1 \leq i \leq n_1$. □

Example 11.2.5 Let n be a positive integer and let $X = \{x_1, \dots, x_n\}$ be a set of cardinal n. Let $\mathrm{FAM}(X)$ be the free abelian multiplicative monoid on X, that is $\mathrm{FAM}(X)$ is the monoid with presentation

$$\mathrm{FAM}(X) = \langle X : x_i x_j = x_j x_i \text{ for all } 1 \leq i, j \leq n \rangle.$$

Let $\Delta = x_1 \cdots x_n \in \mathrm{FAM}(X)$. Then $(\mathrm{FAM}(X), \Delta)$ is a Garside monoid and its group of fractions is the free abelian multiplicative group on X, which is isomorphic to $\mathbb{Z}^n$.

11.3 Finite Involutive Solutions to the YBE

In [47, Theorem 3.3], Chouraqui proved that the structure group of a finite involutive solution to the YBE is a Garside group. We shall prove this result using the natural structure of the skew brace of the structure group of a solution to the YBE.

Lemma 11.3.1 *Let (X, r) be a finite involutive solution to the YBE. Let*

$$n = [G(X, r) : \mathrm{Soc}(G(X, r))].$$

Let z be an integer. Let $\Delta_z = \sum_{x \in X} zx$. Then $g\Delta_z = g + \Delta_z$ for all $g \in G(X, r)$. In particular, $\Delta_z^m = m\Delta_z$ for all integer m, and $n\Delta_z$ is a central element of the structure group $G(X, r)$.

Proof Let $g \in G(X, r)$. We have

$$g\Delta_z = g + \lambda_g(\Delta_z) = g + \lambda_g\left(\sum_{x \in X} zx\right) = g + \sum_{x \in X} z\lambda_g(x) = g + \Delta_z.$$

The second part is a consequence of the first and the fact that $n\Delta_z \in \mathrm{Soc}(G(X, r))$. □

Proposition 11.3.2 *Let (X, r) be an involutive solution to the YBE. Then the structure monoid $M(X, r)$ of the solution (X, r) is isomorphic to the submonoid of $G(X, r)$ generated by X.*

Proof By Proposition 9.2.4, $M(X, r)$ is a cancellative monoid. Furthermore, in the proof of Proposition 9.2.4, we have seen that the derived structure monoid

$$A(X, r) = \langle X : xy = yx \text{ for all } x, y \in X\rangle$$

is the free abelian monoid on X. By Proposition 9.2.3, there exist a unique monoid homomorphism $\theta\colon M(X, r) \to \mathrm{Aut}(A(X, r))$ and a unique bijective 1-cocycle $\varphi\colon M(X, r) \to A(X, r)$ such that $\theta(x)(y) = \sigma_x(y)$ and $\varphi(x) = x$ for all $x, y \in X$. We shall prove that $M(X, r)$ is a right Ore monoid. Let $a, b \in M(X, r)$. Let

$$c = \varphi^{-1}(\theta(a)^{-1}(\varphi(b))) \text{ and } d = \varphi^{-1}(\theta(b)^{-1}(\varphi(a))).$$

Now we have that

$$\begin{aligned}
\varphi(ac) &= \varphi(a)\theta(a)(\varphi(c)) \\
&= \varphi(a)\varphi(b) \\
&= \varphi(b)\varphi(a) \\
&= \varphi(b)\theta(b)(\varphi(d)) \\
&= \varphi(bd).
\end{aligned}$$

Hence $ac = bd \in aM(X, r) \cap bM(X, r)$. Therefore $M(X, r)$ is a right Ore monoid. Let G be its right group of fractions. We may assume that $M(X, r)$ is a submonoid of G. By the definitions of $M(X, r)$ and $G(X, r)$, the natural map $i\colon X \to G(X, r)$ induces a unique monoid homomorphism $f\colon M(X, r) \to G(X, r)$ such that $f(x) =$

$i(x)$ for all $x \in X$. By Proposition 11.1.3, there exists a unique group homomorphism $g\colon G \to G(X,r)$ such that $g(a) = f(a)$ for all $a \in M(X,r)$. On the other hand, since in the group G

$$xy = \sigma_x(y)\tau_y(x)$$

for all $x, y \in X$, there exists a unique group homomorphism

$$h\colon G(X,r) \to G$$

such that $h(i(x)) = x$ for all $x \in X$. Now it is easy to check that $gh = \mathrm{id}_{G(X,r)}$ and $hg = \mathrm{id}_G$. Hence $G(X,r) \cong G$. In particular, the natural map $i\colon X \to G(X,r)$ is injective and we may identify $i(x)$ with x for each $x \in X$. Hence

$$M(X,r) \cong g(M(X,r)) = f(M(X,r)),$$

the submonoid of $G(X,r)$ generated by X. Therefore the result follows. □

By Proposition 11.3.2, for every involutive solution (X,r) to the YBE, we identify $M(X,r)$ with the submonoid of $G(X,r)$ generated by X.

Lemma 11.3.3 *Let (X,r) be a finite involutive solution to the YBE. Then*

$$M(X,r) = \left\{ \sum_{x\in X} z_x x : z_x \in \mathbb{N} \right\}$$

and $M(X,r)^{-1} = -M(X,r)$.

Proof Let $M = \{\sum_{x\in X} z_x x : z_x \in \mathbb{N}\}$. Note that $\lambda_g(a) \in M$ for all $g \in G(X,r)$ and all $a \in M$.

Every element of $M(X,r)$ is of the form $x_1 \cdots x_m$, with $x_1, \dots, x_m \in X$. We shall prove that $x_1 \cdots x_m \in M$ by induction on m. For $m = 1$ it is clear. Suppose that $m > 1$ and that $y_1 \cdots y_{m-1} \in M$ for all $y_1, \dots, y_{m-1} \in X$. By the inductive hypothesis, $x_2 \cdots x_m \in M$, and thus

$$x_1 \cdots x_m = x_1 + \lambda_{x_1}(x_2 \cdots x_m) \in M.$$

Hence $M(X,r) \subseteq M$. Let $\sum_{x\in X} z_x x \in M$. We shall prove that $\sum_{x\in X} z_x x \in M(X,r)$ by induction on $t = \sum_{x\in X} z_x$. For $t = 1$ it is clear. Suppose that $t > 1$ and that $\sum_{x\in X} a_x x \in M(X,r)$ for all $\sum_{x\in X} a_x x \in M$ with $\sum_{x\in X} a_x < t$. Since $t > 1$ there exists $x_0 \in X$ such that $z_{x_0} > 0$. Let $a = -\lambda_{x_0}^{-1}(x_0) + \sum_{x\in X} z_x \lambda_{x_0}^{-1}(x)$. Now we have that $a \in M$ and, by the inductive hypothesis, $a \in M(X,r)$. Thus

$$\sum_{x\in X} z_x x = x_0 - x_0 + \sum_{x\in X} z_x x = x_0 \lambda_{x_0}^{-1}\left(-x_0 + \sum_{x\in X} z_x x\right) = x_0 a \in M(X,r).$$

Hence $M = M(X, r)$.

Note that $g^{-1} = -\lambda_{g^{-1}}(g)$ and $-g = \lambda_{(-g)^{-1}}(g)^{-1}$ for all $g \in G(X, r)$. Therefore $M^{-1} = -M$, and the result follows. □

Note that $M(X, r)$ has a degree function

$$\deg\colon M(X, r) \longrightarrow \mathbb{N}$$

defined by $\deg(x_1 \cdots x_m) = m$ for all $x_1, \dots, x_m \in X$.

Lemma 11.3.4 *Let (X, r) be a finite involutive solution to the YBE. Let $a \in M(X, r)$. By Lemma 11.3.3, $a = \sum_{x \in X} a_x x$ for some $a_x \in \mathbb{N}$. Then* $\deg(a) = \sum_{x \in X} a_x$.

Proof We will prove the result by induction on $\deg(a)$. If $\deg(a) = 1$, then $a = x$ for some $x \in X$, and the result is clear. Suppose that $\deg(a) = m > 1$ and that the result is true for all $b \in M(X, r)$ with $\deg(b) < m$. We have that $a = x_1 \cdots x_m$, for some $x_1, \dots, x_m \in X$. By the induction hypothesis,

$$x_2 \cdots x_m = \sum_{x \in X} b_x x$$

for some $b_x \in \mathbb{N}$ such that $\sum_{x \in X} b_x = m - 1$. We have

$$a = x_1 + \lambda_{x_1}(x_2 \cdots x_m) = x_1 + \lambda_{x_1}\left(\sum_{x \in X} b_x x\right) = x_1 + \sum_{x \in X} b_x \lambda_{x_1}(x).$$

Since $\lambda_{x_1}(x) \in X$ for all $x_1, x \in X$, the result follows. □

Lemma 11.3.5 *Let (X, r) be a finite involutive solution to the YBE, $a, b \in M(X, r)$ and $c, d \in M(X, r)^{-1}$. By Lemma 11.3.3,*

$$a = \sum_{x \in X} a_x x,\ b = \sum_{x \in X} b_x x,\ c = \sum_{x \in X} c_x x \ \text{ and } \ d = \sum_{x \in X} d_x x,$$

for some $a_x, b_x, -c_x, -d_x \in \mathbb{N}$. Then the following statements hold:

(1) $b^{-1}a \in M(X, r)$ if and only if $b_x \leq a_x$ for all $x \in X$.
(2) $c^{-1}d \in M(X, r)^{-1}$ if and only if $c_x \geq d_x$ for all $x \in X$.

Proof (1) Note that

$$b^{-1}a = b^{-1} + \lambda_{b^{-1}}(a) = -\lambda_{b^{-1}}(b) + \lambda_{b^{-1}}(a) = \lambda_{b^{-1}}(a - b).$$

Therefore $b^{-1}a \in M(X, r)$ if and only if $a - b \in M(X, r)$ and the result follows by Lemma 11.3.3.

(2) As above $c^{-1}d = \lambda_{c^{-1}}(d - c)$. Therefore $c^{-1}d \in M(X, r)^{-1}$ if and only if $d - c \in M(X, r)^{-1}$ and the result follows by Lemma 11.3.3. □

Lemma 11.3.6 *Let (X, r) be a finite involutive solution to the YBE. Any two elements of $M(X, r)$ have a left- and a right-lcm and a left- and a right-gcd.*

Proof Let $a, b \in M(X, r)$. By Lemma 11.3.3

$$a = \sum_{x \in X} a_x x \quad \text{and} \quad b = \sum_{x \in X} b_x x$$

for some $a_x, b_x \in \mathbb{N}$. By Lemma 11.3.5, it is clear that the left-gcd of a and b is

$$\operatorname{lgcd}(a, b) = \sum_{x \in X} \min(a_x, b_x)x,$$

and that its right-lcm is

$$\operatorname{rlcm}(a, b) = \sum_{x \in X} \max(a_x, b_x)x.$$

Now consider the monoid $M(X, r)^{-1}$. Since $a^{-1}, b^{-1} \in M(X, r)^{-1}$, by Lemma 11.3.3

$$a^{-1} = \sum_{x \in X} a'_x x \quad \text{and} \quad b^{-1} = \sum_{x \in X} b'_x x,$$

for some non-positive integers a'_x, b'_x. By Lemma 11.3.5 it is clear that the left-gcd of a^{-1} and b^{-1} in $M(X, r)^{-1}$ is

$$\operatorname{lgcd}(a^{-1}, b^{-1}) = \sum_{x \in X} \max(a'_x, b'_x)x,$$

and that its right-lcm is

$$\operatorname{rlcm}(a^{-1}, b^{-1}) = \sum_{x \in X} \min(a'_x, b'_x)x.$$

Note that if $d^{-1} = \operatorname{lgcd}(a^{-1}, b^{-1})$, then d is the right-gcd of a and b in $M(X, r)$, and if $m^{-1} = \operatorname{rlcm}(a^{-1}, b^{-1})$, then m is the left-lcm of a and b in $M(X, r)$. Thus the result follows. □

Theorem 11.3.7 (Chouraqui) *Let (X, r) be a finite involutive solution to the YBE. Let m be a positive integer. Then $(M(X, r), \Delta_m)$ is a Garside monoid and thus $G(X, r)$ is a Garside group.*

Proof It is clear that the submonoid $M(X, r)$ of $G(X, r)$ is cancellative. We have seen that $M(X, r)$ has a degree function deg. In particular condition (1) in the definition of Garside monoid is satisfied.

By Lemma 11.3.6, condition (2) in the definition of Garside monoid is satisfied by $M(X, r)$.

By Lemma 11.3.3, $\Delta_m \in M(X, r)$ and, by Lemma 11.3.1, $\Delta_m^{-1} = -\Delta_m$. By Lemma 11.3.5, the set of left-divisors of Δ_m in $M(X, r)$ is

$$D_1 = \left\{ \sum_{x \in X} z_x x : z_x \in \mathbb{Z},\ 0 \le z_x \le m,\ \text{for all } x \in X \right\},$$

and the set of left-divisors of Δ_m^{-1} in the monoid $M(X, r)^{-1}$ is

$$D_2 = \left\{ -\sum_{x \in X} z_x x : z_x \in \mathbb{Z},\ 0 \le z_x \le m,\ \text{for all } x \in X \right\}.$$

Note that

$$\left(\sum_{x \in X} z_x x \right)^{-1} = -\lambda_{(\sum_{x \in X} z_x x)^{-1}} \left(\sum_{x \in X} z_x x \right) = -\sum_{x \in X} z_x \lambda_{(\sum_{x \in X} z_x x)^{-1}}(x)$$

for all $z_x \in \mathbb{Z}$. Hence $D_2^{-1} = D_1$. Thus the set of right-divisors of Δ_m in $M(X, r)$ also is D_1, which is finite and contains X. Therefore Δ_m is a Garside element of $M(X, r)$. Hence $(M(X, r), \Delta_m)$ is a Garside monoid, and clearly $G(X, r)$ is its left (right) group of fractions. □

Definition 11.3.8 If (M, Δ) is a Garside monoid and G is its group of fractions, a surjective homomorphism $\pi : G \longrightarrow \overline{G}$ is said to provide a *Garside germ* for (G, M, Δ) if there exists a map $\sigma : \overline{G} \longrightarrow M$ such that $\pi\sigma$ is the identity, the image of σ is the family of all divisors of Δ in M, and M admits a presentation

$$\langle \sigma(\overline{G}) : \{\sigma(f)\sigma(g) = \sigma(fg) : f, g \in \overline{G} \text{ and } \|f\|_{\overline{S}} + \|g\|_{\overline{S}} = \|fg\|_{\overline{S}}\} \rangle,$$

where $\overline{S}$ is the image under π of the atom set of M, and $\|f\|_{\overline{S}}$ is the length of a shortest decomposition of f as a product of elements of $\overline{S}$.

Proposition 11.3.9 (Dehornoy) *Let (X, r) be a finite involutive solution to the YBE. Let $|X| = n$, $G = G(X, r)$, $M = M(X, r)$ and $d = [G : \operatorname{Soc}(G)]$. Let k be a positive integer such that $kd - 1 \ge 1$ and $\Delta = \Delta_{kd-1}$. Then $N = \{kdg : g \in G\}$ is a normal subgroup of G and the natural map $\pi : G \longrightarrow G/N$ provides a Garside germ for (G, M, Δ). The group G/N has order $(kd)^n$ and the kernel of π is isomorphic to $\mathbb{Z}^n$.*

Proof By Theorem 11.3.7, (M, Δ) is a Garside monoid. We also know that the set of left-divisors of Δ is

$$D_1 = \left\{ \sum_{x \in X} z_x x : z_x \in \mathbb{Z},\ 0 \le z_x \le kd - 1 \right\}.$$

It is easy to see that N is a left ideal of the skew brace G. Let $g, h \in G$. Then, since $d = [G : \mathrm{Soc}(G)]$, $dg \in \mathrm{Soc}(G)$ and

$$\begin{aligned} h^{-1}(kdg)h &= h^{-1}(kdg + h) \\ &= h^{-1} + \lambda_{h^{-1}}(kdg + h) \\ &= h^{-1} + \lambda_{h^{-1}}(h + kdg) \\ &= h^{-1} + \lambda_{h^{-1}}(h) + kd\lambda_{h^{-1}}(g) \\ &= kd\lambda_{h^{-1}}(g) \in N. \end{aligned}$$

Hence N is an ideal of the skew brace G. In particular, the multiplicative group of N is a normal subgroup of the structure group G. The natural homomorphism $\pi : G \longrightarrow G/N$ is in fact a homomorphism of skew braces. Since the additive group of the skew brace G is free abelian with basis X, the restriction of π to D_1 is a bijection $\pi|_{D_1} : D_1 \longrightarrow G/N$. Let $\sigma = (\pi|_{D_1})^{-1}$. We have that $\pi(\sigma(f)) = f$ for all $f \in G/N$. By Proposition 11.3.2, M admits a presentation as a monoid

$$\langle X : xy = zt \text{ whenever } r(x, y) = (z, t)\rangle.$$

The set of atoms of M is X. Let $\overline{X} = \pi(X)$. Let $\|a\|_{\overline{X}}$ denote the length of a shortest decomposition of $a \in G/N$ as a product of elements of $\overline{X}$. Let $a \in G/N$ and

$$\sigma(a) = \sum_{x \in X} a_x x$$

for some $a_x \in \mathbb{N}$. By Lemma 11.3.4, $\|a\|_{\overline{X}} = \sum_{x \in X} a_x$. Hence, for $f, g \in G/N$,

$$\|f\|_{\overline{X}} + \|g\|_{\overline{X}} = \|fg\|_{\overline{X}}$$

if and only if

$$\sigma(f)\sigma(g) = \sigma(fg).$$

Since $kd > 1$, for $x, y \in X$, $xy \notin D_1$ if and only if $kd = 2$ and $xy = x + \lambda_x(y) = 2x$, that is if and only if $kd = 2$ and $\lambda_x(y) = x$. In this case,

$$r(x, y) = (\lambda_x(y), \lambda^{-1}_{\lambda_x(y)}(x)) = (x, y).$$

Now it is easy to see that M admits a presentation

$$\langle \sigma(G/N) : \{\sigma(f)\sigma(g) = \sigma(fg) : f, g \in G/N \text{ and } \|f\|_{\overline{X}} + \|g\|_{\overline{X}} = \|fg\|_{\overline{X}}\}\rangle.$$

Therefore the result follows. □

11.4 Exercises

11.4.1 Complete the proof of Proposition 11.1.3.

11.5 Notes

Theorem 11.3.7 was proved by Chouraqui in [47]. Our proof is based on the work of Dehornoy [52] and the presentation of Cedó's survey [36].

In [52] Dehornoy developed a sort of right-cyclic calculus to describe finite quotients of $G(X, r)$ that play a role similar to the role that Coxeter groups play for Artin–Tits groups. Thus he proved Proposition 11.3.9. Here we have proved this using the skew brace structure of $G(X, r)$.

Chapter 12
Left Nilpotent Skew Braces

In the first section of this chapter, we study classical results on invariant subgroups. These results are useful for studying the left nilpotency of skew braces in the second section.

12.1 Invariant Subgroups

We say that a group G acts on a group K by automorphisms if the (left) action

$$G \times K \to K, \quad (g, x) \mapsto g \cdot x,$$

satisfies

$$g \cdot (xy) = (g \cdot x)(g \cdot y)$$

for all $g \in G$ and $x, y \in K$. In this case, the map $\alpha \colon G \to \operatorname{Aut}(K), g \mapsto \alpha(g)$, where $\alpha(g)(k) = g \cdot k$ for all $g \in G$ and $k \in K$, is a homomorphism of groups. Consider the semidirect product $\Gamma = K \rtimes_\alpha G$. We will identify K with $K \times \{1\}$ and G with $\{1\} \times G$ in Γ. With this identification we have that $g \cdot k = gkg^{-1}$ for all $g \in G$ and $k \in K$. The group

$$C_K(G) = \{x \in K : g \cdot x = x \text{ for all } g \in G\}$$

acts on the set of G-orbits by left multiplication. Indeed, if $x \in K$ and $c \in C_K(G)$, then $g \cdot c = c$ for all $g \in G$. Thus

$$\begin{aligned} c(G \cdot x) &= \{c(g \cdot x) : g \in G\} \\ &= \{(g \cdot c)(g \cdot x) : g \in G\} = \{g \cdot (cx) : g \in G\} = G \cdot (cx). \end{aligned}$$

F. Cedó and L. Vendramin, *Groups, Radical Rings, and the Yang-Baxter Equation*, Progress in Mathematics 361, https://doi.org/10.1007/978-3-032-01524-2_12

The following theorem goes back to Deaconescu and Walls [51]. Our proof is that of Isaacs, see [69].

Theorem 12.1.1 (Deaconescu–Walls) *Let G be a group acting by automorphisms on a finite group K. Let $C = C_K(G)$ and $N = C \cap [G, K]$, where $[G, K]$ is the subgroup of K generated by $[g, x] = (g \cdot x)x^{-1}$ for all $g \in G$ and $x \in K$. Then the index $(C : N)$ divides the number of G-orbits of K.*

Proof The group C acts by left multiplication on the set Ω of G-orbits on K. Let $X = G \cdot x \in \Omega$ be an orbit and C_X be the stabilizer of X in C. If $c \in C_X$, then $cX = X$. In particular, if $c \in C_X$ and $x \in X$, then $cx = g \cdot x$ for some $g \in G$. Thus

$$c = (g \cdot x)x^{-1} = [g, x] \in [G, K]$$

and hence $C_X \subseteq N$.

To prove that $(C : N)$ divides the size of Ω, decompose Ω as a disjoint union of C-orbits. Then it is enough to show that $(C : N)$ divides the size of each C-orbit. If $X \in \Omega$, then $C \cdot X$ has size

$$(C : C_X) = (C : N)(N : C_X).$$

Thus $(C : N)$ divides the size of $C \cdot X$. □

Corollary 12.1.2 *Let K be a non-trivial finite group with k conjugacy classes. If $|Z(K)|$ and k are coprime, then $Z(K) \subseteq [K, K]$.*

Proof Let the group K act on K by conjugation, which is an action by automorphisms. Deaconescu–Walls theorem implies that $(Z(K) : Z(K) \cap [K, K])$ divides k. Since k and $|Z(K)|$ are coprime, it follows that $Z(K) = Z(K) \cap [K, K] \subseteq [K, K]$. □

Let K be a group and $f \in \operatorname{Aut}(K)$. Then f is *central* if $f(x)x^{-1} \in Z(K)$ for all $x \in K$. Note that $f \in \operatorname{Aut}(K)$ is central if and only if $f \in C_{\operatorname{Aut}(K)}(\operatorname{Inn}(K))$.

Corollary 12.1.3 *Let K be a finite group with k conjugacy classes and c central automorphisms. If $\gcd(|K|, kc) = 1$, then $[K, K] = Z(K)$.*

Proof By Corollary 12.1.2, $Z(K) \subseteq [K, K]$.

Let us prove that $Z(K) \supseteq [K, K]$. Let $G = C_{\operatorname{Aut}(K)}(\operatorname{Inn}(K))$. Since $\gcd(|K|, kc) = 1$ and, by Deaconescu–Walls' theorem, $(C_K(G) : C_K(G) \cap [G, K])$ divides c, it follows that $C_K(G) = C_K(G) \cap [G, K]$. Since $[K, K] \subseteq C_K(G)$, as

$$a \cdot [x, y] = [(a \cdot x)x^{-1}x, (a \cdot y)y^{-1}y] = [x, y]$$

for all $a \in G$, $x, y \in K$ and $[G, K] \subseteq Z(K)$, we conclude that

$$[K, K] \subseteq C_K(G) = C_K(G) \cap [G, K] \subseteq [G, K] \subseteq Z(K).$$

□

Corollary 12.1.4 *Let p be a prime number. If K is a group with p conjugacy classes, then $Z(K) \subseteq [K, K]$ or $|K| = p$.*

Proof Let K acts on K by conjugation. Since every element of $C = Z(K)$ form a conjugacy class, $|C| \leq p$. If $|C| = p$, then $K = C = Z(K)$ has p elements. Otherwise, $\gcd(|C|, p) = 1$ and, by Corollary 12.1.2, $Z(K) = C \subseteq [K, K]$. □

Let G and K be groups such that G acts on K by automorphisms. A subgroup H of K is *G-invariant* if $g \cdot H \subseteq H$ for all $g \in G$.

Now we will develop Sylow theory for invariant subgroups.

Lemma 12.1.5 (Glauberman) *Let G and K be finite groups of coprime order, where at least one of G or K is solvable. Assume that G acts on K by automorphisms and that G acts on a set Ω, K acts transitively on Ω and*

$$(g \cdot x) \cdot (g \cdot \omega) = g \cdot (x \cdot \omega)$$

for all $g \in G$, $x \in K$ and $\omega \in \Omega$. Then the following statements hold:

(1) There exists a G-invariant element of Ω.
(2) If $\omega, \omega_1 \in \Omega$ are G-invariant elements, then $c \cdot \omega = \omega_1$ for some $c \in C_K(G)$.

Proof We demonstrate the first claim. Let $\Gamma = K \rtimes G$. We identify K and G with the subgroups $K \times \{1\}$ and $\{1\} \times G$ of Γ respectively. Each $\gamma \in \Gamma$ can be written uniquely as $\gamma = xg$ for $x \in K$ and $g \in G$. Thus Γ acts on Ω by

$$(xg) \cdot \omega = x \cdot (g \cdot \omega), \quad x \in K,\ g \in G,\ \omega \in \Omega.$$

To prove that this is an action we use the compatibility condition to compute

$$\begin{aligned}(xg) \cdot ((x_1 g_1) \cdot \omega) &= (xg) \cdot (x_1 \cdot (g_1 \cdot \omega)) \\ &= x \cdot (g \cdot (x_1 \cdot (g_1 \cdot \omega))) = x \cdot ((g \cdot x_1) \cdot (g \cdot (g_1 \cdot \omega)) \\ &= (x(g \cdot x_1)) \cdot ((gg_1) \cdot \omega) = (x(g \cdot x_1)(gg_1)) \cdot \omega \\ &= ((xg)(x_1 g_1)) \cdot \omega.\end{aligned}$$

Let $\omega \in \Omega$ and $U = \Gamma_\omega$ be the stabilizer of ω in Γ. Let $\gamma \in \Gamma$. Since K acts transitively on Ω, there exists $x \in K$ such that $\gamma \cdot \omega = x \cdot \omega$. Thus $x^{-1}\gamma \in U$ and hence $\gamma \in xU \subseteq KU$, and therefore $KU = \Gamma$.

Since K is normal in Γ, it follows that $U \cap K$ is normal in U. Moreover,

$$(U : U \cap K) = (KU : K) = (\Gamma : K) = |G|$$

is coprime with $|U \cap K|$. By Schur–Zassenhaus theorem (Theorem 7.2.4), there exists a complement H of $U \cap K$ in U. Since

$$|H| = (U : U \cap K) = |G|,$$

it follows that H is also a complement of K in Γ. Since G is a complement of K in Γ, by Theorem 7.2.5, H and G are conjugate in Γ, this means $G = \gamma H \gamma^{-1}$ for some $\gamma \in \Gamma$. Since $H \subseteq U$, it follows that H stabilizes ω and hence $\gamma \cdot \omega$ is a G-invariant element.

Let us prove the second claim. Let $\omega, \omega_1 \in \Omega$ be G-invariant elements. By assumption, K acts transitively on Ω and thus the set

$$X = \{x \in K : x \cdot \omega = \omega_1\}$$

is non-empty.

Claim. The group G acts on X.

If $x \in X$, then $x \cdot \omega = \omega_1$. By applying $g \in G$ to this equality we obtain that $g \cdot (x \cdot \omega) = g \cdot \omega_1$. By the compatibility condition and using that ω and ω_1 are both G-invariant elements,

$$(g \cdot x) \cdot \omega = (g \cdot x) \cdot (g \cdot \omega) = g \cdot (x \cdot \omega) = g \cdot \omega_1 = \omega_1.$$

This proves that $G \cdot X \subseteq X$ and hence G acts on X.

Claim. There exists a G-invariant element of X. This completes the proof, as if $x \in X$ is G-invariant, then $x \in K$ is such that $x \cdot \omega = \omega_1$ and $g \cdot x = x$ for all $g \in G$, i.e., $x \in C_K(G)$.

Let $H = K_{\omega_1}$ be the stabilizer of ω_1 in K. Note that H is a subgroup of K. The group H acts transitively on X by left multiplication: if $h \in H$ and $x \in X$, then

$$(hx) \cdot \omega = h \cdot (x \cdot \omega) = h \cdot \omega_1 = \omega_1.$$

Note also that $(g \cdot h) \cdot \omega_1 = (g \cdot h) \cdot (g \cdot \omega_1) = g \cdot (h \cdot \omega_1) = g \cdot \omega_1 = \omega_1$. Thus the action of G on K restricts to H and hence G acts on H by automorphisms. The orders of G and H are coprime, and either G or H is solvable. The group G acts on X and the compatibility condition holds:

$$g \cdot (h \cdot x) = g \cdot (hx) = (g \cdot h)(g \cdot x) = (g \cdot h) \cdot (g \cdot x),$$

as G acts on K by automorphisms. Hence the first part of the lemma implies the existence of a G-invariant element of X. □

Note that by the Feit–Thompson theorem, the solvability of G or K is not really needed since at least one of $|G|$ or $|K|$ is of odd order.

We now prove Sylow's theorems for invariant subgroups.

Theorem 12.1.6 *Let G and K be finite groups of coprime order. Assume that G acts by automorphisms on K, and that at least one of G or K is solvable. Let p be a prime number.*

(1) There exists a Sylow p-group of K which is G-invariant.

(2) If S and T are G-invariant Sylow p-subgroups, then there exists $c \in C_K(G)$ such that $cSc^{-1} = T$.

Proof Let us prove the first claim. By Sylow's theorems, the group K acts transitively by conjugation on the non-empty set $\Omega = \mathrm{Syl}_p(K)$. Since G acts on K by automorphisms, G acts on Ω. Let us check the compatibility condition. Since $g \cdot P \in \Omega$ for all $g \in G$ and $P \in \mathrm{Syl}_p(K)$, K acts by conjugation on Ω and G acts by automorphisms on K,

$$(g \cdot x) \cdot (g \cdot P) = (g \cdot x)(g \cdot P)(g \cdot x)^{-1} = g \cdot (xPx^{-1}) = g \cdot (x \cdot P),$$

for all $g \in G, x \in K$ and $P \in \mathrm{Syl}_p(K)$. The first part of Glauberman's lemma implies that there exists a Sylow p-subgroup of K that is G-invariant.

Let us now prove the second claim. Let S and T be G-invariant Sylow p-subgroups of K, so S and T are G-invariant elements of Ω. By the second part of Glauberman's lemma, there exits $c \in C_K(G)$ such that $c \cdot S = cSc^{-1} = T$. □

Corollary 12.1.7 *Let G and K be finite groups of coprime orders. Assume that G acts by automorphisms on K and at least one of K or G is solvable. Let P be a G-invariant p-subgroup of K for some prime number p. Then P is contained in some G-invariant Sylow p-subgroup of K.*

Proof We may assume that P is a maximal G-invariant p-subgroup of K. Let $N = N_K(P)$. Note that N is G-invariant: if $n \in N$ and $g \in G$, then

$$(g \cdot n)P(g \cdot n)^{-1} = g \cdot (nPn^{-1}) = g \cdot P = P$$

since P is G-invariant. We claim that $P \in \mathrm{Syl}_p(K)$. By the first part of Theorem 12.1.6, there exists $S \in \mathrm{Syl}_p(N)$ that is G-invariant. By Sylow's theorem, some conjugate of P is contained in S, say $xPx^{-1} \subseteq S$ for some $x \in N$, and hence

$$P = xPx^{-1} \subseteq S$$

since P is normal in N. Since P is a maximal G-invariant p-subgroup of K, it follows that $P = S$ and thus $P \in \mathrm{Syl}_p(N)$. Let $Q \in \mathrm{Syl}_p(K)$ be such that $P \subseteq Q$. If $P \neq Q$, then $Q \cap N$ is a p-subgroup of N such that $P \subsetneq Q \cap N$, by the normalizer condition (Lemma 5.1.10), a contradiction because $P \in \mathrm{Syl}_p(N)$. Hence $P = Q$ and therefore $P \in \mathrm{Syl}_p(K)$. □

Theorem 12.1.8 *Let G and K be finite groups of coprime order. Assume that G acts by automorphisms on K and that at least one of K or G is solvable. Then $K = C_K(G)[G, K]$.*

Proof Note that $[G, K]$ is a normal subgroup of K and it is G-invariant. In fact, for $k_1, k_2 \in K$ and $g_1, g_2 \in G$, we have

$$k_1[g_1, k_2]k_1^{-1} = k_1(g_1 \cdot k_2)k_2^{-1}k_1^{-1}$$

$$
\begin{aligned}
&= (g_1 \cdot (g_1^{-1} \cdot k_1))(g_1 \cdot k_2)k_2^{-1}(g_1^{-1} \cdot k_1^{-1})(g_1^{-1} \cdot k_1)k_1^{-1} \\
&= (g_1 \cdot ((g_1^{-1} \cdot k_1)k_2))k_2^{-1}(g_1^{-1} \cdot k_1)^{-1}(g_1^{-1} \cdot k_1)k_1^{-1} \\
&= [g_1, (g_1^{-1} \cdot k_1)k_2][g_1^{-1}, k_1] \in [G, K],
\end{aligned}
$$

and

$$(g_1 \cdot [g_2, k_1])[g_2, k_1]^{-1} = [g_1, [g_2, k_1]] \in [G, K].$$

Hence G acts on $K/[G, K]$ and K acts transitively on $K/[G, K]$ by left multiplication. Furthermore,

$$
\begin{aligned}
g_1 \cdot (k_1 \cdot k_2[G, K]) &= g_1 \cdot (k_1k_2[G, K]) = (g_1 \cdot (k_1k_2))[G, K] \\
&= (g_1 \cdot k_1)(g_1 \cdot k_2)[G, K] = (g_1 \cdot k_1) \cdot (g_1 \cdot (k_2[G, K])).
\end{aligned}
$$

Note that for every $g \in G$ and $k \in K$, we have that

$$g \cdot (k[G, K]) = (g \cdot k)[G, K] = kk^{-1}(g \cdot k)k^{-1}k[G, K] = k[G, K].$$

By Glauberman's lemma, for every $k \in K$, there exists $c \in C_K(G)$ such that

$$k[G, K] = c[G, K] \subseteq C_K(G)[G, K],$$

and thus $K = C_K(G)[G, K]$. □

Corollary 12.1.9 *Let G and K be finite groups of coprime order. Assume that G acts by automorphisms on K. Then $[G, [G, K]] = [G, K]$.*

Proof It is clear that $[G, [G, K]] \subseteq [G, K]$. Let $g \in G$ and $k \in K$. We shall prove that $[g, k] \in [G, [G, K]]$. Consider the group $A = \langle g \rangle$. Since A is abelian, by Theorem 12.1.8, we have that $K = C_K(A)[A, K]$. Hence there exists $c \in C_K(A)$ and $x \in [A, K]$ such that $k = xc$. Now we have that

$$
\begin{aligned}
[g, k] &= (g \cdot k)k^{-1} = (g \cdot x)(g \cdot c)c^{-1}x^{-1} \\
&= (g \cdot x)x^{-1} = [g, x] \in [G, [G, K]].
\end{aligned}
$$

Hence $[G, [G, K]] = [G, K]$. □

An action of a group G on a group K by automorphisms is said to be *faithful* if, for $g \in G$, $g \cdot x = x$ for all $x \in K$ implies that $g = 1$. Note that if the action is faithful, then $C_G(K) = \{1\}$.

The *kernel* of an action of a group G on a group K by automorphisms is the (normal) subgroup N of elements of G that act trivially. Let $\pi : G \to G/N$ be the canonical map. In this case, the quotient group G/N acts by $\pi(g) \cdot x = g \cdot x$ and the action is faithful. Note that the kernel of the action is the subgroup $C_G(K)$ of $K \rtimes G$. Note that $C_G(K)$ is isomorphic to the largest subgroup N of G such that

$[N, K] = \{1\}$. If $N = C_G(K)$, then N is normal in G. In this case, the quotient group G/N acts on K by $g \cdot x = \pi(g) \cdot x$.

It will be convenient to introduce the following notation. Let

$$[G, \ldots, G, K]_1 = [G, K] \qquad \text{and}$$
$$[G, \ldots, G, K]_{m+1} = [G, \ldots, G, [G, K]]_m \qquad \text{for } m \geq 1.$$

Then

$$[G, \ldots, G, K]_m = [\underbrace{G, \ldots, G}_{m-\text{copies}}, K]$$

for $m \geq 1$.

Theorem 12.1.10 *Let G and K be groups such that G acts by automorphisms on K. If $[G, \ldots, G, K]_m = \{1\}$, then $G^{(m-1)} \subseteq C_G(K)$. In particular, if the action is faithful, then G is solvable and its derived series has length $\leq m - 1$.*

Proof We claim that $G^{(m-1)} \subseteq C_G(K)$. We proceed by induction on m. If $m = 1$, then $[G, K] = \{1\}$ by assumption and thus $G^{(0)} = G = C_G(K)$. Assume now that $m > 1$ and the result holds for $m - 1$. Let $L = [G, K]$. Then

$$\{1\} = [G, \ldots, G, K]_m = [G, \ldots, G, L]_{m-1}.$$

Since L is G-invariant, G acts on L. By the inductive hypothesis, $G^{(m-2)} \subseteq C_G(L)$, which implies that $[G^{(m-2)}, L] = \{1\}$. Thus

$$[G^{(m-2)}, [G^{(m-2)}, K]] \subseteq [G^{(m-2)}, [G, K]] = [G^{(m-2)}, L] = \{1\}.$$

Moreover, since $[G^{(m-2)}, K] = [K, G^{(m-2)}]$, it follows that

$$[G^{(m-2)}, [K, G^{(m-2)}]] = \{1\}.$$

The three-subgroups Lemma with $X = Y = G^{(m-2)}$ and $Z = K$ implies that

$$\{1\} = [K, [G^{(m-2)}, G^{(m-2)}]] = [K, G^{(m-1)}].$$

Hence $G^{(m-1)} \subseteq C_G(K)$ and the claim follows by induction. In particular, if the action is faithful, then $C_G(K) = \{1\}$ and the result follows. □

Theorem 12.1.11 (Hall) *Let G be a finite group that acts on a finite group K by automorphisms. If $[G, \ldots, G, K]_m = \{1\}$ for some m, then there exists a positive integer n such that $[[G, \ldots, G, G]_n, K] = \{1\}$. Furthermore, if the action of G on K is faithful, then G is nilpotent.*

Proof We proceed by induction on $|K|$. We may assume that K is non-trivial. Let

$$G = \gamma_1(G) \supseteq \gamma_2(G) \supseteq \cdots \supseteq \gamma_n(G) \supseteq \cdots$$

be the lower central series of G. Since G is finite, this sequence stabilizes. Let $n \in \mathbb{N}$ be the smallest positive integer such that $\gamma_n(G) = \gamma_{n+k}(G)$ for all k.

Claim. $[G, K] \neq K$.

In fact, if $[G, K] = K$, then

$$\{1\} = [G, \dots, G, K]_m = [G, \dots, G, K]_{m-1} = \cdots = [G, K] = K,$$

a contradiction.

Claim. $\gamma_n(G)$ acts trivially on K.

By assumption,

$$\{1\} = [G, \dots, G, K]_m = [G, \dots, G, [G, K]]_{m-1}.$$

Since $[G, K]$ is a proper G-invariant subgroup of K, the inductive hypothesis implies that $\gamma_n(G)$ acts trivially on $[G, K]$ and hence

$$[\gamma_n(G), [G, K]] = [\gamma_n(G), [K, G]] = \{1\}.$$

To use the three-subgroups lemma, we need to show that

$$[G, [\gamma_n(G), K]] = \{1\}.$$

Suppose that $[\gamma_n(G), K] \neq \{1\}$. Since $[\gamma_n(G), K] \subseteq K$ and $[G, \dots, G, K]_m = \{1\}$, we have that there exists a positive integer $l \leq m$ such that $[G, \dots, G, [\gamma_n(G), K]]_l = \{1\}$ and $[G, \dots, G, [\gamma_n(G), K]]_{l-1} \neq \{1\}$. Since $[\gamma_n(G), [G, K]] = \{1\}$, we have that

$$[K, [\gamma_n(G), [G, K]]] = [K, \{1\}] = \{1\}$$

and hence, since $[G, K]$ is normal in K, it follows that

$$[\gamma_n(G), [[G, K], K]] \subseteq [\gamma_n(G), [G, K]] = \{1\}.$$

By the three-subgroups lemma with $X = K$, $Y = \gamma_n(G)$ and $Z = [G, K]$,

$$[[G, K], [\gamma_n(G), K]] = [[G, K], [K, \gamma_n(G)]] = \{1\}. \tag{12.1}$$

Let $C = C_{[\gamma_n(G), K]}(G)$. Then

$$[G, \dots, G, [\gamma_n(G), K]]_{l-1} \subseteq C$$

and thus C is a non-trivial subgroup of K and G acts trivially on C. By definition, $C \subseteq [\gamma_n(G), K]$ and $[C, G] = \{1\}$. Thus, by (12.1),

$$[C, [G, K]] \subseteq [[\gamma_n(G), K], [G, K]] = \{1\}$$

and

$$[K, [C, G]] = [K, \{1\}] = \{1\}.$$

By using the three-subgroups lemma with $X = K, Y = C$ and $Z = G$, it follows that $[G, [C, K]] = \{1\}$, so G centralizes $[C, K]$. Since $C \subseteq [\gamma_n(G), K]$ and $[\gamma_n(G), K]$ is a normal subgroup of K, it follows that $[C, K] \subseteq [\gamma_n(G), K]$. But

$$[C, K] \subseteq C_{[\gamma_n(G),K]}(G) = C$$

and therefore C is normal in K.

Let $\pi : K \to K/C$ be the canonical map. Since $|K/C| < |K|$ and

$$[G, \dots, G, \pi(K)]_m = \pi([G, \dots, G, K]_m) = \{1\},$$

the inductive hypothesis implies that

$$\{1\} = [\gamma_n(G), \pi(K)] = \pi([\gamma_n(G), K]).$$

Thus $[\gamma_n(G), K] \subseteq C$. In particular,

$$[G, [\gamma_n(G), K]] = \{1\},$$

as G acts trivially on C. Then by the three-subgroups lemma,

$$\{1\} = [K, [G, \gamma_n(G)]] = [K, \gamma_{n+1}(G)] = [K, \gamma_n(G)],$$

a contradiction. Therefore $[\gamma_n(G), K] = \{1\}$ and the result follows by induction. □

For the next result we need a lemma.

Lemma 12.1.12 *Let G and K be groups such that G acts on K by automorphisms. If $[G, [G, K]] = \{1\}$, then $[G, K]$ is abelian.*

Proof Recall that we consider K and G as subgroups of $\Gamma = K \rtimes G$. Since $[G, [G, K]] = \{1\}$, it follows that $[K, [[G, [G, K]]] = \{1\}$. Moreover, $[G, K]$ is normal in K and thus

$$[G, [[G, K], K]] \subseteq [G, [G, K]] = \{1\}.$$

The three-subgroups lemma with $X = K, Y = G$ and $Z = [G, K]$ implies that

$$\{1\} = [[G, K], [K, G]] = [[G, K], [G, K]],$$

so the commutator subgroup of the group $[G, K]$ is trivial. This means that $[G, K]$ is abelian. □

Theorem 12.1.13 *Let p be a prime number and G be a finite p-group that acts by automorphisms on a finite group K. If $[G, \dots, G, K]_m = \{1\}$ for some m, then $[G, K]$ is a p-group.*

Proof We proceed by induction on $|K|$. We may assume that $K \neq \{1\}$. Then, since $[G, \dots, G, K]_m = \{1\}$, it follows that $[G, K] \neq K$. Let $L = [G, K]$. Then L is a G-invariant proper normal subgroup of K. Since

$$[G, \dots, G, L]_{m-1} = [G, \dots, G, K]_m = \{1\},$$

the inductive hypothesis on L implies that $[G, L]$ is p-group. Note that $[G, L]$ is a normal subgroup of L. Then $[G, L] \subseteq O_p(L)$ and since $O_p(L)$ is characteristic in L and L is normal in K, it follows that $O_p(L)$ is normal in K. Since L is G-invariant and $O_p(L)$ is characteristic in L, we have that $O_p(L)$ is G-invariant. Let $\pi : K \to K/O_p(L)$ be the canonical map. Then G acts by automorphisms on $\pi(K)$. Since $[G, L] \subseteq O_p(L)$, we have that $[G, [G, \pi(K)]] = \{1\}$. By Lemma 12.1.12, $[G, \pi(K)]$ is abelian. Let $g \in G \setminus \{1\}$ and $k \in K$. We have that g has order p^n for some positive integer n. Since $[G, [G, \pi(K)]] = \{1\}$, we have that

$$\begin{aligned}
1 &= [\pi(k), g^{p^n}] \\
&= [\pi(k), g^{p^n-1}] g^{p^n-1} [\pi(k), g] (g^{p^n-1})^{-1} \\
&= [\pi(k), g^{p^n-1}][\pi(k), g] \\
&\;\;\vdots \\
&= [\pi(k), g]^{p^n}.
\end{aligned}$$

Hence $[G, \pi(K)] \cong ([G, K]O_p(L))/O_p(L) = L/O_p(L)$ is an abelian p-group. Therefore $[G, K]$ is a p-group, and the result follows by induction. □

Theorem 12.1.14 *Let G and K be finite groups such that G acts on K by automorphisms. If $[G, \dots, G, K]_m = \{1\}$ for some m, then $[G, K]$ is nilpotent.*

Proof We proceed by induction on $|G|$. If $G = \{1\}$, then $[G, K] = \{1\}$ and there is nothing to demonstrate. So we may assume that $G \neq \{1\}$. Let G_1 be a proper subgroup of G. Since

$$[G_1, \dots, G_1, K]_m \subseteq [G, \dots, G, K]_m = \{1\},$$

the inductive hypothesis implies that $[G_1, K]$ is nilpotent. Since $[G_1, K]$ is a normal subgroup of K, by Fitting's theorem, $[G_1, K] \subseteq F(K)$, the Fitting subgroup of K. Let $g \in G$. If the order of g is a power of a prime p, then by Theorem 12.1.13, $[\langle g \rangle, K]$ is a p-group and it is normal in K. By Fitting's theorem $[\langle g \rangle, K] \subseteq F(K)$. If the order of g is not a power of any prime, then the order of g is mn for some coprime integers $m, n > 1$. There exist integers a, b such that $am + bn = 1$. Hence $g = g^{am}g^{bn}$ and $\langle g^{am}\rangle$ and $\langle g^{bn}\rangle$ are proper subgroups of G. By the inductive hypothesis, $[\langle g^{am}\rangle, K]$ and $[\langle g^{bn}, K\rangle]$ are nilpotent. Hence

$$[\langle g\rangle, K] = [\langle g^{am}\rangle, K][\langle g^{bn}\rangle, K] \subseteq F(K).$$

Therefore $[G, K] \subseteq F(K)$ and the result follows by induction and Fitting's theorem (Theorem 5.4.5). □

12.2 Left Nilpotent Skew Braces

Definition 12.2.1 Let A be a skew brace. One defines $A^1 = A$ and for $n \geq 1$

$$A^{n+1} = A * A^n = \langle a * x : a \in A,\ x \in A^n\rangle_+.$$

The sequence $A^1 \supseteq A^2 \supseteq A^3 \supseteq \cdots \supseteq A^n \supseteq \cdots$ is the *left series* of A.

Proposition 12.2.2 *Let A be a skew brace. Each A^n is a left ideal of A.*

Proof We proceed by induction on n. The case $n = 1$ is trivial, so we may assume that the result is true for some $n \geq 1$. Let $a, b \in A$ and $x \in A^n$. By the inductive hypothesis, $\lambda_a(x) \in A^n$ and hence

$$\begin{aligned}\lambda_a(b * x) &= \lambda_a(\lambda_b(x) - x) = \lambda_a\lambda_b(x) - \lambda_a(x)\\ &= \lambda_{a\circ b\circ a'}(\lambda_a(x)) - \lambda_a(x) = (a \circ b \circ a') * \lambda_a(x) \in A^{n+1}.\end{aligned} \tag{12.2}$$

This implies that $\lambda_a(A^{n+1}) \subseteq A^{n+1}$. Thus the result follows. □

Definition 12.2.3 A skew brace A is said to be *left nilpotent* if $A^m = \{0\}$ for some $m \geq 1$.

Some basic properties of left nilpotent skew braces appear in Exercises 12.3.3–12.3.5.

Proposition 12.2.4 *Let A be a nonzero left nilpotent skew brace and I be a nonzero left ideal of A. Then $I \cap \mathrm{Fix}(A) \neq \{0\}$. In particular, $\mathrm{Fix}(A) \neq \{0\}$.*

Proof Let $m = \max\{k : I \cap A^k \neq \{0\}\}$. Since $A * (I \cap A^m) \subseteq I \cap A^{m+1} = \{0\}$, it follows that there exists a nonzero $x \in I \cap A^m$ such that $a * x = 0$ for all $a \in A$. Thus $0 \neq x \in \mathrm{Fix}(A) \cap I$. For the second claim, apply the first case with $I = A$. □

We first prove that if both groups of a finite skew brace A are nilpotent, then A can be decomposed as a direct product of skew braces of prime-power size.

Lemma 12.2.5 *Let A be a skew brace such that the additive group is a direct sum of ideals I_1, I_2, that is $A = I_1 + I_2$ and $I_1 \cap I_2 = \{0\}$. Then the map $f : A \to I_1 \times I_2$ defined by $f(a_1 + a_2) = (a_1, a_2)$, for all $a_1 \in I_1$ and $a_2 \in I_2$, is an isomorphism of skew braces.*

Proof The operations of the skew brace $I_1 \times I_2$ are defined component-wise. Clearly, f is an isomorphism of the additive groups of A and $I_1 \times I_2$. Let $a_1 \in I_1$ and $a_2 \in I_2$. Since I_1 and I_2 are ideals we have that

$$a_1 + a_2 - a_1 - a_2, a_1 * a_2, a_2 * a_1 \in I_1 \cap I_2 = \{0\},$$

thus $a_1 + a_2 = a_2 + a_1$ and $a_1 \circ a_2 = a_1 + a_2 = a_2 \circ a_1$. Hence

$$\begin{aligned} f((a_1 + a_2) \circ (b_1 + b_2)) &= f(a_1 \circ a_2 \circ b_1 \circ b_2) \\ &= f(a_1 \circ b_1 \circ a_2 \circ b_2) \\ &= f(a_1 \circ b_1 + a_2 \circ b_2) \\ &= (a_1 \circ b_1, a_2 \circ b_2) \\ &= (a_1, a_2) \circ (b_1, b_2) \\ &= f(a_1 + a_2) \circ f(b_1 + b_2) \end{aligned}$$

for all $a_1, b_1 \in I_1$ and $a_2, b_2 \in I_2$. □

Theorem 12.2.6 *Let n be a positive integer. Let A be a skew brace such that the additive group is a direct sum of ideals $I_1, \dots, I_n$, that is every element $a \in A$ is uniquely written as $a = a_1 + \cdots + a_n$, with $a_j \in I_j$ for all j. Then the map*

$$f : A \to I_1 \times \cdots \times I_n, \quad f(a_1 + \cdots + a_n) = (a_1, \dots, a_n),$$

for all $a_j \in I_j$, is an isomorphism of skew braces.

Proof We shall prove the result by induction on n. For $n = 1$, it is clear. Suppose that $n > 1$ and that the result is true for $n - 1$. Let $A_1 = I_1 + \cdots + I_{n-1}$. Then A_1 is an ideal of A and A is the direct sum of the ideals A_1 and I_n. By Lemma 12.2.5, the map $f_1 : A \to A_1 \times I_n$ defined by $f(a + a_n) = (a, a_n)$, for all $a \in A_1$ and $a_n \in I_n$, is an isomorphism of skew braces. By the inductive hypothesis, the map

$$f_2 : A_1 \to I_1 \times \cdots \times I_{n-1}, \quad f_2(a_1 + \cdots + a_{n-1}) = (a_1, \dots, a_{n-1}),$$

is an isomorphism of skew braces. Therefore $f = (f_2 \times \mathrm{id}) f_1 : A \to I_1 \times \cdots \times I_n$ is an isomorphism of skew braces and $f(a_1 + \cdots + a_n) = (a_1, \dots, a_n)$, for all $a_j \in I_j$. The result then follows. □

Corollary 12.2.7 *Let A be a finite skew brace such that $(A, +)$ and $(A, \circ)$ are nilpotent. Let $I_1, \dots, I_n$ be the distinct Sylow subgroups of the additive group of A. Then $I_1, \dots, I_n$ are ideals of A and the map*

$$f : A \to I_1 \times \cdots \times I_n, \quad f(a_1 + \cdots + a_n) = (a_1, \dots, a_n),$$

for all $a_j \in I_j$, *is an isomorphism of skew braces.*

Proof Since $(A, +)$ is nilpotent, for every prime divisor p of the order of A, there is a unique Sylow p-subgroup I of $(A, +)$. Hence I is a normal subgroup of $(A, +)$, and $\lambda_a(b) \in I$ for all $a \in A$ and $b \in I$. Thus I is a left ideal of A and thus it is a Sylow p-subgroup of $(A, \circ)$. Since $(A, \circ)$ is nilpotent, I is the unique Sylow p-subgroup of $(A, \circ)$ and, thus, it is normal in $(A, \circ)$. Therefore I is an ideal of A. Hence $I_1, \dots, I_n$ are ideals of A and clearly the additive group of A is the direct sum of $I_1, \dots, I_n$. The result follows by Theorem 12.2.6. □

Let A be a skew brace. Let G be the multiplicative group of A and K be the additive group of A. The group G acts on K by automorphisms. Let Γ be the semidirect product $\Gamma = K \rtimes G$. The operation of Γ is

$$(x, g)(y, h) = (x + \lambda_g(y), g \circ h).$$

Identifying each $g \in G$ with $(0, g) \in \Gamma$ and each $x \in K$ with $(x, 0) \in \Gamma$,

$$\begin{aligned}[g, x] &= gxg^{-1}x^{-1} = (0, g)(x, 0)(0, g')(-x, 0)\\ &= (\lambda_g(x), g)(-\lambda_g^{-1}(x), g') = (\lambda_g(x) - x, 0) = \lambda_g(x) - x = g * x.\end{aligned}$$

Note that $A^2 = [G, K]$ and $A^{n+1} = [G, \dots, G, K]_n$ for $n > 0$.

Proposition 12.2.8 *Let p be a prime and A be a skew brace of size p^m. Then A is left nilpotent.*

Proof Let G be the multiplicative group of A and K be the additive group of A. Since the semidirect product $\Gamma = K \rtimes G$ is a p-group, it is nilpotent. Thus there exists a positive integer k such that $[\Gamma, \Gamma, \dots, \Gamma]_k = \{(0, 1)\}$. Since

$$A^{k+1} = [G, \dots, G, K]_k \subseteq [\Gamma, \dots, \Gamma]_k,$$

it follows that A is left nilpotent. □

Theorem 12.2.9 *Let A be a finite left nilpotent skew brace. Then the following statements hold:*

(1) The additive group of A^2 is nilpotent.
(2) The multiplicative group of $A/\ker\lambda$ is nilpotent.

Proof Let G be the multiplicative group of A and K be the additive group of A. As above we consider G and K as subgroups of $\Gamma = K \rtimes G$. Since A is left nilpotent, there exists a positive integer n such that $A^n = \{0\}$. Hence $[G, \dots, G, K]_{n-1} = A^n = \{0\}$. By Theorem 12.1.14, $[G, K]$ is nilpotent, i.e., the additive group of A^2 is nilpotent.

Let C be the centralizer of K in G. Note that C is a normal subgroup of G. Since $[G, \dots, G, K]_m = \{1\}$, by Theorem 12.1.11, there exists a positive integer n such that $[[G, \dots, G]_n, K] = \{1\}$. Hence $[G, \dots, G]_n \subseteq C$. Now G/C is nilpotent. Note that

$$\begin{aligned} C &= \{g \in G \mid gxg^{-1} = x, \text{ for all } x \in K\} \\ &= \{g \in A \mid \lambda_g(x) = x, \text{ for all } x \in A\} = \ker \lambda. \end{aligned}$$

Thus G/C is the multiplicative group of $A/\ker\lambda$, and the result follows. □

Lemma 12.2.10 *Let A be a finite skew brace of nilpotent type. Let p and q distinct prime numbers and let P and Q be Sylow subgroups of $(A, +)$ of sizes p^n and q^m, respectively. Then P, Q and $P + Q$ are left ideals of A.*

Proof Let us first prove that P is a left ideal. Since $(A, +)$ is nilpotent, P is a normal subgroup of $(A, +)$. Let $a \in A$ and $x \in P$. Then $\lambda_a(x) \in P$ since λ_a is a group homomorphism. Similarly one proves that Q is a left ideal. From this, it follows that $P + Q$ is a left ideal. □

The following is based on [104, Theorem 5(1)]. However, the proof is completely different.

Theorem 12.2.11 *Let A be a finite skew brace of nilpotent type. Let p and q distinct prime numbers and let A_p and A_q be Sylow subgroups of $(A, +)$ of sizes p^n and q^m, respectively. If p does not divide $q^t - 1$ for all $t \in \{1, \dots, m\}$, then $A_p * A_q = \{0\}$. In particular, $\lambda_x(y) = y$ for all $x \in A_p$ and $y \in A_q$.*

Proof By Lemma 12.2.10 A_p, A_q and $A_p + A_q$ are left ideals of A. In particular, $A_p + A_q$ is a subbrace of A and A_p and A_q are Sylow subgroups of $(A_p + A_q, \circ)$. By Sylow's theorem, the number n_p of Sylow p-subgroups of the multiplicative group of $A_p + A_q$ is

$$n_p = [A_p + A_q : N] \equiv 1 \bmod p,$$

where $N = \{g \in A_p + A_q : g \circ A_p \circ g' = A_p\}$ is the normalizer of A_p in the multiplicative group of $A_p + A_q$. Since $[A_p + A_q : N] = q^s$ for some $s \in \{0, \dots, m\}$ and p does not divide $q^t - 1$ for all $t \in \{1, \dots, m\}$, it follows that $s = 0$ and hence A_p is a normal subgroup of the multiplicative group of $A_p + A_q$. Thus A_p is an ideal of the skew brace $A_p + A_q$. Since A_p is an ideal of $A_p + A_q$ and A_q is a left ideal, we have that $A_p * A_q \subseteq A_p \cap A_q = \{0\}$, and the result follows. □

Corollary 12.2.12 *Let A be a skew brace of size $p_1^{\alpha_1} \cdots p_k^{\alpha_k}$, where $p_1 < \cdots < p_k$ are prime numbers and $\alpha_1, \dots, \alpha_k$ are positive integers. Assume that the additive group of A is nilpotent. Let A_j be the Sylow p_j- subgroups of the additive group of A. Assume that, for some $j \le k$, p_j does not divide $p_i^{t_i} - 1$ for all $t_i \in \{1, \dots \alpha_i\}$ for all $i \ne j$. Then $\operatorname{Soc}(A_j) \subseteq \operatorname{Soc}(A)$.*

Proof Write $A = A_1 + \cdots + A_k$. Let $a \in \mathrm{Soc}(A_j)$ and $b \in A$. Hence there exist elements $b_k \in A_k$ such that $b = b_1 + \cdots + b_k$. By Theorem 12.2.11, $\lambda_a(b_i) = b_i$, for all $i \neq j$. Then $\lambda_a(b) = \lambda_a(b_1) + \cdots + \lambda_a(b_k) = b_1 + \cdots + b_k = b$ and hence $a \in \mathrm{Soc}(A)$. Thus the result follows. □

A finite group G is said to be *p-nilpotent* for a prime p if G has a normal p-complement. It is an easy exercise to show that a finite group is nilpotent if and only if it is p-nilpotent for every prime p.

Let A be a skew brace. For subsets X and Y of A we define inductively $L_0(X, Y) = Y$ and $L_{n+1}(X, Y) = X * L_n(X, Y)$ for $n \geq 0$.

Definition 12.2.13 Let p be a prime number. A finite skew brace A of nilpotent type is said to be *left p-nilpotent* if there exists $n \geq 1$ such that $L_n(A, A_p) = \{0\}$, where A_p is the Sylow p-subgroup of $(A, +)$.

Lemma 12.2.14 *Let A be a skew brace such that its additive group is the direct product of the left ideals B and C. Then $A * (B + C) = A * B + A * C$. Moreover, if $A = \oplus_{i=1}^n B_i$ where the B_i are left ideals, then*

$$A * \sum_{i=1}^{n} B_i = \sum_{i=1}^{n} A * B_i.$$

Proof Let $a \in A$, $b \in B$ and $c \in C$. Then

$$a * (b + c) = a * b + b + a * c - b = a * b + a * c$$

holds for all $a \in A$, $b \in B$ and $c \in C$. The second part follows by induction. □

Proposition 12.2.15 *Let A be a finite skew brace of nilpotent type. Then A is left nilpotent if and only if A is left p-nilpotent for all prime divisors p of $|A|$.*

Proof Let $\pi(A)$ be the set of all prime divisors of $|A|$. For each $p \in \pi(A)$ there exists $n(p) \in \mathbb{N}$ such that $L_{n(p)}(A, A_p) = \{0\}$. Let $n = \max\{n(p) : p \in \pi(A)\}$. Then $L_n(A, A_p) = \{0\}$ for all $p \in \pi(A)$. Since A is of nilpotent type, the additive group $(A, +)$ is isomorphic to the direct sum of the A_p for $p \in \pi(A)$. Then Lemma 12.2.14 implies that

$$L_n(A, A) = \sum_{p \in \pi(A)} L_n(A, A_p) = \{0\}.$$

The other implication is trivial. □

Let A be a skew brace. Recall that the group $(A, \circ)$ acts on $(A, +)$ by automorphisms. If, in the semidirect product $(A, +) \rtimes (A, \circ)$, we identify a with $(0, a)$ and b with $(b, 0)$, then $[a, b] = [(0, a), (b, 0)] = (a * b, 0) = a * b$. Under this identification, we write $[X, Y] = X * Y$ for any pair of subsets $X, Y \subseteq A$. Then the iterated commutator satisfies

$$[X, \ldots, X, Y] = [X, [X, \ldots, [X, Y] \ldots]] = L_n(X, Y),$$

where the subset X appears n times.

In a finite skew brace A of nilpotent type we denote by $A_{p'}$ the p-complement of $(A, +)$ for every prime p.

Theorem 12.2.16 *Let A be a finite skew brace of nilpotent type. The following statements are equivalent:*

(1) A is left p-nilpotent.
(2) $A_{p'} * A_p = \{0\}$.
(3) The group $(A, \circ)$ is p-nilpotent.

Proof We first prove that (1) implies (2). Since A is left p-nilpotent, there exists $n \in \mathbb{N}$ such that $L_n(A_{p'}, A_p) \subseteq L_n(A, A_p) = \{0\}$. Since $(A_{p'}, \circ)$ acts by automorphisms on $(A_p, +)$ and this is a coprime action, it follows from Corollary 12.1.9 that

$$L_1(A_{p'}, A_p) = A_{p'} * A_p = A_{p'} * (A_{p'} * A_p) = L_2(A_{p'}, A_p).$$

By induction, one then proves that $A_{p'} * A_p = L_n(A_{p'}, A_p) = \{0\}$.

We now prove that (2) implies (3). It is enough to prove that $(A_{p'}, \circ)$ is a normal subgroup of $(A, \circ)$. By using Lemma 12.2.14,

$$A_{p'} * A = A_{p'} * (A_p + A_{p'}) = (A_{p'} * A_p) + (A_{p'} * A_{p'}) \subseteq A_{p'},$$

since $A_{p'}$ is a left ideal of A and $A_{p'} * A_p = \{0\}$. Then $A_{p'}$ is an ideal of A by Proposition 8.4.9. In particular, $(A_{p'}, \circ)$ is a normal subgroup of $(A, \circ)$.

Finally, we prove that (3) implies (1). We need to prove that $L_n(A, A_p) = \{0\}$ for some n. Since $(A, \circ)$ is p-nilpotent, there exists a normal p-complement that is a characteristic subgroup of $(A, \circ)$. This group is $A_{p'}$ and hence $A_{p'}$ is an ideal of A. Then $A_{p'} * A_p \subseteq A_{p'} \cap A_p = \{0\}$. We now prove that $L_n(A, A_p) = L_n(A_p, A_p)$ for all $n \geq 0$. The case where $n = 0$ is trivial, so assume that the result holds for some $n \geq 0$. By the inductive hypothesis,

$$L_{n+1}(A, A_p) = A * L_n(A, A_p) = A * L_n(A_p, A_p).$$

Thus it is enough to prove that $A * L_n(A_p, A_p) \subseteq A_p * L_n(A_p, A_p)$. Let $a \in A$ and $b \in L_n(A_p, A_p)$. Write $a = x \circ y$ for $x \in A_p$ and $y \in A_{p'}$. Then

$$a * b = (x \circ y) * b = x * (y * b) + y * b + x * b = x * b \in A_p * L_n(A_p, A_p)$$

since $A_{p'} * A_p = \{0\}$. The skew brace A_p is left nilpotent by Proposition 12.2.8, so there exists $n \in \mathbb{N}$ such that $L_n(A_p, A_p) = \{0\}$, and the result follows. □

Theorem 12.2.17 *Let A be a finite skew brace of nilpotent type. Then, A is left nilpotent if and only if the multiplicative group of A is nilpotent.*

Proof By Proposition 12.2.15, A is left nilpotent if and only if A is left p-nilpotent for every prime p. By Theorem 12.2.16, this is equivalent to the p-nilpotency of the multiplicative group of A for every prime p. Therefore the result follows. □

Definition 12.2.18 For a skew brace A let $\ell_1(a) = a$ and $\ell_{k+1}(a) = a * \ell_k(a)$ for $n \geq 1$. The skew brace A is said to be *left nil* if for every $a \in A$ there exists a positive integer $n = n(a)$ such that $\ell_n(a) = 0$.

12.3 Exercises

12.3.1 Let G and K be groups such that G acts on K by automorphisms. Then $[G, K]$ is the unique smallest G-invariant subgroup of K such that the induced action of G on $K/[G, K]$ is trivial.

12.3.2 Let G and K be groups such that G acts faithfully on K by automorphisms. If $[G, [G, K]] = \{1\}$, then G is abelian.

12.3.3 Let $A_1, \dots, A_k$ be left nilpotent skew braces. Prove that $A_1 \times \cdots \times A_k$ is left nilpotent.

12.3.4 Let $f : A \to B$ be a surjective homomorphism of skew braces. Prove that if A is left nilpotent, then B is left nilpotent.

12.3.5 Let A be a left nilpotent skew brace and $B \subseteq A$ be a subbrace. Prove that B is left nilpotent.

12.3.6 Let A be a finite skew brace such that $A^3 = \{0\}$. Prove that A^2 is a trivial skew brace of abelian type.

12.4 Open Problems

Open problem 12.4.1 Let A be a finite left nil skew brace. Is A left nilpotent? And if furthermore A is of nilpotent type, is A left nilpotent?

12.5 Notes

The left series of a skew brace was defined by Rump [99] in the context of skew braces of abelian type. In that paper, he proved Proposition 12.2.8 by a different method in the case of skew braces of abelian type.

Theorem 12.2.6 was proved by Byott in the context of Hopf–Galois structures [28].

Theorem 12.2.16 was proved by Ballester-Bolinches, Meng, and Esteban-Romero in [88] for skew braces of abelian type.

Theorem 12.2.17 was proved by Smoktunowicz in [105, Theorem 1.1] for skew braces of abelian type. The generalization to skew braces of nilpotent type appeared in [46, Theorem 4.8]. The proof presented in this chapter appeared in [1] and it is heavily based on the ideas of Ballester-Bolinches, Meng, and Esteban-Romero [88].

Open problem 12.4 was proved in the case of skew braces of abelian type by Smoktunowicz [104].

Chapter 13
Right Nilpotent Skew Braces

In this chapter, we study right nilpotency of skew braces. We will see later that this algebraic notion translates into an important combinatorial property of solutions.

13.1 Right Series

Definition 13.1.1 Let A be a skew brace. One defines $A^{(1)} = A$ and for $n \geq 1$

$$A^{(n+1)} = A^{(n)} * A = \langle x * a : x \in A^{(n)},\ a \in A \rangle_+,$$

where $\langle X \rangle_+$ denotes the subgroup of the additive group of A generated by the subset X. The sequence

$$A^{(1)} \supseteq A^{(2)} \supseteq A^{(3)} \supseteq \cdots \supseteq A^{(n)} \supseteq \cdots$$

is the *right series* of A.

Proposition 13.1.2 *Let A be a skew brace. Each $A^{(n)}$ is an ideal of A.*

Proof We want to prove that for each positive integer n, $A^{(n)}$ is a normal subgroup of $(A, +)$, that $\lambda_a(A^{(n)}) \subseteq A^{(n)}$ for all $a \in A$ and that $A^{(n)}$ is a normal subgroup of $(A, \circ)$. We proceed by induction on n. The case $n = 1$ is trivial. We assume that the claim is true for some $n \geq 1$. We first prove that $A^{(n+1)}$ is a normal subgroup of $(A, +)$. Let $a, b \in A$ and $x \in A^{(n)}$. Then $a + x * b - a \in A^{(n+1)}$ since

$$\begin{aligned} a + x * b - a &= a + \lambda_x(b) - b - a \\ &= a + \lambda_x(b) - (a + b) = a + \lambda_x(-a + a + b) - (a + b) \\ &= a + \lambda_x(-a) + \lambda_x(a + b) - (a + b) = -x * a + x * (a + b). \end{aligned}$$

Now we prove that $A^{(n+1)}$ is an ideal. Let $a, b \in A$ and $x \in A^{(n)}$. Then, by (12.2),

F. Cedó and L. Vendramin, *Groups, Radical Rings, and the Yang-Baxter Equation*, Progress in Mathematics 361, https://doi.org/10.1007/978-3-032-01524-2_13

$$\lambda_a(x * b) = (a \circ x \circ a') * \lambda_a(b) \in A^{(n+1)}$$

since $a \circ x \circ a' \in A^{(n)}$ by the inductive hypothesis. From this, it immediately follows that $\lambda_a(A^{(n+1)}) \subseteq A^{(n+1)}$. Now let $y \in A^{(n+1)}$. By using the formula (8.13) of Proposition 8.4.9 one obtains that $a \circ y \circ a' = a + \lambda_a(y + y * a') - a \in A^{(n+1)}$. Thus the result follows by induction. □

The second term of the right series is particularly important.

Proposition 13.1.3 *Let A be a skew brace. Then $A^{(2)}$ is the smallest ideal of A such that $A/A^{(2)}$ is a trivial skew brace.*

Proof Proposition 13.1.2 shows that $A^{(2)}$ is an ideal of A. Let I be an ideal of A and $\pi : A \to A/I$ be the canonical map. Then A/I is trivial as a skew brace if and only if $\lambda_a(b) - b \in I$ for all $a, b \in A$. Since this condition is equivalent to $A^{(2)} \subseteq I$, the claim follows. □

Corollary 13.1.4 *Let A be a finite skew brace of size p^n for some prime number p and some positive integer n. Then either A is the trivial skew brace of order p or it has a nonzero proper ideal.*

Proof By Proposition 12.2.8, A is left nilpotent. In particular, if $A \neq \{0\}$, then $A^{(2)} = A^2 \neq A$. Since A^2 is an ideal, we may assume that $A^2 = \{0\}$. In this case, $a \circ b = a + b$ for all $a, b \in A$. Therefore the derived subgroup $[A, A]_+$ of the additive group of A is a proper ideal of A. Hence we may assume that $A^2 = [A, A]_+ = \{0\}$. In this case, A is a trivial skew brace of abelian type and the result follows. □

Definition 13.1.5 A skew brace A is said to be *right nilpotent* if $A^{(m)} = \{0\}$ for some $m \geq 1$.

Let us review some basic properties of right nilpotent skew braces.

Lemma 13.1.6 *Let A and B be skew braces.*

(1) If $f : A \to B$ is a surjective homomorphism of skew braces and A is right nilpotent, then B is right nilpotent.
(2) If B is a subbrace of A and A is right nilpotent, then B is right nilpotent.

Proof To prove the first claim, we shall prove that $f(A^{(k)}) = B^{(k)}$ by induction on k. The case $k = 1$ is trivial. Let us assume that the result is valid for some $k \geq 1$. Since $f(A^{(k)}) = B^{(k)}$,

$$f(A^{(k+1)}) = f(A^{(k)} * A) = f(A^{(k)}) * f(A) = B^{(k)} * B = B^{(k+1)}.$$

The second claim also follows by induction, as $B^{(k)} \subseteq A^{(k)}$ for all k. □

Theorem 13.1.7 *Let A be a nonzero right nilpotent skew brace of nilpotent type and I be a nonzero ideal of A. Then $I \cap \mathrm{Soc}(A) \neq \{0\}$. In particular, $\mathrm{Soc}(A) \neq \{0\}$.*

Proof Since $(A, +)$ is nilpotent and each $I \cap A^{(k)}$ is a normal subgroup of $(A, +)$, it follows from Theorem 5.1.15 that $I \cap A^{(k)} \cap Z(A, +) \neq \{0\}$ whenever $I \cap A^{(k)} \neq \{0\}$. Let $m = \max\{k \in \mathbb{N} : I \cap A^{(k)} \cap Z(A, +) \neq \{0\}\}$. Since

$$(I \cap A^{(m)} \cap Z(A, +)) * A \subseteq I \cap (A^{(m)} * A) = I \cap A^{(m+1)} = \{0\},$$

it follows that $I \cap A^{(m)} \cap Z(A, +) \subseteq I \cap \mathrm{Soc}(A)$. For the second claim just take $I = A$ in the first claim. □

Let A be a skew brace. Let $A^{[1]} = A$ and for $n \geq 1$ let $A^{[n+1]}$ be the additive subgroup of A generated by $\bigcup_{1 \leq i \leq n} A^{[i]} * A^{[n+1-i]}$. One easily proves by induction that $A^{[k]} \supseteq A^{[k+1]}$ for all $k \geq 1$.

Definition 13.1.8 A skew brace A is said to be *strongly nilpotent* if there is a positive integer n such that $A^{[n]} = \{0\}$.

Definition 13.1.9 A skew brace A is said to be *strongly nil* if for every $a \in A$ there is a positive integer $n = n(a)$ such that any $*$-product of n copies of a is zero.

Proposition 13.1.10 *Let A be a skew brace. Each $A^{[n]}$ is a left ideal of A.*

Proof Each $A^{[n]}$ is a subgroup of $(A, +)$. Since $A * A^{[n]} \subseteq A^{[n+1]} \subseteq A^{[n]}$, the claim follows from Proposition 8.4.8. □

There exists a skew brace A such that $A^{[n]} = A^{[n+1]} \neq \{0\}$ for some positive integer n and $A^{[n+2]} = \{0\}$.

Example 13.1.11 Let

$$G = \langle r, s : r^8 = s^2 = 1,\ srs = r^7 \rangle \cong \mathbb{D}_{16},$$
$$K = \langle a, b : 8a = 2b = 0,\ a + b = b + a \rangle \cong \mathbb{Z}/(8) \times \mathbb{Z}/(2).$$

The group G acts by automorphisms on K via

$$r \cdot a = a + b, \quad r \cdot b = 4a + b, \quad s \cdot a = 3a, \quad s \cdot b = 4a + b.$$

A direct calculation shows that the map $\pi : G \to K$ given by

$$\begin{array}{cccc}
1 \mapsto 0, & r \mapsto a, & r^2 \mapsto 2a + b, & r^3 \mapsto 7a + b, \\
r^4 \mapsto 4a, & r^5 \mapsto 5a, & r^6 \mapsto 6a + b, & r^7 \mapsto 3a + b, \\
rs \mapsto 6a, & r^2 s \mapsto 7a, & r^3 s \mapsto b, & r^4 s \mapsto 5a + b, \\
r^5 s \mapsto 2a, & r^6 s \mapsto 3a, & r^7 s \mapsto 4a + b, & s \mapsto a + b
\end{array}$$

is a bijective 1-cocycle. Therefore there exists a skew brace A with additive group isomorphic to K and multiplicative group isomorphic to G. The addition of A is that of K and the multiplication is given by

$$x \circ y = \pi(\pi^{-1}(x)\pi^{-1}(y)), \quad x, y \in K.$$

Since

$$a * a = -a + a \circ a - a = -a + (2a + b) - a = b,$$
$$(5a + b) * a = -(5a + b) + (5a + b) \circ a - a = -(5a + b) + b - a = 2a,$$

it follows that $A^{[2]}$ contains $\langle 2a, b\rangle_+ = \{0, 2a, 4a, 6a, b, 2a + b, 4a + b, 6a + b\}$, the additive subgroup of $(A, +)$ generated by $2a$ and b. Therefore $A^{[2]} = \langle 2a, b\rangle_+$ since $A^{[2]} \neq A$. Routine calculations prove that

$$A^{[3]} = \{0, 2a + b, 4a, 6a + b\}, A^{[4]} = A^{[5]} = \{0, 4a\}, A^{[6]} = \{0\}.$$

The relation between the sequence of the $A^{[n]}$ and the left and right series is given in the following theorem.

Theorem 13.1.12 *Let A be a skew brace. The following statements are equivalent:*

(1) $A^{[\alpha]} = \{0\}$ *for some positive integer* α.
(2) $A^{(\beta)} = \{0\}$ *and* $A^{\gamma} = \{0\}$ *for some positive integers* β *and* γ.

Proof To prove that (1) $\implies$ (2) one proves by induction that $A^n \subseteq A^{[n]}$ and $A^{(n)} \subseteq A^{[n]}$ for all positive integers n. Let us prove that (2) $\implies$ (1). We proceed by induction on β. If $\beta \in \{1, 2\}$, the result is true because $\{0\} = A^{(2)} = A^2 = A^{[2]}$. Fix a positive integer $\beta \geq 2$ and suppose that the result holds for this β, so for every γ there exists $\alpha = \alpha(\gamma)$ depending on γ such that $A^{[\alpha]} = \{0\}$. We need to show that $A^{(\beta+1)} = \{0\}$ and $A^{\gamma} = \{0\}$ imply that $A^{[n]} = \{0\}$ for some n. Let $n > \alpha(\gamma)$. Every element of $A^{[n]}$ is a sum of elements from $A^{[i]} * A^{[j]}$, where $i + j = n$ and $1 \leq i \leq n - 1$. Note that if $\alpha(\gamma) \leq i \leq n - 1$, $a_i \in A^{[i]}$ and $a_{n-i} \in A^{[n-i]}$, then by the inductive hypothesis applied to the quotient $A/A^{(\beta)}$, $a_i \in A^{(\beta)}$ and thus $a_i * a_{n-i} \in A^{(\beta+1)} = \{0\}$. Hence we may assume that the elements of $A^{[n]}$ are sums of elements from $A^{[i]} * A^{[j]}$ for $1 \leq i < \alpha(\gamma)$ and $j \geq n - \alpha(\gamma)$ such that $i + j = n$. Then

$$A^{[n]} \subseteq A * A^{[n-\alpha(\gamma)]} \subseteq A^2.$$

Applying the same argument for $n' = n - \alpha(\gamma)$ we obtain that $A^{[n']} \subseteq A * A^{[n'-\alpha(\gamma)]}$ provided that $n' > \alpha(\gamma)$. Therefore

$$A^{[n]} \subseteq A * (A * A^{[n-2\alpha(\gamma)]}) \subseteq A^3$$

provided that $n > 2\alpha(\gamma)$. Continuing in this way we obtain that $A^{[n]} \subseteq A^k$ provided that $n > (k - 1)\alpha(\gamma)$. Then it follows that $A^{[(\gamma-1)\alpha(\gamma)+1]} \subseteq A^{\gamma} = \{0\}$. □

13.2 Right Nilpotent Skew Braces

We now present some characterizations of right nilpotent skew braces. We first need some definitions.

Definition 13.2.1 Let A be a skew brace. An *s-series* of A is a sequence

$$A = I_0 \supseteq I_1 \supseteq I_2 \supseteq \cdots \supseteq I_n = \{0\}$$

of ideals of A such that $I_{j-1}/I_j \subseteq \operatorname{Soc}(A/I_j)$ for each $j \in \{1, \dots, n\}$.

Definition 13.2.2 Let A be a skew brace. We define $\operatorname{Soc}_0(A) = \{0\}$ and, for $n \geq 1$, $\operatorname{Soc}_n(A)$ is the ideal of A containing $\operatorname{Soc}_{n-1}(A)$ such that

$$Soc_n(A)/\operatorname{Soc}_{n-1}(A) = \operatorname{Soc}(A/\operatorname{Soc}_{n-1}(A)).$$

Hence

$$\{0\} = \operatorname{Soc}_0(A) \subseteq \operatorname{Soc}_1(A) \subseteq \cdots \subseteq \operatorname{Soc}_n(A) \subseteq \dots$$

is a series of ideals of the skew brace A. It is called the *socle series* of A.

Lemma 13.2.3 *Let A be a skew brace and let $A = I_0 \supseteq I_1 \supseteq I_2 \supseteq \cdots \supseteq I_n = \{0\}$ be an s-series for A. Then $A^{(i+1)} \subseteq I_i$ for all i.*

Proof We proceed by induction on i. The case $i = 0$ is trivial, so let us assume that the result holds for some $i \geq 0$. Let $\pi : A \to A/I_{i+1}$ be the canonical map. From $\pi(I_i) \subseteq \operatorname{Soc}(A/I_{i+1})$ and $\pi(I_i * A) = \pi(I_i) * \pi(A) = \{0\}$ it follows that $I_i * A \subseteq I_{i+1}$. The inductive hypothesis then implies that $A^{(i+2)} = A^{(i+1)} * A \subseteq I_i * A \subseteq I_{i+1}$. Thus the result follows by induction. □

Lemma 13.2.4 *Let A be a skew brace. Then A admits an s-series if and only if there exists a positive integer n such that $A = \operatorname{Soc}_n(A)$.*

Proof Suppose that there exists a positive integer n such that $A = \operatorname{Soc}_n(A)$. Then

$$A = \operatorname{Soc}_n(A) \supseteq \operatorname{Soc}_{n-1}(A) \supseteq \cdots \supseteq \operatorname{Soc}_0(A) = \{0\},$$

is an s-series. Conversely, suppose that A admits an s-series. Let

$$A = I_0 \supseteq I_1 \supseteq I_2 \supseteq \cdots \supseteq I_n = \{0\}$$

be an s-series of A. We shall prove that $I_{n-j} \subseteq \operatorname{Soc}_j(A)$ by induction on j. For $j = 0$, $I_n = \{0\} = \operatorname{Soc}_0(A)$. Suppose that $j > 0$ and $I_{n-j+1} \subseteq \operatorname{Soc}_{j-1}(A)$. Since $I_{n-j}/I_{n-j+1} \subseteq \operatorname{Soc}(A/I_{n-j+1})$, $I_{n-j} * A \subseteq I_{n-j+1} \subseteq \operatorname{Soc}_{j-1}(A)$, by the inductive hypothesis. Furthermore, for all $x \in A$ and all $y \in I_{n-j}$,

$$x + y - x - y \in I_{n-j+1} \subseteq \operatorname{Soc}_{j-1}(A).$$

Therefore $I_{n-j} \subseteq \operatorname{Soc}_j(A)$. Hence $A = I_0 = \operatorname{Soc}_n(A)$ and the result follows. □

Proposition 13.2.5 *Let A be a skew brace of abelian type. Then* $\operatorname{Soc}_n(A) = A$ *if and only if* $A^{(n+1)} = \{0\}$.

Proof Suppose that $\operatorname{Soc}_n(A) = A$. Then

$$A = \operatorname{Soc}_n(A) \supseteq \operatorname{Soc}_{n-1}(A) \supseteq \cdots \supseteq \operatorname{Soc}_0(A) = \{0\}$$

is an s-series. By Lemma 13.2.3, $A^{(n+1)} = \{0\}$.

Conversely, suppose that $A^{(n+1)} = \{0\}$. We shall prove that $\operatorname{Soc}_n(A) = A$ by induction on n. For $n = 0$, $\operatorname{Soc}_0(A) = \{0\} = A^{(1)} = A$. Suppose that $n \geq 1$ and that, for every $k < n$, $\operatorname{Soc}_k(B) = B$ for all skew brace B of abelian type such that $B^{(k+1)} = \{0\}$. Note that $A^{(n)} \subseteq \ker \lambda = \operatorname{Soc}(A)$. Let $\overline{A} = A/\operatorname{Soc}(A)$. Then $\overline{A}^{(n)} = \{0\}$. By the inductive hypothesis, $\operatorname{Soc}_{n-1}(\overline{A}) = \overline{A}$. Since $\operatorname{Soc}_{n-1}(\overline{A}) = \operatorname{Soc}_n(A)/\operatorname{Soc}(A)$, we have that $\operatorname{Soc}_n(A) = A$. Therefore the result follows by induction. □

Proposition 13.2.6 *A skew brace of nilpotent type is right nilpotent if and only if it admits an s-series.*

Proof Let A be a skew brace of nilpotent type. If A admits an s-series, then A is right nilpotent by Lemma 13.2.3. Conversely, suppose that A is right nilpotent. There exists a positive integer such that $A^{(m)} = \{0\}$. We shall prove that A admits an s-series by induction on m. For $m = 1$, $A = A^{(1)} = \{0\}$ is an s-series. Suppose that $m > 1$ and that the result is true for $m - 1$. Consider $\bar{A} = A/A^{(m-1)}$. Since $\bar{A}^{(m-1)} = \{0\}$, by the induction hypothesis $\bar{A}$ admits an s-series. Thus there is a sequence

$$A = I_0 \supseteq I_1 \supseteq I_2 \supseteq \cdots \supseteq I_n = A^{(m-1)}$$

of ideals of A such that $I_{j-1}/I_j \subseteq \operatorname{Soc}(A/I_j)$ for each $j \in \{1, \ldots, n\}$. Since $A^{(m)} = \{0\}$, we have that $A^{(m-1)} \subseteq \ker(\lambda)$. Since A is of nilpotent type, there exists a positive integer s such that $\gamma_s^+(A) = \{0\}$, where $\gamma_i^+(A)$ denotes the lower central series of the additive group of A, that is, $\gamma_1^+(A) = A$ and $\gamma_{i+1}^+(A) = [A, \gamma_i^+(A)]_+$, for all positive integers i. Let $I_{n+j-1} = A^{(m-1)} \cap \gamma_j^+(A)$ for $j = 1, \ldots, s$. Note that I_{n+j-1} is a normal subgroup of the additive group of A invariant by λ_x, for all $x \in A$, and $I_{n+j-1} * A = \{0\}$, for all $j = 1, \ldots, s$, because $A^{(m-1)} \subseteq \ker(\lambda)$. By Proposition 8.4.9, I_{n+j-1} is an ideal of A, for all $j = 1, \ldots, s$. Note that $I_{n+j-1}/I_{n+j} \subseteq Z(A/I_{n+j}, +)$, for all $j = 1, \ldots, s-1$. Therefore, since $I_{n+j-1} \subseteq \ker(\lambda)$, we have that $I_{n+j-1}/I_{n+j} \subseteq \operatorname{Soc}(A/I_{n+j})$, for all $j = 1, \ldots, s-1$. Hence

$$A = I_0 \supseteq I_1 \supseteq I_2 \supseteq \cdots \supseteq I_n = A^{(m-1)} \supseteq I_{n+1} \supseteq \cdots \supseteq I_{n+s-1} = \{0\}$$

is an s-series of A, and the result follows by induction. □

Proposition 13.2.7 *Let A be a skew brace such that* $A/\operatorname{Soc}(A)$ *is right nilpotent. Then A is right nilpotent.*

Proof Note that $(A/\operatorname{Soc}(A))^{(k)} = \{0\}$ if and only if $A^{(k)} \subseteq \operatorname{Soc}(A)$ by the definition of the quotient skew brace. Then $A^{(k+1)} = A^{(k)} * A \subseteq \operatorname{Soc}(A) * A = \{0\}$ as required.
□

Definition 13.2.8 A skew brace A has *finite multipermutation level* if the sequence S_n defined as $S_1 = A$ and $S_{n+1} = S_n/\operatorname{Soc}(S_n)$ for $n \geq 1$ reaches zero.

Proposition 13.2.9 *Let A be a skew brace. Then A has finite multipermutation level if and only if A admits an s-series.*

Proof Let $S_1 = A$ and $S_{n+1} = S_n/\operatorname{Soc}(S_n)$ for $n \geq 1$. By induction on n, we prove that $S_{n+1} \cong A/\operatorname{Soc}_n(A)$. For the cases $n \in \{0, 1\}$, the result is clear since $\operatorname{Soc}_0(A) = \{0\}$ and $\operatorname{Soc}_1(A) = \operatorname{Soc}(A)$. Suppose that $n > 1$ and the result is true for $n - 1$. Hence, by the inductive hypothesis,

$$\begin{aligned}
S_n &= S_{n-1}/\operatorname{Soc}(S_{n-1}) \\
&\cong (A/\operatorname{Soc}_{n-2}(A))/\operatorname{Soc}(A/\operatorname{Soc}_{n-2}(A)) \\
&= (A/\operatorname{Soc}_{n-2}(A))/(\operatorname{Soc}_{n-1}(A)/\operatorname{Soc}_{n-2}(A)) \cong A/\operatorname{Soc}_{n-1}(A).
\end{aligned}$$

Therefore $S_n = \{0\}$ if and only if $A = \operatorname{Soc}_{n-1}(A)$. Now Lemma 13.2.4 applies. □

Theorem 13.2.10 *Let A be a skew brace. Then A has finite multipermutation level, if and only if A is right nilpotent and $(A, +)$ is nilpotent.*

Proof Suppose that A has finite multipermutation level. We proceed by induction on the multipermutation level n. The case $n = 1$ is trivial. Let A be a skew brace of finite multipermutation level $n + 1$. Since $A/\operatorname{Soc}(A)$ has multipermutation level n, the inductive hypothesis implies that $(A/\operatorname{Soc}(A))^{(m)} = \{0\}$ for some m and $(A/\operatorname{Soc}(A), +)$ is nilpotent. This implies that $A^{(m)} \subseteq \operatorname{Soc}(A)$ and hence $A^{(m+1)} = \{0\}$, furthermore, since $\operatorname{Soc}(A)$ is central in $(A, +)$, we have that $(A, +)$ is nilpotent. Conversely, suppose that A is right nilpotent and $(A, +)$ is nilpotent. By Proposition 13.2.6, A admits an s-series. Thus the result follows from Proposition 13.2.9. □

The following example shows that the assumption on the nilpotency of the additive group of the skew brace is needed in Theorem 13.2.10.

Example 13.2.11 Let A be a nonzero trivial skew brace, that is, $a \circ b = a + b$ for all $a, b \in A$. Then $A^{(2)} = \{0\}$, thus A is right nilpotent. If $Z(A, +) = \{0\}$, then $\operatorname{Soc}(A) = \{0\}$ and A does not have finite multipermutation level. For example, we can take $(A, +) = (A, \circ)$ any non-abelian simple group.

Let A be a skew brace. For subsets X and Y of A we define inductively $R_0(X, Y) = X$ and $R_{n+1}(X, Y)$ as the additive subgroup generated by $R_n(X, Y) * Y$ and $[R_n(X, Y), Y]_+$ for $n \geq 0$.

Lemma 13.2.12 *Let I be an ideal of a skew brace A. Then $R_{n+1}(I, A) \subseteq R_n(I, A)$ for all $n \geq 0$.*

Proof We proceed by induction on n. The case $n = 0$ is trivial as I is an ideal of A. Let us assume that the claim holds for some $n \geq 0$. Since by the inductive hypothesis $R_n(I, A) * A \subseteq R_{n-1}(I, A) * A \subseteq R_n(I, A)$ and

$$[R_n(I, A), A]_+ \subseteq [R_{n-1}(I, A), A]_+ \subseteq R_n(I, A),$$

it follows that $R_{n+1}(I, A) \subseteq R_n(I, A)$. □

Proposition 13.2.13 *Let I be an ideal of a skew brace A. Then each $R_n(I, A)$ is an ideal of A.*

Proof We proceed by induction on n. The case where $n = 0$ follows from the fact that I is an ideal of A. So assume that the result holds for some $n \geq 0$. We first prove that $R_{n+1}(I, A)$ is a normal subgroup of $(A, +)$. Let $a, b \in A$ and $x \in R_n(I, A)$. Then

$$a + x * b - a = -x * a + x * (a + b) \in R_{n+1}(I, A),$$

by definition. Since, moreover,

$$\begin{aligned} &a + (x + b - x - b) - a \\ &\quad = (a + x - a) + (a + b - a) - (a + x - a) - (a + b - a) \in R_{n+1}(I, A) \end{aligned}$$

by the inductive hypothesis, it follows that $R_{n+1}(I, A)$ is a normal subgroup of $(A, +)$.

We now prove that

$$\lambda_a(R_{n+1}(I, A)) \subseteq R_{n+1}(I, A) \tag{13.1}$$

for all $a \in A$. Using the inductive hypothesis and that each $\lambda_a \in \mathrm{Aut}(A, +)$, by (12.2), we have that

$$\lambda_a(x * b) = (a \circ x \circ a') * \lambda_a(b) \in R_{n+1}(I, A)$$

for all $x \in R_n(I, A)$ and $b \in A$, and

$$\begin{aligned} \lambda_a([R_n(I, A), A]_+) &\subseteq [\lambda_a(R_n(I, A)), \lambda_a(A)]_+ \\ &\subseteq [R_n(I, A), A]_+ \subseteq R_{n+1}(I, A), \end{aligned}$$

and thus (13.1) follows.

Since $R_{n+1}(I, A) \subseteq R_n(I, A)$ by Lemma 13.2.12,

$$R_{n+1}(I, A) * A \subseteq R_n(I, A) * A \subseteq R_{n+1}(I, A).$$

Hence the claim follows from Proposition 8.4.9. □

Lemma 13.2.14 *Let A be a skew brace, X be a subset of A and $n, m \in \mathbb{N}$. Then $R_m(X, A) \subseteq \mathrm{Soc}_n(A)$ if and only if $X \subseteq \mathrm{Soc}_{m+n}(A)$.*

Proof We proceed by induction on m. The case where $m = 0$ is trivial, so assume that the result is valid for some $m \geq 0$. Note that $R_{m+1}(X, A) \subseteq \mathrm{Soc}_n(A)$ is equivalent to $R_m(X, A) * A \subseteq \mathrm{Soc}_n(A)$ and $[R_m(X, A), A]_+ \subseteq \mathrm{Soc}_n(A)$. By the definition of the socle series, this is equivalent to $R_m(X, A) \subseteq \mathrm{Soc}_{n+1}(A)$, which is equivalent to $X \subseteq \mathrm{Soc}_{m+n+1}(A)$ by the inductive hypothesis. □

Lemma 13.2.15 *A skew brace A of nilpotent type is right nilpotent if and only if $R_n(A, A) = \{0\}$ for some $n \in \mathbb{N}$.*

Proof By Lemma 13.2.14, $R_n(A, A) = \{0\}$ if and only if $A = \mathrm{Soc}_n(A)$. By Lemma 13.2.4 and Proposition 13.2.6, the latter is equivalent to A being right nilpotent. □

Recall that a finite group G is said to be *p-nilpotent* if G has a normal p-complement for a prime number p. One proves that this subgroup is characteristic in G.

Definition 13.2.16 Let p be a prime number. A finite skew brace A of nilpotent type is said to be *right p-nilpotent* if there exists $n \geq 1$ such that $R_n(A_p, A) = \{0\}$, where A_p is the Sylow p-subgroup of $(A, +)$.

Proposition 13.2.17 *Let A be a finite skew brace of nilpotent type and p be a prime divisor of $|A|$. Then $A_p \subseteq \mathrm{Soc}_n(A)$ for some $n \geq 1$ if and only if A is right p-nilpotent.*

Proof By Lemma 13.2.14, $R_n(A_p, A) = \{0\}$ if and only if $A_p \subseteq \mathrm{Soc}_n(A)$. □

Proposition 13.2.18 *A finite skew brace A of nilpotent type is right nilpotent if and only if A is right p-nilpotent for all prime divisors p of $|A|$.*

Proof Let $\pi(A)$ be the set of all prime divisors of $|A|$. Assume first that A is right nilpotent. By Lemma 13.2.4 and Proposition 13.2.6, there exists a positive integer n such that $A_p \subseteq A = \mathrm{Soc}_n(A)$ for all $p \in \pi(A)$. Hence the claim follows from Proposition 13.2.17. Assume now that A is right p-nilpotent for all $p \in \pi(A)$. This means that for each $p \in \pi(A)$ there exists a positive integer $n(p)$ such that $A_p \subseteq \mathrm{Soc}_{n(p)}(A)$. Let

$$n = \max\{n(p) : p \in \pi(A)\}.$$

Then $A_p \subseteq \mathrm{Soc}_n(A)$ for all $p \in \pi(A)$. Since $\mathrm{Soc}_n(A)$ is an ideal of A and A is of nilpotent type, $A = \bigoplus_{p \in \pi(A)} A_p \subseteq \mathrm{Soc}_n(A)$. Hence A is right nilpotent by Lemma 13.2.4 and Proposition 13.2.6. □

Lemma 13.2.19 *Let A be a finite skew brace of nilpotent type. If $(A, \circ)$ has a normal Sylow p-subgroup for some prime number p, then A_p is an ideal of A.*

Proof Since the group $(A, +)$ is nilpotent, there exists a unique normal Sylow p-subgroup A_p of $(A, +)$. Thus A_p is a left ideal of A. Then A_p is a Sylow p-subgroup of $(A, \circ)$ and it is normal by hypothesis. Hence A_p is an ideal of A. □

Lemma 13.2.20 *Let A be a finite skew brace of nilpotent type. If $(A, \circ)$ has a normal Sylow p-subgroup for some prime number p, then* $\operatorname{Soc}(A_p) = \operatorname{Soc}(A) \cap A_p$. *In particular,* $\operatorname{Soc}(A_p)$ *is an ideal of A.*

Proof By Lemma 13.2.19, A_p is an ideal of A. Since $\operatorname{Soc}(A_p) \supseteq \operatorname{Soc}(A) \cap A_p$, we only need to prove that

$$\operatorname{Soc}(A_p) \subseteq \operatorname{Soc}(A) \cap A_p.$$

If $a \in \operatorname{Soc}(A_p)$, then $a \in Z(A_p, +) = Z(A, +) \cap A_p$, since $(A, +)$ is nilpotent, and $a * b = 0$ for all $b \in A_p$. Let $c \in A$ and write $c = x + y$, where $x \in A_p$ and $y \in A_{p'}$, where $A_{p'}$ is the p-complement of $(A, +)$. Note that $a * y \in A_{p'}$, because $A_{p'}$ is a left ideal, and $a * y \in A_p$, because A_p is an ideal. Hence $a * y \in A_{p'} \cap A_p = \{0\}$. Since

$$a * c = a * (x + y) = a * x + x + a * y - x = 0$$

and $a \in Z(A, +)$, the lemma is proved. □

Theorem 13.2.21 *Let A be a finite skew brace of nilpotent type. If $(A, \circ)$ has an abelian normal Sylow p-subgroup for some prime number p, then A is right p-nilpotent.*

Proof Let us assume that the result does not hold and let A be a counterexample of minimal size. We may assume that A is non-trivial, i.e., $\operatorname{Soc}(A) \neq A$. By Lemma 13.2.19, A_p is an ideal of A.

Since $\lambda_a \in \operatorname{Aut}(A_p, +)$, $\lambda_a(Z(A_p, +)) \subseteq Z(A_p, +)$ and hence $Z(A_p, +)$ is a left ideal of A_p.

By Lemma 13.2.20, $\operatorname{Soc}(A_p)$ is an ideal of A. Furthermore, since $(A_p, \circ)$ is abelian,

$$\begin{aligned}\operatorname{Soc}(A_p) &= \{a \in A_p : a * b = 0 \text{ for all } b \in A_p\} \cap Z(A_p, +)\\ &= \{a \in A_p : a \circ b = a + b \text{ for all } b \in A_p\} \cap Z(A_p, +)\\ &= \{a \in A_p : b \circ a = b + a \text{ for all } b \in A_p\} \cap Z(A_p, +)\\ &= \operatorname{Fix}(A_p) \cap Z(A_p, +).\end{aligned}$$

Since $|A_p| = p^m$ for some $m \geq 1$, the skew brace A_p is left nilpotent by Proposition 12.2.8, and, moreover, $Z(A_p, +)$ is a nonzero characteristic subgroup of $(A_p, +)$. Thus $Z(A_p, +)$ is a left ideal of A_p. Then

$$\operatorname{Soc}(A_p) = \operatorname{Fix}(A_p) \cap Z(A_p, +) \neq \{0\}$$

by Proposition 12.2.4. In particular, by Lemma 13.2.20,

$$\{0\} \neq \operatorname{Soc}(A_p) \subseteq \operatorname{Soc}(A)$$

and $I = \operatorname{Soc}(A_p)$ is a non-trivial ideal of A. Then A/I is a skew brace of nilpotent type such that $0 < |A/I| < |A|$. The minimality of $|A|$ implies that A/I is right

p-nilpotent and therefore $R_n(A_p/I, A/I) = \{0\}$ for some n. That is $R_n(A_p, A) \subseteq I \subseteq \mathrm{Soc}(A)$. Hence $R_{n+1}(A_p, A) = \{0\}$. Then A is right p-nilpotent, a contradiction. Therefore the result follows. □

A finite group G has the *Sylow tower property* if there exists a normal series

$$\{1\} = G_0 \subseteq G_1 \subseteq \cdots \subseteq G_n = G$$

such that each quotient G_i/G_{i-1} is isomorphic to a Sylow subgroup of G, in particular, G_1 is a normal Sylow subgroup of G. Note that if a finite group G has the Sylow tower property, then G is solvable. We also recall that A-*groups* are finite groups whose Sylow subgroups are abelian.

Corollary 13.2.22 *Let A be a finite skew brace of nilpotent type. Assume that $(A, \circ)$ has the Sylow tower property and that all Sylow subgroups of $(A, \circ)$ are abelian. Then A is right nilpotent.*

Proof Assume that the result is not true and let A be a counterexample of minimal size. Note that A is a non-trivial skew brace. Since $(A, \circ)$ has the Sylow tower property, there exists a normal Sylow p-subgroup A_p of $(A, \circ)$ for some prime divisor p of $|A|$. As in the proof of Theorem 13.2.21, one proves that A_p is a nonzero ideal of A and

$$\{0\} \neq \mathrm{Soc}(A_p) = \mathrm{Soc}(A) \cap A_p \subseteq \mathrm{Soc}(A).$$

The group $(A/\mathrm{Soc}(A), \circ)$ has abelian Sylow subgroups and has the Sylow tower property. Since A is a non-trivial skew brace, $0 < |A/\mathrm{Soc}(A)| < |A|$, and therefore $A/\mathrm{Soc}(A)$ is right nilpotent by the minimality of $|A|$. By Proposition 13.2.7, A is right nilpotent, a contradiction. Therefore the result follows. □

There are examples of right nilpotent skew braces of abelian type where the multiplicative group contains a non-abelian Sylow subgroup or does not have the Sylow tower property:

Example 13.2.23 The operation $a \circ b = a + 3^a b$ turns $\mathbb{Z}/8$ into a right nilpotent skew brace of abelian type with multiplicative group isomorphic to the quaternion group. This example appears in [7].

Example 13.2.24 Let $G = \mathbb{A}_4 \times \mathbb{S}_3$. Each Sylow subgroup of G is abelian, so it follows from [[40], Theorem 2.1] that there exists a skew brace of abelian type with multiplicative group isomorphic to G. The group G does not have the Sylow tower property. The database of skew braces of abelian type of [64] shows that there are only four skew braces of abelian type with multiplicative group isomorphic to G, all with additive group isomorphic to $C_6 \times C_6 \times C_2$. However, only one of these four skew braces is not right nilpotent.

A group G is *supersolvable* if it has a normal series

$$\{1\} = G_0 \subseteq G_1 \subseteq \cdots \subseteq G_n = G$$

such that each quotient G_i/G_{i-1} is cyclic. Note that subgroups and homomorphic images of supersolvable groups are supersolvable.

Proposition 13.2.25 *Every finite supersolvable group has the Sylow tower property.*

Proof Let G be a finite supersolvable group. We may assume that G is non-trivial. Let p be the largest prime divisor of $|G|$. We claim that G has a normal non-trivial p-subgroup. We shall prove the claim by induction on $|G|$. If $|G|$ is prime, then the claim holds. Suppose that $|G|$ is not prime and that the claim holds for every finite non-trivial supersolvable group H with $|H| < |G|$. Since G is supersolvable, there exists a normal cyclic subgroup N of G. Let q be a prime divisor of $|N|$. There exists a unique subgroup M of N of order q. Clearly, M is normal in G. We may assume that $q \neq p$. Let $\pi : G \to G/M$ be the canonical map. Since $|G/M| < |G|$ and p is the largest prime divisor of $|G/M|$, by the inductive hypothesis, G/M has a normal non-trivial p-subgroup T. Let P be a Sylow p-subgroup of $\pi^{-1}(T)$. Clearly $\pi^{-1}(T) = MP$ and it is normal in G. Let $\alpha\colon P \to \operatorname{Aut}(M)$ be the map defined by $\alpha(x)(y) = xyx^{-1}$ for all $x \in P$ and all $y \in M$. Since $|\operatorname{Aut}(M)| = q - 1 < p$, we have that $xyx^{-1} = y$ for all $x \in P$ and all $y \in M$. Hence P is a characteristic subgroup of $\pi^{-1}(T)$. Thus P is normal in G and the claim follows by induction. Now it is easy to prove by induction on $|G|$ that if p is the largest prime divisor of $|G|$, then the Sylow p-subgroup of G is normal in G. Therefore G has the Sylow tower property. □

Corollary 13.2.26 *Let A be a finite skew brace of nilpotent type. If all Sylow subgroups of the multiplicative group of A are cyclic, then A is right nilpotent.*

Proof By Theorem 9.5.2, the multiplicative group of A is solvable. We may assume that $A \neq \{0\}$. By Lemma 6.1.1, every minimal normal subgroup of $(A, \circ)$ is an elementary abelian p-group for some prime p. Since all Sylow subgroups of $(A, \circ)$ are cyclic, the group $(A, \circ)$ has a normal subgroup of order p for some prime divisor p of $|A|$. Now it is easy to see that $(A, \circ)$ is supersolvable. By Proposition 13.2.25, $(A, \circ)$ has the Sylow tower property. Then the claim follows from Corollary 13.2.22. □

13.3 Exercises

13.3.1 Let $A_1, \dots, A_k$ be right nilpotent skew braces. Prove that $A_1 \times \cdots \times A_k$ is right nilpotent.

13.3.2 Let A be a right nilpotent skew brace of nilpotent type and I be a minimal ideal of A. Prove that $I \subseteq \operatorname{Soc}(A)$.

13.3.3 Let I be an ideal of a skew brace A such that $I \cap A^2 = \{0\}$. Prove that I is a trivial skew brace.

13.4 Notes

The right series first appeared in the work of Rump [99], where it was proved Proposition 13.1.2 for skew braces of abelian type. The proof of the general case appeared in [46].

The socle series was introduced by Rump [99] for skew braces of abelian type. For skew braces, it was studied in [46].

Strongly nilpotent skew braces of abelian type were defined by Smoktunowicz in [105]. These definitions extend to skew braces, see [46].

Theorem 13.1.12 was proved by Smoktunowicz in [105] for skew braces of abelian type and it was extended to nilpotent type in [46].

Theorem 13.2.21 was proved by Meng, Ballester-Bolinches, and Esteban-Romero in the case of skew braces of abelian type [88]. For general skew braces, the proof is due to Acri, Łutowski and Vendramin [1].

For a skew brace A let $\rho_1(a) = a$ and $\rho_{k+1}(a) = \rho_k(a) * a$ for $n \geq 1$. The skew brace A is said to be *right nil* if for every $a \in A$ there exists a positive integer $n = n(a)$ such that $\rho_n(a) = 0$.

Regarding nil skew braces, two natural questions arises: If A is a finite right nil (resp. a strongly nil) skew brace, is A right nilpotent (resp. strongly nilpotent)? Recently, Ballester-Bolinches, Ferrara, Pérez-Calabuig and Trombetti in [16] answered these questions in the negative. They constructed a strongly nil skew brace of abelian type which is not right nilpotent. Compare with Open problem 12.4.

Chapter 14
Multipermutation Solutions

One important tool in the study of solutions to the YBE is the retraction relation. This allows us to associate to each solution another, often simpler, solution called the retraction solution. In this chapter, we study the class of multipermutation solutions, which exhibit optimal behavior with respect to the retraction relation. For finite involutive solutions to the YBE, multipermutation solutions are characterized by their structure groups.

14.1 The Retraction of a Solution

Let (X, r) be a solution to the YBE. We write

$$r(x, y) = (\sigma_x(y), \tau_y(x)) \quad \text{and} \quad r^{-1}(x, y) = (\widehat{\sigma}_x(y), \widehat{\tau}_y(x))$$

for all $x, y \in X$. Since $\sigma_x\sigma_y = \sigma_{\sigma_x(y)}\sigma_{\tau_y(x)}$ for all $x, y \in X$, the map $\sigma: X \to \mathbb{S}_X$, defined by $\sigma(x) = \sigma_x$ for all $x \in X$, induces a unique homomorphism

$$\sigma: G(X, r) \to \mathbb{S}_X$$

such that $\sigma(a) = \sigma_a$ and $\sigma_{i(x)}(y) = \sigma_x(y)$ for all $x, y \in X$ and $a \in G(X, r)$, where $i: X \to G(X, r)$ is the natural map. Similarly, the map $\widehat{\sigma}: X \to \mathbb{S}_X$, defined by $\widehat{\sigma}(x) = \widehat{\sigma}_x$ for all $x \in X$, induces a unique homomorphism

$$\widehat{\sigma}: G(X, r) \to \mathbb{S}_X$$

such that $\widehat{\sigma}(a) = \widehat{\sigma}_a$ and $\widehat{\sigma}_{i(x)}(y) = \widehat{\sigma}_x(y)$ for all $x, y \in X$ and $a \in G(X, r)$.

The map $\tau: X \to \mathbb{S}_X$, defined by $\tau(x) = \tau_x$ for all $x \in X$, induces a unique anti-homomorphism

F. Cedó and L. Vendramin, *Groups, Radical Rings, and the Yang-Baxter Equation*, Progress in Mathematics 361, https://doi.org/10.1007/978-3-032-01524-2_14

$$\tau\colon G(X,r) \to \mathbb{S}_X$$

such that $\tau(a) = \tau_a$ and $\tau_{i(x)}(y) = \tau_x(y)$ for all $x, y \in X$ and $a \in G(X,r)$, and the map $\widehat{\tau}\colon X \to \mathbb{S}_X$, defined by $\widehat{\tau}(x) = \widehat{\tau}_x$ for all $x \in X$, induces a unique anti-homomorphism

$$\widehat{\tau}\colon G(X,r) \to \mathbb{S}_X$$

such that $\widehat{\tau}(a) = \widehat{\tau}_a$ and $\widehat{\tau}_{i(x)}(y) = \widehat{\tau}_x(y)$ for all $x, y \in X$ and $a \in G(X,r)$.

By Theorem 9.3.5, there exists a unique braiding operator r_G on $G(X,r)$ such that $r_G(i \times i) = (i \times i)r$. We write

$$r_G(a,b) = (\lambda_a(b), \mu_b(a)) \quad \text{and} \quad r_G^{-1}(a,b) = (\widehat{\lambda}_a(b), \widehat{\mu}_b(a))$$

for all $a, b \in G(X,r)$.

Lemma 14.1.1 *Let (X,r) be a solution to the YBE. Then*

$$\sigma_a^{-1}(x) = \widehat{\tau}_{\widehat{\lambda}_{i(x)}^{-1}(a)}(x), \quad \widehat{\sigma}_a^{-1}(x) = \tau_{\lambda_{i(x)}^{-1}(a)}(x),$$
$$\tau_a^{-1}(x) = \widehat{\sigma}_{\widehat{\mu}_{i(x)}^{-1}(a)}(x), \quad \hat{\tau}_a^{-1}(x) = \sigma_{\mu_{i(x)}^{-1}(a)}(x)$$

for all $a \in G(X,r)$ and $x \in X$.

Proof Let $\tilde{g}\colon G(X,r) \to \mathbb{S}_X$ be the map defined by $\tilde{g}(a) = \tilde{g}_a$ and $\tilde{g}_a(x) = \tau_{\lambda_{i(x)}^{-1}(a)}(x)$ for all $a \in G(X,r)$ and $x \in X$. We shall prove that $\tilde{g}$ is an anti-homomorphism of groups. Let $b_1, b_2 \in G(X,r)$ and $x \in X$. We have that

$$\begin{aligned}
\tilde{g}_{b_2b_1}(x) &= \tau_{\lambda_{i(x)}^{-1}(b_2b_1)}(x) = \tau_{\lambda_{i(x)^{-1}}(b_2b_1)}(x)\\
&= \tau_{\lambda_{i(x)^{-1}}(b_2)\lambda_{\mu_{b_2}(i(x)^{-1})}(b_1)}(x)\\
&= \tau_{\lambda_{\mu_{b_2}(i(x)^{-1})}(b_1)}\tau_{\lambda_{i(x)^{-1}}(b_2)}(x)\\
&= \tau_{\lambda_{\mu_{\lambda_{i(x)^{-1}}(b_2)}(i(x))^{-1}}(b_1)}\tau_{\lambda_{i(x)^{-1}}(b_2)}(x)\\
&= \tau_{\lambda^{-1}_{\mu_{\lambda_{i(x)}^{-1}(b_2)}(i(x))}(b_1)}\tau_{\lambda_{i(x)}^{-1}(b_2)}(x)\\
&= \tau_{\lambda^{-1}_{i(\tau_{\lambda_{i(x)}^{-1}(b_2)}(x))}(b_1)}\tau_{\lambda_{i(x)}^{-1}(b_2)}(x)\\
&= \tilde{g}_{b_1}(\tau_{\lambda_{i(x)}^{-1}(b_2)}(x))\\
&= \tilde{g}_{b_1}\tilde{g}_{b_2}(x).
\end{aligned}$$

Hence $\tilde{g}$ is an anti-homomorphism of groups.

Similarly, one verifies that the map $\tilde{f}\colon G(X,r) \to \mathbb{S}_X$, defined by $\tilde{f}(a) = \tilde{f}_a$ for all $a \in G(X,r)$, where $\tilde{f}_a(x) = \sigma_{\mu_{i(x)}^{-1}(a)}(x)$ for all $a \in G(X,r)$ and $x \in X$, is a homomorphism of groups. Note that the map $\widehat{\sigma}^{-1}\colon G(X,r) \to \mathbb{S}(X)$ defined by $\widehat{\sigma}^{-1}(a) = \widehat{\sigma}_a^{-1}$ is an anti-homomorphism of groups. Since

$$\widehat{\sigma}_{i(x)}^{-1}(y) = \widehat{\sigma}_x^{-1}(y) = \tau_{\sigma_y^{-1}(x)}(y) = \tau_{i(\sigma_y^{-1}(x))}(y) = \tau_{\lambda_{i(y)}^{-1}(i(x))}(y) = \tilde{g}_{i(x)}(y)$$

for all $x, y \in X$, we have that $\widehat{\sigma}^{-1} = \tilde{g}$, i.e.:

$$\widehat{\sigma}_a^{-1}(x) = \tau_{\lambda_{i(x)}^{-1}(a)}(x)$$

for all $a \in G(X,r)$ and $x \in X$. We also have that the map $\widehat{\tau}^{-1}\colon G(X,r) \to \mathbb{S}(X)$, defined by $\widehat{\tau}^{-1}(a) = \widehat{\tau}_a^{-1}$ for all $a \in G(X,r)$, is a homomorphism of groups. Since

$$\widehat{\tau}_{i(x)}^{-1}(y) = \widehat{\tau}_x^{-1}(y) = \sigma_{i(\tau_y^{-1}(x))}(y) = \sigma_{\mu_{i(y)}^{-1}(i(x))}(y) = \tilde{f}_{i(x)}(y)$$

for all $x, y \in X$, we have that $\widehat{\tau}^{-1} = \tilde{f}$, i.e.:

$$\widehat{\tau}_a^{-1}(x) = \sigma_{\mu_{i(x)}^{-1}(a)}(x)$$

for all $a \in G(X,r)$ and $x \in X$. This proves two of the equalities in the statement of the result. The other two equalities follow similarly. □

Lemma 14.1.2 *Let (X,r) be a solution to the YBE. Then*

$$\ker\sigma \cap \ker\tau = \ker\sigma \cap \ker\widehat{\sigma} = \ker\widehat{\sigma} \cap \ker\widehat{\tau} = \ker\tau \cap \ker\widehat{\tau}.$$

Proof By the proof of Lemma 9.4.1, $\ker\sigma \subseteq \ker\lambda$ and $\ker\tau \subseteq \ker\mu$.

First we shall see that $\mu_{i(x)}(\ker\sigma) = \ker\sigma$ for all $x \in X$. To do so, let us fix $a \in \ker\sigma$ and $x \in X$. We have

$$\sigma_{i(x)} = \sigma_a\sigma_{i(x)} = \sigma_{\lambda_a(i(x))}\sigma_{\mu_{i(x)}(a)} = \sigma_{i(x)}\sigma_{\mu_{i(x)}(a)}$$

and

$$\sigma_{i(x)^{-1}} = \sigma_a\sigma_{i(x)^{-1}} = \sigma_{\lambda_a(i(x)^{-1})}\sigma_{\mu_{i(x)^{-1}}(a)} = \sigma_{i(x)^{-1}}\sigma_{\mu_{i(x)^{-1}}(a)}.$$

Hence $\sigma_{\mu_{i(x)}(a)} = \mathrm{id}$ and $\sigma_{(\mu_i(x))^{-1}(a)} = \mathrm{id}$, and thus $\mu_{i(x)}(\ker\sigma) = \ker\sigma$.

Similarly one proves that

$$\lambda_{i(x)}(\ker\tau) = \ker\tau, \quad \widehat{\mu}_{i(x)}(\ker\widehat{\sigma}) = \ker\widehat{\sigma}, \quad \widehat{\lambda}_{i(x)}(\ker\widehat{\tau}) = \ker\widehat{\tau}$$

for all $x \in X$.

Now, we shall see that

$$\ker\sigma \cap \ker\widehat{\sigma} = \ker\sigma \cap \ker\tau = \ker\tau \cap \ker\widehat{\tau}.$$

Indeed, let $a \in \ker\sigma \cap \ker\widehat{\sigma}$. For every $x \in X$, we have $\widehat{\sigma}_{\widehat{\mu}_{i(x)}^{-1}(a)} = \mathrm{id}$ and $\sigma_{\mu_{i(x)}^{-1}(a)} = \mathrm{id}$. Thus by Lemma 14.1.1,

$$\tau_a^{-1}(x) = \widehat{\sigma}_{\widehat{\mu}_{i(x)}^{-1}(a)}(x) = x \quad \text{and} \quad \widehat{\tau}_a^{-1}(x) = \sigma_{\mu_{i(x)}^{-1}(a)}(x) = x$$

for all $x \in X$. This shows that $\ker \sigma \cap \ker \widehat{\sigma} \subseteq \ker \tau \cap \ker \widehat{\tau}$. The other inclusion follows by a symmetric argument. Hence $\ker \sigma \cap \ker \widehat{\sigma} = \ker \tau \cap \ker \widehat{\tau}$.

Note that we have also proven that

$$\ker \sigma \cap \ker \widehat{\sigma} \subseteq \ker \sigma \cap \ker \tau.$$

Let $b \in \ker \sigma \cap \ker \tau$. As $\ker \tau$ is $\lambda_{i(x)}$-invariant, for every $x \in X$, we have that $\tau_{\lambda^{-1}_{i(x)}(b)} = \mathrm{id}$. Therefore $\widehat{\sigma}_b^{-1}(x) = \tau_{\lambda^{-1}_{i(x)}(b)}(x) = x$ for all $x \in X$. This shows that $\ker \sigma \cap \ker \tau \subseteq \ker \sigma \cap \ker \widehat{\sigma}$, and thus $\ker \sigma \cap \ker \tau = \ker \sigma \cap \ker \widehat{\sigma}$. The other equality follows from the symmetry between (X, r) and (X, r^{-1}). □

Note that in case the solution (X, r) is involutive, we have $\sigma = \widehat{\sigma}$ and $\tau = \widehat{\tau}$ and therefore

$$\ker \sigma = \ker \sigma \cap \ker \tau = \ker \tau.$$

Let (X, r) be a solution to the YBE. We consider over X the relation

$$x \sim y \iff \sigma_x = \sigma_y \text{ and } \tau_x = \tau_y.$$

Then $\sim$ is an equivalence relation. Let $\overline{X}$ be the set of equivalence classes and $[x]$ denote the equivalence class of x.

Proposition 14.1.3 *Let (X, r) be a solution to the YBE. Then r induces a solution $\overline{r}$ to the YBE on $\overline{X}$ by*

$$\overline{r}([x], [y]) = ([\sigma_x(y)], [\tau_y(x)])$$

for all $x, y \in X$.

Proof Let $x, y, z \in X$ be elements such that $x \sim y$. Note that

$$\begin{aligned}
\sigma_{\sigma_x(z)}\sigma_{\tau_z(x)} &= \sigma_x\sigma_z = \sigma_y\sigma_z = \sigma_{\sigma_y(z)}\sigma_{\tau_z(y)} = \sigma_{\sigma_x(z)}\sigma_{\tau_z(y)},\\
\tau_{\tau_x(z)}\tau_{\sigma_z(x)} &= \tau_x\tau_z = \tau_y\tau_z = \tau_{\tau_y(z)}\tau_{\sigma_z(y)} = \tau_{\tau_x(z)}\tau_{\sigma_z(y)},\\
\sigma_{\sigma_z(x)}\sigma_{\tau_x(z)} &= \sigma_z\sigma_x = \sigma_z\sigma_y = \sigma_{\sigma_z(y)}\sigma_{\tau_y(z)} = \sigma_{\sigma_z(y)}\sigma_{\tau_x(z)},\\
\tau_{\tau_z(x)}\tau_{\sigma_x(z)} &= \tau_z\tau_x = \tau_z\tau_y = \tau_{\tau_z(y)}\tau_{\sigma_y(z)} = \tau_{\tau_z(y)}\tau_{\sigma_x(z)}.
\end{aligned}$$

Hence $\sigma_z(x) \sim \sigma_z(y)$ and $\tau_z(x) \sim \tau_z(y)$. Therefore $\overline{r}$ is well defined. Note that $x \sim y$ if and only if $i(x)i(y)^{-1} \in \ker \sigma \cap \ker \tau \subseteq G(X, r)$. It follows from Lemma 14.1.2 that $\widehat{\sigma}_z(x) \sim \widehat{\sigma}_z(y)$ and $\widehat{\tau}_z(x) \sim \widehat{\tau}_z(y)$, and so the map $s : \overline{X}^2 \to \overline{X}^2$, defined by $s([x], [y]) = ([\widehat{\sigma}_x(y)], [\widehat{\tau}_y(x)])$ for all $x, y \in X$, is well defined. Clearly $s = \overline{r}^{-1}$. It is clear that $\overline{r}$ is a bijective set-theoretic solution to the Yang–Baxter equation. We will prove that it is non-degenerate. Define $\sigma_{[x]} \colon \overline{X} \to \overline{X}$ and $\tau_{[x]} \colon \overline{X} \to \overline{X}$ by $\sigma_{[x]}([y]) = [\sigma_x(y)]$ and $\tau_{[x]}([y]) = [\tau_x(y)]$ for all $x, y \in X$. Clearly, in order to prove that $\sigma_{[x]}$ is bijective, it is enough to prove that it is injective (surjectivity of $\sigma_{[x]}$ is an immediate consequence of surjectivity of σ_x).

Let $x, y, z \in X$ be elements such that $\sigma_z(x) \sim \sigma_z(y)$. We have that

$$x = \sigma_z^{-1}\sigma_z(x) = \widehat{\tau}_{\widehat{\sigma}^{-1}_{\sigma_z(x)}(z)}(\sigma_z(x)) \sim \widehat{\tau}_{\widehat{\sigma}^{-1}_{\sigma_z(x)}(z)}(\sigma_z(y)) = \widehat{\tau}_{\widehat{\sigma}^{-1}_{\sigma_z(y)}(z)}(\sigma_z(y)) = y.$$

Hence $\sigma_{[x]}$ is injective and thus it is bijective. Similarly one can prove that $\tau_{[x]}$ is bijective. Hence $\overline{r}$ is non-degenerate, and the result follows. □

In the case of involutive solutions, it follows from Proposition 4.2.6 that $\sigma_x = \sigma_y$ if and only if $\tau_x = \tau_y$. In particular, if (X, r) is involutive, then

$$\begin{aligned} \mathcal{G}(X, r) &\cong \langle \sigma_x : x \in X \rangle = \{\sigma_a : a \in G(X, r)\} \\ &\cong \{\lambda_a : a \in G(X, r)\} \cong G(X, r)/\operatorname{Soc}(G(X, r)). \end{aligned}$$

Definition 14.1.4 Let (X, r) be a solution. The solution $\operatorname{Ret}(X, r) = (\overline{X}, \overline{r})$ induced by the equivalence relation $\sim$ is the *retraction* of (X, r).

We define inductively $\operatorname{Ret}^0(X, r) = (X, r)$, $\operatorname{Ret}^1(X, r) = \operatorname{Ret}(X, r)$ and

$$\operatorname{Ret}^{n+1}(X, r) = \operatorname{Ret}(\operatorname{Ret}^n(X, r)) \quad n \geq 1.$$

Definition 14.1.5 A solution (X, r) is said to be *multipermutation* solution of *level* n if n is the smallest non-negative integer such that $|\operatorname{Ret}^n(X, r)| = 1$. The solution (X, r) is said to be *irretractable* if $\operatorname{Ret}(X, r) = (X, r)$.

The trivial solution over the set with one element is a multipermutation solution of level zero. Permutation solutions over sets with more than one element are multipermutation solutions of level one.

Example 14.1.6 Let $X = \{1, 2, 3, 4\}$ and $r(x, y) = (\sigma_x(y), \tau_y(x))$, where $\sigma_1 = (234)$, $\sigma_2 = \sigma_3 = \sigma_4 = (243)$, $\tau_1 = (243)$ and $\tau_2 = \tau_3 = \tau_4 = (234)$. Then (X, r) is a multipermutation solution of level two.

Example 14.1.7 Let $X = \{1, \dots, 6\}$ and $r(x, y) = (\sigma_x(y), \tau_y(x))$, where

$$\sigma_1 = \sigma_2 = (56), \quad \sigma_3 = (12)(34)(56), \sigma_4 = (15)(26)(34), \quad \sigma_5 = \sigma_6 = (12)$$

and $\tau_x = \sigma_x$ for all $x \in X$. Then (X, r) is a multipermutation solution of level four.

For size ≤ 7 the numbers of Table 14.1 coincide with those in [56] but there are some differences for solutions of size eight.

Table 14.1 Involutive solutions of size ≤ 10

n	2	3	4	5	6	7	8	9	10
Solutions	2	5	23	88	595	3456	34530	321931	4895272
Multipermutation	2	5	21	84	554	3295	32155	305916	4606440
Irretractable	0	0	2	4	9	13	191	685	3590

Table 14.2 Non-involutive solutions of size ≤ 8

n	2	3	4	5	6	7	8
Solutions	2	21	230	3519	100071	4602720	422449480
Multipermutation		15	206	3165	95517	4461805	416725250
Irretractable		6	24	98	514	2659	17370

Theorem 14.1.8 *Let (X, r) be a finite multipermutation solution to the YBE. If $|X| > 1$, then r has even order.*

Proof Since $(X, r) \to \operatorname{Ret}(X, r) = (\overline{X}, \overline{r})$, $x \mapsto [x]$ is a homomorphism of solutions, the order of the solution $\overline{r}$ divides the order of r. Assume that (X, r) has multipermutation level n. There exists a homomorphism of solutions $(X, r) \to \operatorname{Ret}^{n-1}(X, r)$, thus it is enough to prove the theorem in the case where $r(x, y) = (\sigma(y), \tau(x))$ for commuting permutations σ and τ, i.e., multipermutation solutions of level one. If r has order $2k + 1$, then

$$(x, y) = r^{2k+1}(x, y) = (\sigma^{k+1}\tau^k(y), \sigma^k\tau^{k+1}(x)).$$

This implies that $\sigma^{k+1}\tau^k(y) = x$ for all $x, y \in X$. This equality in particular implies that $x = y$ because $\sigma^{k+1}\tau^k$ is a permutation, a contradiction. □

The connection between the socle of a skew brace and the retract of a solution was discovered by Rump in the case of involutive solutions and skew braces of abelian type, see [99].

Proposition 14.1.9 *Let A be a skew brace and (A, r) be its associated solution. Then the retraction $\operatorname{Ret}(A, r)$ is the canonical solution associated with the quotient brace $A/\operatorname{Soc}(A)$.*

Proof The equivalence relation $\sim$ on A is defined as $a \sim b$ if and only if $\lambda_a = \lambda_b$ and $\mu_a = \mu_b$. Let $\overline{A}$ be the set of equivalence classes. The equivalence class of an element a is then

$$\begin{aligned}[a] &= \{b \in A : a \sim b\} = \{b \in A : \lambda_a = \lambda_b,\ \mu_a = \mu_b\} \\ &= \{b \in A : a' \circ b \in \ker\lambda \cap \ker\mu\} = \{b \in A : a' \circ b \in \operatorname{Soc}(A)\},\end{aligned}$$

by Proposition 8.4.14. This means that $[a] = [b]$ if and only if $\pi(a) = \pi(b)$, where $\pi \colon A \to A/\operatorname{Soc}(A)$, $x \mapsto x \circ \operatorname{Soc}(A)$ is the canonical skew brace homomorphism. Moreover, $A/\operatorname{Soc}(A) = \overline{A}$ as sets. Now we compute the retraction of (A, r):

$$\begin{aligned}\overline{r}([a], [b]) &= ([\lambda_a(b)], [\mu_b(a)]) = (\pi(\lambda_a(b)), \pi(\mu_b(a))) \\ &= \left(\lambda_{\pi(a)}(\pi(b)), \mu_{\pi(b)}(\pi(a))\right) = \left(\lambda_{[a]}([b]), \mu_{[b]}([a])\right).\end{aligned}$$

Therefore $\operatorname{Ret}(A, r) = (A/\operatorname{Soc}(A), \overline{r})$, and the result follows. □

Note that a skew brace A has finite multipermutation level (see Definition 13.2.8) if and only if its associated solution (A, r_A) is a multipermutation solution.

Proposition 14.1.10 *Let A be a skew brace of nilpotent type. Then its associated solution (A, r_A) is a multipermutation solution if and only if A is right nilpotent.*

Proof This follows by Theorem 13.2.10. □

14.2 Multipermutation Solutions

In this section, we study some properties of multipermutation solutions and its structure group.

Proposition 14.2.1 *Let (X, r) and (Y, s) be solutions to the YBE. Each surjective homomorphism of solutions $f: (X, r) \to (Y, s)$ induces a surjective homomorphism of solutions* $\operatorname{Ret}(X, r) \to \operatorname{Ret}(Y, s)$. *In particular, if (X, r) is a multipermutation solution of level m and (Y, s) is an homomorphic image of (X, r), then (Y, s) is a multipermutation solution of level $\leq m$.*

Proof Write $r(x, y) = (\sigma_x(y), \tau_y(x))$ and $s(z, t) = (\sigma'_z(t), \tau'_t(z))$. Let $x, x_1 \in X$ be such that $x \sim x_1$. If $y \in X$, then

$$\sigma'_{f(x)} f(y) = f(\sigma_x(y)) = f(\sigma_{x_1}(y)) = \sigma'_{f(x_1)} f(y).$$

Since f is surjective, it follows that $\sigma'_{f(x)} = \sigma'_{f(x_1)}$. A similar calculation proves that $\tau'_{f(x)} = \tau'_{f(x_1)}$. If $\pi: (Y, s) \to \operatorname{Ret}(Y, r)$, $y \mapsto [y]$, is the canonical map, the composition $\pi f: (X, r) \to \operatorname{Ret}(Y, s)$ is a surjective homomorphism of solutions. Therefore the map $\operatorname{Ret}(X, r) \to \operatorname{Ret}(Y, s)$, $[x] \mapsto \pi(f(x))$ is then a well-defined surjective homomorphism of solutions. This proves the first part of the result. The second part follows easily from the first part and induction on m. □

Proposition 14.2.2 *Let (X, r) be a solution to the YBE of finite multipermutation level m and $Y \subseteq X$ be such that $r(Y \times Y) \subseteq Y \times Y$. Then the* subsolution $(Y, r|_{Y \times Y})$ *is of finite multipermutation level $\leq m$.*

Proof We shall prove the result by induction on m. For $m = 0$, $|X| = 1$ and thus $Y = X$ and the result holds in this case. Suppose that $m \geq 1$ and the result holds for all solutions to the YBE of multipermutation level $< m$. Let $\operatorname{Ret}(X, r) = (\overline{X}, \overline{r})$. Let $\pi: (X, r) \to \operatorname{Ret}(X, r)$ be the natural map. Clearly $(\pi(Y), \overline{r}|_{\pi(Y)^2})$ is a subsolution of $\operatorname{Ret}(X, r)$. Since $\operatorname{Ret}(X, r)$ has multipermutation level $m - 1$, by the inductive hypothesis, $(\pi(Y), \overline{r}|_{\pi(Y)^2})$ has multipermutation level $\leq m - 1$. Let $\sim_X$ be the retract relation on X and $\sim_Y$ the retract relation on Y. For $y \in Y$, we denote by $[y]_X$ the $\sim_X$-class of y and by $[y]_Y$ the $\sim_Y$-class of y, i.e.:

$$[y]_X = \{x \in X : x \sim_X y\} \quad \text{and} \quad [y]_Y = \{z \in Y : z \sim_Y y\}.$$

Note that if $x, y \in Y$ and $x \sim_X y$, then $x \sim_Y y$. Thus the map

$$f\colon (\pi(Y), \overline{r}|_{\pi(Y)^2}) \to \operatorname{Ret}(Y, r|_{Y^2}), \quad f([y]_X) = [y]_Y$$

is a surjective homomorphism of solutions. By Proposition 14.2.1, $\operatorname{Ret}(Y, r|_{y^2})$ has multipermutation level $\leq m - 1$. Hence $(Y, r|_{Y^2})$ has multipermutation level $\leq m$ and the result follows by induction. □

Theorem 14.2.3 *Let (X, r) be a solution. The following statements are equivalent:*

(1) *(X, r) has finite multipermutation level.*
(2) *$\big(\mathcal{G}(X, r), r_{\mathcal{G}(X,r)}\big)$ has finite multipermutation level.*
(3) *$\big(G(X, r), r_{G(X,r)}\big)$ has finite multipermutation level.*

Proof Let us first prove that (2) implies (1). The map $X \to \mathcal{G}(X, r), x \mapsto (\sigma_x, \tau_x^{-1})$, is a homomorphism of solutions that induces an injective homomorphism of solutions $j\colon \operatorname{Ret}(X, r) \to \big(\mathcal{G}(X, r), r_{\mathcal{G}(X,r)}\big)$. Since $\big(\mathcal{G}(X, r), r_{\mathcal{G}(X,r)}\big)$ has finite multipermutation level, (X, r) has finite multipermutation level by Proposition 14.2.2.

Let us now prove that (3) implies (2). The canonical map $G(X, r) \to \mathcal{G}(X, r)$ yields a surjective homomorphism of solutions. Then Proposition 14.2.1 applies.

Finally, we prove that (1) implies (3). We shall prove that if (X, r) is a multipermutation solution to the YBE of level m, then $(G(X, r), r_{G(X,r)})$ is a multipermutation solution of level $\leq m + 1$, by induction on m. For $m = 0$, we have that $|X| = 1$ and $G(X, r)$ is the trivial skew brace of abelian type $(\mathbb{Z}, +)$. Hence $\big(G(X, r), r_{G(X,r)}\big)$ has multipermutation level 1 in this case. Suppose that $m \geq 1$ and $\big(G(Y, s), r_{G(Y,s)}\big)$ has finite multipermutation level $\leq k + 1$ for every multipermutation solution (Y, s) to the YBE of level $k < m$. By the inductive hypothesis, $\big(G(\operatorname{Ret}(X, r)), r_{G(\operatorname{Ret}(X,r))}\big)$ has finite multipermutation level $\leq m$. By Lemmas 9.4.1 and 9.4.2, there exists a surjective homomorphism of skew braces $\pi\colon \mathcal{G}(X, r) \to G(X, r)/\operatorname{Soc}(G(X, r))$. By Proposition 14.1.9,

$$\operatorname{Ret}(G(X, r), r_{G(X,r)}) = \big(G(X, r)/\operatorname{Soc}(G(X, r)), r_{G(X,r)/\operatorname{Soc}(G(X,r))}\big).$$

Hence π also is a surjective homomorphism of solutions

$$\pi\colon \big(\mathcal{G}(X, r), r_{\mathcal{G}(X,r)}\big) \to \operatorname{Ret}(G(X, r)).$$

By Theorem 9.3.5, the injective homomorphism of solutions

$$j\colon \operatorname{Ret}(X, r) \to \big(\mathcal{G}(X, r), r_{\mathcal{G}(X,r)}\big)$$

induces a homomorphism of braided groups

$$\varphi\colon \big(G(\operatorname{Ret}(X, r)), r_{G(\operatorname{Ret}(X,r))}\big) \to \big(\mathcal{G}(X, r), r_{\mathcal{G}(X,r)}\big),$$

such that $\varphi i([x]) = j([x]) = (\sigma_x, \tau_x^{-1})$ for all $x \in X$, where

$$i\colon \operatorname{Ret}(X,r) \to G(\operatorname{Ret}(X,r))$$

is the natural map. Clearly, φ is surjective and thus $\pi\varphi$ is a surjective homomorphism of solutions. Since

$$\big(G(\operatorname{Ret}(X,r)), r_{G(\operatorname{Ret}(X,r))}\big)$$

has finite multipermutation level $\leq m$, by Proposition 14.2.1, $\operatorname{Ret}(G(X,r))$ has finite multipermutation level $\leq m$. Hence $\big(G(X,r), r_{G(X,r)}\big)$ has finite multipermutation level $\leq m+1$. Therefore the result follows by induction. □

Proposition 14.2.4 *Let B be a skew brace such that its associated solution (B, r_B) to the YBE has finite multipermutation level m. Then $(B, \circ)$ is solvable of derived length $\leq m$ and $(B, +)$ is nilpotent of nilpotency class $\leq m$.*

Proof By Proposition 14.1.9, $\operatorname{Ret}(B, r_B) = (B/\operatorname{Soc}(B), r_{B/\operatorname{Soc}(B)})$. Since $\operatorname{Soc}(B)$ is a trivial skew brace an $(\operatorname{Soc}(B), +)$ is central in $(B, +)$, the result follows easily by induction on m. □

An easy consequence of the proof of Theorem 14.2.3 and Proposition 14.2.4 is the following result.

Theorem 14.2.5 *Let (X, r) be a multipermutation solution to the YBE of level m. Then the multiplicative group of the skew brace $G(X, r)$ is solvable of derived length $\leq m+1$ and $(G(X, r), +)$ is nilpotent of nilpotency class $\leq m+1$.*

Proof By the proof of Theorem 14.2.3, $\big(G(X,r), r_{G(X,r)}\big)$ is a multipermutation solution of level $\leq m+1$. The result follows by Proposition 14.2.4. □

14.3 Finite Multipermutation Solutions

The following result appeared in [56].

Proposition 14.3.1 *Let (X, r) be a finite involutive solution to the YBE. If the additive group of the skew brace $\mathcal{G}(X, r)$ is cyclic, then (X, r) is multipermutation.*

Proof Let (X, r) be a counterexample of minimal cardinality. If K is the additive group of $\mathcal{G}(X, r)$, then K is finite and cyclic. Write G for the multiplicative group of $\mathcal{G}(X, r)$. Since $|\operatorname{Aut}(K)| = \varphi(|K|) < |K|$, where φ is the Euler function, the group homomorphism $\lambda\colon G \to \operatorname{Aut}(K)$ has a non-trivial kernel, so $\operatorname{Soc}(\mathcal{G}(X, r))$ is non-zero. This implies that (X, r) is retractable. By Theorem 9.3.2, the canonical surjective homomorphism of solutions $(X, r) \to \operatorname{Ret}(X, r)$ induces a surjective homomorphism of skew braces $f\colon G(X, r) \to G(\operatorname{Ret}(X, r))$. Since $f(\operatorname{Soc}(G(X, r))) \subseteq \operatorname{Soc}(G(\operatorname{Ret}(X, r)))$, f induces a surjective homomorphism of skew braces $\mathcal{G}(X, r) \to \mathcal{G}(\operatorname{Ret}(X, r))$. Hence the additive group of $\mathcal{G}(\operatorname{Ret}(X, r))$ is cyclic and $\operatorname{Ret}(X, r)$ is an involutive solution. The minimality of $|X|$ implies that $\operatorname{Ret}(X, r)$ is a multipermutation solution, and hence so is (X, r), a contradiction. □

Example 14.3.2 Let $X = \{1, 2, 3, 4\}$ and $r(x, y) = (\varphi_x(y), \varphi_y(x))$, where

$$\varphi_1 = \varphi_2 = (34), \quad \varphi_3 = \varphi_4 = (12).$$

Then (X, r) is an involutive multipermutation solution. One easily checks that $\mathcal{G}(X, r) \cong C_2 \times C_2$.

The previous result does not hold in the case of arbitrary solutions.

Example 14.3.3 Let $X = \{1, 2, 3, 4, 5, 6\}$ and $r(x, y) = (\sigma_x(y), \tau_y(x))$, where

$$\begin{aligned}
&\sigma_1 = \text{id}, && \sigma_2 = \text{id}, && \sigma_3 = \text{id},\\
&\sigma_4 = (23)(56), && \sigma_5 = (23)(56), && \sigma_6 = (23)(56),\\
&\tau_1 = \text{id}, && \tau_2 = (456), && \tau_3 = (465),\\
&\tau_4 = \text{id}, && \tau_5 = (465), && \tau_6 = (456).
\end{aligned}$$

Then (X, r) is an irretractable solution to the YBE. The brace $\mathcal{G}(X, r)$ has multiplicative group isomorphic to $\mathbb{S}_3$ and additive group isomorphic to the cyclic group of order six.

The following result appears in the work of Rump [99].

Proposition 14.3.4 *Let A be a finite non-trivial radical ring. Then* $\text{Soc}(A) \neq \{0\}$ *and (A, r_A) is an involutive multipermutation solution.*

Proof Let A be a counterexample of minimal size. This means that $\text{Soc}(A) = \{0\}$. Since A is finite, there exists a nonzero minimal left ideal I of A. Recall that A is a radical ring with product $a * b = \lambda_a(b) - b$. Since A is a radical ring, A is a nil ring, which implies by Nakayama's lemma that $I * A = \{0\}$. This means that if $x \in I$, then $x \in \text{Soc}(A)$, as $0 = x * a = \lambda_x(a) - a$ for all $a \in A$. In particular, $\text{Soc}(A) \neq \{0\}$, a contradiction. □

Proposition 14.3.4 has a nice application. The results appeared first in [39].

Theorem 14.3.5 *Let (X, r) be a finite involutive solution to the YBE. If the multiplicative group of the skew brace $\mathcal{G}(X, r)$ is abelian, then (X, r) is a multipermutation solution.*

Proof Note that $(\mathcal{G}(X, r), +, *)$ is a finite radical ring. By Proposition 14.3.4, $\mathcal{G}(X, r)$ either is zero or has nonzero socle. By Proposition 14.1.9, it follows easily that $\big(\mathcal{G}(X, r), r_{\mathcal{G}(X,r)}\big)$ is a multipermutation solution. Therefore, the result follows by Theorem 14.2.3. □

The following theorem characterizes finite multipermutation involutive solutions to the YBE in terms of its structure group.

Theorem 14.3.6 *Let (X, r) be a finite involutive solution to the YBE. The following statements are equivalent:*

(1) *(X, r) is a multipermutation solution.*
(2) *Every non-trivial subgroup of $G(X, r)$ has non-trivial center.*
(3) *$G(X, r)$ is poly-$\mathbb{Z}$.*
(4) *$G(X, r)$ is left orderable.*
(5) *$G(X, r)$ is locally indicable.*
(6) *$G(X, r)$ is diffuse.*

Proof Let $G = G(X, r)$ and $\mathcal{G} = \mathcal{G}(X, r)$. By Theorem 10.4.9, G is a Bieberbach group. By Theorem 10.4.8, the statements (2)–(6) are equivalent.

(1) $\implies$ (3). Suppose that (X, r) has multipermutation level m. We shall prove (3) by induction on m. For $m = 0$, $G \cong \mathbb{Z}$ and thus (3) holds. Suppose that $m \geq 1$ and that $G(Y, s)$ is poly-$\mathbb{Z}$ for every finite involutive solution (Y, s) to the YBE of multipermutation level $< m$. Hence, by the inductive hypothesis, $G(\mathrm{Ret}(X, r))$ is poly-$\mathbb{Z}$. Let $\pi : X \to \overline{X}$ be the map defined by $f(x) = [x]$ for all $x \in X$, where $\mathrm{Ret}(X, r) = (\overline{X}, \overline{r})$. By Theorem 9.3.2, there is a unique homomorphism of skew braces $\overline{f} : G \to G(\overline{X}.\overline{r})$ such that $\overline{f}(x) = [x]$ for all $x \in X$. Clearly $\overline{f}$ is surjective. Note that $xy^{-1} \in \ker \overline{f}$ for all $x, y \in X$ such that $\sigma_x = \sigma_y$, and in this case $\lambda_x = \lambda_y$. Hence $\{xy^{-1} : x, y \in X \text{ and } [x] = [y]\} \subseteq \mathrm{Soc}(G)$. Let I be the ideal of G generated by $\{xy^{-1} : x, y \in X \text{ and } [x] = [y]\}$. Let $\pi : G \to G/I$ be the natural map. Since $I \subseteq \ker \overline{f}$, there exists a homomorphism of skew braces $\tilde{f} : G/I \to G(\overline{X}, \overline{r})$ such that $\overline{f} = \tilde{f}\pi$. Note also that the map $\overline{X} \to G/I$ defined by $[x] \mapsto xI$ is a homomorphism of solutions $(\overline{X}, \overline{r}) \to (G/I, r_{G/I})$. By Theorem 9.3.2, there exists a unique homomorphism of skew braces $g : G(\overline{X}, \overline{r}) \to G/I$ such that $g([x]) = xI$ for all $x \in X$. Since $\tilde{f}(xI) = [x]$ for all $x \in X$, we have that g is bijective and $g^{-1} = \tilde{f}$. Since $I \subseteq \mathrm{Soc}(G)$ and $\mathrm{Soc}(G)$ is a finitely generated free abelian group, I also is a finitely generated free abelian group. Hence G is poly-$\mathbb{Z}$. Therefore (3) holds by induction.

(2) $\implies$ (1). Assume that (X, r) is not a multipermutation solution. By Theorem 14.2.3, (G, r_G) and $(\mathcal{G}, r_{\mathcal{G}})$ are not multipermutation solutions. Since G and $\mathcal{G}$ are skew braces of abelian type, by Proposition 14.1.10, the skew braces G and $\mathcal{G}$ are not right nilpotent, i.e., $G^{(n)} \neq \{0\}$ and $\mathcal{G}^{(n)} \neq \{0\}$ for all positive integer n. Since $\mathcal{G}$ is finite, there exists a positive integer n such that $\mathcal{G}^{(n+1)} = \mathcal{G}^{(n)} \neq \{0\}$. Let $H = G^{(n+1)}$. We shall prove that H has trivial center. Let $z \in Z(H)$. Since $\mathrm{Soc}(G)$ has finite index in G and G is torsion free, without loss of generality we may assume that $z \in \mathrm{Soc}(G)$. Notice that if $h \in H$, then

$$\lambda_h(z) = -h + hz = zh - h = z + h - h = z. \tag{14.1}$$

Let $X_1, \ldots, X_k$ be the orbits of X under the action of $\mathcal{G}^{(n)}$. These are the orbits of X under the action of $G^{(n)}$ though the map λ. By Theorem 9.3.4, $(G, +)$ is free abelian with basis X. Hence the element z can be uniquely written as

$$z = z_1 + \cdots + z_k,$$

where each $z_i \in \langle X_i \rangle_+$. From the uniqueness of the decomposition of z and (14.1), one obtains that $\lambda_h(z_i) = z_i$ for all $i \in \{1, \dots, k\}$ and $h \in H$. Now we write each z_i as

$$z_i = \sum_{t \in X_i} n_t t,$$

where $n_t \in \mathbb{Z}$. Note that

$$\begin{aligned} H = G^{(n+1)} &= G^{(n)} * G = \langle a * b : a \in G^{(n)},\ b \in G \rangle_+ \\ &= \langle \lambda_a(b) - b : a \in G^{(n)},\ b \in G \rangle_+ \\ &= \langle \lambda_a(x) - x : a \in G^{(n)},\ x \in X \rangle_+ \\ &= \langle y - x : x, y \in X_i,\ i \in \{1, \dots, k\} \rangle_+. \end{aligned}$$

Hence $\sum_{t \in X_i} n_t = 0$ for all $i \in \{1, \dots, k\}$. Let $x, y \in X_i$. Then there exists $g \in G^{(n)}$ such that $\lambda_g(x) = y$. From $\mathcal{G}^{(n+1)} = \mathcal{G}^{(n)}$ it follows that

$$G^{(n)} = G^{(n+1)} + (\operatorname{Soc}(G) \cap G^{(n)}) = H + (\operatorname{Soc}(G) \cap G^{(n)}).$$

Thus $g = g_1 + g_2$, where $g_1 \in H$ and $g_2 \in \operatorname{Soc}(G) \cap G^{(n)}$. Since $g_2 \in \operatorname{Soc}(G)$, $g = g_2 g_1$. Therefore

$$y = \lambda_g(x) = \lambda_{g_2 g_1}(x) = \lambda_{g_2} \lambda_{g_1}(x) = \lambda_{g_1}(x).$$

Since $z_i = \lambda_{g_1}(z_i) = \sum_{t \in X_i} n_t \lambda_{g_1}(t)$, we conclude that $n_x = n_y$. Hence $n_x = n_y$ for all $x, y \in X_i$. Since $\sum_{t \in X_i} n_t = 0$, it follows that $n_t = 0$ for all $t \in X_i$ and all $i \in \{1, \dots, k\}$. Therefore $z = 0 = 1$ and the result follows. □

Recall that a group G has the *unique product property* if for all finite non-empty subsets A and B of G there exists $x \in G$ that can be written uniquely as $x = ab$ with $a \in A$ and $b \in B$.

It is natural to ask when $G(X, r)$ has the unique product property. By Theorem 14.3.6, if (X, r) is an involutive multipermutation solution, then $G(X, r)$ has the unique product property since $G(X, r)$ is left orderable.

Example 14.3.7 Let $X = \{1, 2, 3, 4\}$ and $r(x, y) = (\sigma_x(y), \tau_y(x))$ be the irretractable involutive solution given by

$$\begin{aligned} &\sigma_1 = (12), && \sigma_2 = (1324), && \sigma_3 = (34), && \sigma_4 = (1423), \\ &\tau_1 = (14), && \tau_2 = (1243), && \tau_3 = (23), && \tau_4 = (1342). \end{aligned}$$

We claim that the group $G(X, r)$ with generators x_1, x_2, x_3, x_4 and relations

$$\begin{aligned} &x_1^2 = x_2 x_4, x_1 x_3 = x_3 x_1, x_1 x_4 = x_4 x_3, \\ &x_2 x_1 = x_3 x_2, x_2^2 = x_4^2, x_3^2 = x_4 x_2 \end{aligned}$$

does not have the unique product property. Let $x = x_1x_2^{-1}$ and $y = x_1x_3^{-1}$ and

$$S = \{x^2y, y^2x, xyx^{-1}, (y^2x)^{-1}, (xy)^{-2}, y, (xy)^2x, (xy)^2, (xyx)^{-1}, yxy, y^{-1}, x, xyx, x^{-1}\}. \quad (14.2)$$

To prove that $G(X,r)$ does not have the unique product property it is enough to prove that each $s \in S^2 = \{s_1s_2 : s_1, s_2 \in S\}$ admits at least two different decompositions of the form $s = ab = uv$ for $a, b, u, v \in S$. To perform these calculations we use the injective group homomorphism $G \to \mathbf{GL}(5, \mathbb{Z})$ given by

$$x_1 \mapsto \begin{pmatrix} 0&1&0&0&1\\ 1&0&0&0&0\\ 0&0&1&0&0\\ 0&0&0&1&0\\ 0&0&0&0&1 \end{pmatrix}, \qquad x_2 \mapsto \begin{pmatrix} 0&0&0&1&0\\ 0&0&1&0&1\\ 1&0&0&0&0\\ 0&1&0&0&0\\ 0&0&0&0&1 \end{pmatrix},$$

$$x_3 \mapsto \begin{pmatrix} 1&0&0&0&0\\ 0&1&0&0&0\\ 0&0&0&1&1\\ 0&0&1&0&0\\ 0&0&0&0&1 \end{pmatrix}, \qquad x_4 \mapsto \begin{pmatrix} 0&0&1&0&0\\ 0&0&0&1&0\\ 0&1&0&0&0\\ 1&0&0&0&1\\ 0&0&0&0&1 \end{pmatrix}.$$

This faithful representation of $G(X,r)$ allows us to compute all possible products of the form s_1s_2 for all $s_1, s_2 \in S$. By inspection, each element of S^2 admits at least two different representations.

14.4 Exercises

14.4.1 Let (X,r) be a solution to the YBE. We say that (X,r) is square free if $r(x,x) = (x,x)$ for all $x \in X$. Suppose that (X,r) is a square-free multipermutation solution of level $m > 0$. Prove that $(\mathcal{G}(X,r), r_{\mathcal{G}(X,r)})$ has multipermutation level $m-1$ and $(G(X,r), r_{G(X,r)})$ has multipermutation level $\leq m$.

14.4.2 Let $X = \{1,2,3,4\}$ and $r(x,y) = (\sigma_x(y), \tau_y(x))$ be the irretractable involutive solution given by

$$\sigma_1 = (34), \quad \sigma_2 = (1324), \quad \sigma_3 = (1423), \quad \sigma_4 = (12),$$
$$\tau_1 = (24), \quad \tau_2 = (1432), \quad \tau_3 = (1234), \quad \tau_4 = (13).$$

Prove that $G(X,r)$ does not have the unique product property.

14.5 Open Problems

Open problem 14.5.1 Let (X,r) be a finite involutive solution to the YBE such that $G(X,r)$ has the unique product property. Is (X,r) a multipermutation solution?

Open problem 14.5.2 Estimate the number of (involutive) multipermutation solutions of size n for $n \to \infty$.

Open problem 14.5.3 For $n \geq 1$ let $s(n)$ be the number of isomorphism classes of involutive solutions of size n and $m(n)$ be the number of isomorphism classes of involutive multipermutation solutions of size n. Is it true that

$$\lim_{n\to\infty} \frac{m(n)}{s(n)} = 1?$$

The question, of course, also makes sense for non-involutive solutions.

Open problem 14.5.4 Construct a faithful linear representation of the structure group of a non-involutive solution to the YBE.

14.6 Notes

Multipermutation involutive solutions were introduced in [56]. The notion was extended to the finite non-involutive case in [85], and to the general case in [38].

Theorem 14.1.8 was proved for the associated solution to finite skew braces in [106] and the general case in [38].

Proposition 14.2.1 and Lemma 14.2.2 were proved in [39] for involutive solutions. The general case goes back to [38].

Theorem 14.3.6 combines several results. The implication (1) $\Longrightarrow$ (3) was proved in [75]. The implication (1) $\Longrightarrow$ (4) was proved in [48]. The implication (3) $\Longrightarrow$ (1) was proved in [14].

Non-involutive multipermutation solutions were defined in [85] and later studied in [73].

The set (14.2) appears in the work of Promislow [94].

Exercise 14.4.1 appears in [38]. Exercise 14.4.2 appears in the book of Jespers and Okniński, see [76, Example 8.2.14].

Open problem 14.5.1 appears in [85].

Chapter 15
Factorizations

In Chap. 8, we found that groups with an exact factorization produce skew braces. In this chapter, we will study a different relationship between factorizations of groups and skew braces.

15.1 Factorizations of Groups

A group G is said to be *factorized through subgroups* A and B if $G = AB$. We remark that we do not assume that $A \cap B = \{1\}$.

A group G is *metabelian* if $[G, G]$ is abelian. Thus G is metabelian if and only if there is a normal subgroup K of G such that K and G/K are abelian. The groups $\mathbb{S}_3$ and $\mathbb{A}_4$ are metabelian.

The following theorem is considered the most satisfying result about group factorization. The proof is based on a surprisingly short and smart calculation with commutators.

Theorem 15.1.1 (Itô) *Let $G = AB$ be a factorization of the group G through abelian subgroups A and B. Then G is metabelian.*

Proof Since $G = AB$, it follows that $AB = BA$. Let us prove that $[A, B]$ is a normal subgroup of G. Let $a, a_1, \alpha, \alpha_1 \in A$ and $b, b_1, \beta, \beta_1 \in B$ be such that $\alpha b \alpha^{-1} = b_1 a_1$, $\beta a \beta^{-1} = a_2 b_2$. Since

$$\alpha[a, b]\alpha^{-1} = a(\alpha b \alpha^{-1})a^{-1}(\alpha b^{-1}\alpha^{-1}) = ab_1a_1a^{-1}a_1^{-1}b_1^{-1} = [a, b_1] \in [A, B]$$
$$\beta[a, b]\beta^{-1} = (\beta a \beta^{-1})b(\beta a^{-1}\beta^{-1})b^{-1} = a_2b_2bb_2^{-1}a_2^{-1}b^{-1} = [a_2, b] \in [A, B],$$

it follows that $[A, B]$ is a normal subgroup of G.

Now we prove that $[A, B]$ is abelian. Since

F. Cedó and L. Vendramin, *Groups, Radical Rings, and the Yang-Baxter Equation*, Progress in Mathematics 361, https://doi.org/10.1007/978-3-032-01524-2_15

$$\begin{aligned}\beta\alpha[a,b]\alpha^{-1}\beta^{-1} &= \beta[a,b_1]\beta^{-1} = (\beta a\beta^{-1})b_1(\beta a^{-1}\beta^{-1})b_1^{-1} = [a_2,b_1],\\ \alpha\beta[a,b]\beta^{-1}\alpha^{-1} &= \alpha[a_2,b]\alpha^{-1} = a_2(\alpha b\alpha^{-1})a_2^{-1}(\alpha b\alpha^{-1}) = [a_2,b_1],\end{aligned}$$

a direct calculation shows that

$$[\alpha^{-1},\beta^{-1}][a,b] = [a,b][\alpha^{-1},\beta^{-1}].$$

Since two arbitrary generators of $[A, B]$ commute, the group $[A, B]$ is abelian.

Finally we note that $[G, G] = [A, B]$. Since $[A, B]$ is normal in G, using (5.1) and (5.2) we get

$$[a_1b_1, a_2b_2] = a_1[a_2,b_1]^{-1}a_1^{-1}a_2[a_1,b_2]a_2^{-1} \in [A, B].$$

□

Sysak found the following generalization of Itô's theorem:

Theorem 15.1.2 (Sysak) *If A and B are abelian subgroups of G and H is a subgroup of G contained in the set AB, then H is metabelian.* □

The proof appears in [110].

There are several other interesting results in the theory of factorizable groups. Another important result that is worth mentioning is Kegel–Wielandt (Theorem 6.2.18).

The theorem of Kegel–Wielandt turns out to be the main tool in the proof of the following result on the structure of finite skew braces. This proves a conjecture of Byott, see [29].

Theorem 15.1.3 *Let A be a finite skew brace with nilpotent multiplicative group. Then the additive group of A is solvable.*

Proof Let K be the additive group of A and let G be the multiplicative group of A. The group $\Gamma = K \rtimes G$ has multiplication

$$(k_1, g_1)(k_2, g_2) = (k_1 + \lambda_{g_1}(k_2), g_1 \circ g_2)),$$

for all $k_1, k_1 \in K$ and $g_1, g_2 \in G$. Let $f: G \to \Gamma$, $g \mapsto (g, g)$. Note that

$$\begin{aligned} f(g_1 \circ g_2) &= (g_1 \circ g_2, g_1 \circ g_2) = (g_1 + \lambda_{g_1}(g_2), g_1 \circ g_2)\\ &= (g_1, g_1)(g_2, g_2) = f(g_1)f(g_2).\end{aligned}$$

Hence f is a group homomorphism and $f(G)$ is nilpotent. Since

$$(k, g) = (k, k)(0, k' \circ g) \in f(G)G,$$

for all $k \in K$ and $g \in G$, we have that $\Gamma = f(G)G$ is a product of two nilpotent subgroups. By the theorem of Kegel–Wielandt, Γ is solvable. Since K is a subgroup of Γ, K also is solvable. □

15.2 Factorizations of Skew Braces

Definition 15.2.1 Let A be a skew brace and let B and C be left ideals of A. We say that A admits a *factorization* through B and C if $A = B + C$.

Note that if a skew brace A admits a factorization through B and C, then it follows that

$$A = B + C = C + B = B \circ C = C \circ B.$$

Now we prove an analog of Itô's theorem in the context of skew braces. It turns out that one needs to consider factorizations through strong left ideals. We also need the following definition:

Definition 15.2.2 A skew brace A is said to be *meta-trivial* if $A^{(2)}$ is a trivial skew brace.

A skew brace A is meta-trivial if and only if there is an ideal I of A such that I and A/I are trivial skew braces.

Lemma 15.2.3 *Let A be a skew brace such that $A = B + C$, where B and C are left ideals. If B and C are trivial skew braces then, for any $b, \beta \in B$ and $c, \gamma \in C$, the following statements hold:*

(1) $\lambda_{\beta\circ\gamma} = \lambda_{\gamma\circ\beta}$,
(2) $(c + b) \circ \beta - \beta = c + b + c * \beta$,
(3) $b \circ c \circ b' \circ c' = b \circ c - c \circ b = b + \lambda_b(c) - \lambda_c(b) - c \in \ker \lambda$.

Proof To prove (1) put $c_1 = \lambda_\beta(c) \in C$ and $b_1 = \lambda_\gamma(b) \in B$. As B and C are trivial skew braces, $\lambda_\beta(b + c) = \lambda_\beta(b) + \lambda_\beta(c) = b + c_1$ and similarly $\lambda_\gamma(b + c) = b_1 + c$. Then $\lambda_{\beta\circ\gamma}(b + c) = b_1 + c_1 = \lambda_{\gamma\circ\beta}(b + c)$.

Let us prove (2). As B is a trivial skew brace, it follows from (8.3) that

$$\begin{aligned}(c + b) \circ \beta - \beta &= (c \circ \lambda_{c'}(b)) \circ \beta - \beta \\ &= c \circ (\lambda_{c'}(b) \circ \beta) - \beta \\ &= c \circ (\lambda_{c'}(b) + \beta) - \beta \\ &= c \circ \lambda_{c'}(b) - c + c \circ \beta - \beta \\ &= c + b + c * \beta.\end{aligned}$$

Part (3) follows from the following computation:

$$\begin{aligned}b \circ c \circ b' \circ c' &= (b \circ c) + \lambda_{b\circ c}(b' + \lambda_{b'}(c')) \\ &= b + \lambda_b(c) + \lambda_{b\circ c}(b') + \lambda_{b\circ c\circ b'}(c') \\ &= b + \lambda_b(c) + \lambda_c(b') + \lambda_{b\circ b'}(c') \\ &= b + \lambda_b(c) + \lambda_c(-b) - c\end{aligned}$$

$$= b \circ c - c \circ b.$$

Moreover, by (1) it follows that $b \circ c \circ b' \circ c' \in \ker \lambda$. □

Lemma 15.2.4 *Let A be a skew brace such that $A = B + C$ is a factorization through left ideals B and C. If B and C are trivial skew braces, then*

(1) $B * C$ and $C * B$ are strong left ideals of A,
(2) $B * C$ and $C * B$ are trivial skew braces, and
(3) $A^{(2)} = C * B + B * C = B * C + C * B$.

Proof Since C is a left ideal, it follows that $B * C \subseteq C$. Let $b, \beta \in B$ and $c, \gamma \in C$. As C is trivial, it follows that

$$\begin{aligned}\lambda_{b \circ c}(\beta * \gamma) &= \lambda_b(\beta * \gamma)\\ &= \lambda_b \lambda_\beta(\gamma) - \lambda_b(\gamma)\\ &= \lambda_{b \circ \beta \circ b'} \lambda_b(\gamma) - \lambda_b(\gamma)\\ &= (b \circ \beta \circ b') * \lambda_b(\gamma) \in B * C.\end{aligned}$$

Hence $B * C$ is a left ideal and a trivial skew brace.

Let $a \in A$, $b \in B$ and $c \in C$. Write $a = b_1 + c_1$, with $b_1 \in B$ and $c_1 \in C$. Then

$$\begin{aligned}a + (b * c) - a &= a + \lambda_b(c) - c - a\\ &= a - \lambda_b(a) + \lambda_b(a) + \lambda_b(c) - c - a\\ &= -(b * a) + b * (a + c)\\ &= -(b * (b_1 + c_1)) + b * (b_1 + c_1 + c).\end{aligned} \tag{15.1}$$

As $B + C = C + B$, it follows that for any $\beta \in B$ and $\gamma \in C$, there exist $\beta_1 \in B$ and $\gamma_1 \in C$ such that $\beta + \gamma = \gamma_1 + \beta_1$. Hence, for any $b \in B$ it holds that

$$b * (\beta + \gamma) = b * (\gamma_1 + \beta_1) = b * \gamma_1 + \gamma_1 + b * \beta_1 - \gamma_1 = b * \gamma_1,$$

as B is trivial. Applying this on (15.1) it follows that $B * C$ is a normal subgroup of $(A, +)$. This proves (1) and (2) for $B * C$. The proof for $C * B$ is similar.

Now we show that $A^{(2)} \subseteq C * B + B * C$. Let $b, b_1 \in B$ and $c, c_1 \in C$. Then

$$\begin{aligned}(b \circ c) * (b_1 + c_1) &= (b \circ c) * b_1 + b_1 + (b \circ c) * c_1 - b_1\\ &= \lambda_{b \circ c}(b_1) - b_1 + b_1 + b * (c * c_1) + c * c_1 + b * c_1 - b_1\\ &= \lambda_c(b_1) - b_1 + b_1 + b * c_1 - b_1\\ &= c * b_1 + b_1 + b * c_1 - b_1 \in C * B + B * C.\end{aligned}$$

Clearly $C * B + B * C \subseteq A^{(2)}$ and thus $A^{(2)} = C * B + B * C = B * C + C * B$. □

Theorem 15.2.5 *Let A be a skew brace. If $A = B + C$ is a factorization through strong left ideals B and C that are trivial skew braces, then A is right nilpotent of class at most three. In particular, A is meta-trivial.*

Proof By Lemma 15.2.4, $B * C$ and $C * B$ are strong left ideals of A, and both are trivial skew braces. Furthermore,

$$A^{(2)} = B * C + C * B = (B * C) \circ (C * B).$$

It remains to show that $A^{(2)}$ acts trivially on A, i.e., $\lambda_a(b) = b$, for all $a \in A^{(2)}$ and $b \in A$. Note that this is equivalent to $A^{(3)} = \{0\}$. We first show that $B * C$ acts trivially on A. Since $B * C$ is trivial as skew brace, it is enough to show that $\lambda_{b*c}(a) = a$, for all $b \in B$, $c \in C$ and $a \in A$. For that purpose, let $b \in B$, $c \in C$ and $a \in A$. Write $a = \beta + \gamma$, where $\beta \in B$ and $\gamma \in C$. Then

$$(b * c) * (\beta + \gamma) = (b * c) * \beta + \beta + (b * c) * \gamma - \beta = (b * c) * \beta,$$

as C is a trivial skew brace and $b * c \in C$. By Lemma 15.2.3,

$$(b \circ c - c \circ b) + \beta = (b \circ c - c \circ b) \circ \beta = (b + \lambda_b(c) - \lambda_c(b) - c) \circ \beta.$$

Since $(B, +)$ is a normal subgroup of $(A, +)$,

$$b \circ c - c \circ b = b + \lambda_b(c) - \lambda_c(b) - c = \lambda_b(c) - c + b_1$$

for some $b_1 \in B$. By Lemma 15.2.3,

$$\begin{aligned}(b \circ c - c \circ b) + \beta &= (\lambda_b(c) - c + b_1) \circ \beta \\ &= \lambda_b(c) - c + b_1 + (b * c) * \beta + \beta\end{aligned}$$

and therefore $(b * c) * \beta = 0$. Thus $B * C$ acts trivially on A. As $(C, +)$ also is a normal subgroup of $(A, +)$, it follows by symmetry that $C * B$ acts trivially on A. Hence $A^{(2)} = (B * C) \circ (C * B)$ acts trivially on A. □

Corollary 15.2.6 *Let A be a skew brace. Assume that $A = B + C$, where B and C are (not necessarily strong) left ideals, which are trivial as skew braces. Then A has a meta-trivial ideal I such that A/I is a trivial skew brace.*

Proof By Lemma 15.2.4, the ideal $A^{(2)}$ has a factorization through the strong left ideals $B * C$ and $C * B$, which are trivial skew braces. By Theorem 15.2.5, $A^{(2)}$ is meta-trivial and hence the claim follows. □

Theorem 15.2.5 has application to involutive solutions.

Theorem 15.2.7 *Let (X, r) be an involutive solution to the YBE with $|X| \geq 2$. If the skew brace of abelian type $\mathcal{G}(X, r)$ admits a factorization through left ideals, which are trivial as skew braces, then (X, r) is a multipermutation solution of level at most three.*

Proof Let $\mathcal{G} = \mathcal{G}(X, r)$ and $G = G(X, r)$. Then Theorem 15.2.5 yields $\mathcal{G}^{(m)} = \{0\}$ for some $m \leq 3$. Because $G/\operatorname{Soc}(G) \cong \mathcal{G}$ as skew braces, we get $G^{(m)} \subseteq \operatorname{Soc}(G)$, and thus $G^{(m+1)} = \{0\}$. Hence G is a right nilpotent skew brace of abelian type of class at most four and, by Proposition 13.2.5, $\operatorname{Soc}_m(G) = G$. This means that (G, r_G) is a multipermutation solution of level at most three. Since (X, r) is a subsolution of (G, r_G), by Proposition 14.2.2, (X, r) is a multipermutation solution of level at most three. □

The following examples show that properties of the involutive solutions (X, r) to the YBE are not completely determined by the group theory of the additive and multiplicative groups of the skew brace $\mathcal{G}(X, r)$.

Example 15.2.8 Let $X = \{1, 2, 3, 4\}$ and $r(x, y) = (\sigma_x(y), \tau_y(x))$ be the irretractable involutive solution to the YBE given by

$$\begin{aligned} &\sigma_1 = (34), &&\sigma_2 = (1324), &&\sigma_3 = (1423), &&\sigma_4 = (12),\\ &\tau_1 = (24), &&\tau_2 = (1432), &&\tau_3 = (1234), &&\tau_4 = (13). \end{aligned}$$

The associated skew brace $\mathcal{G}(X, r)$ has additive group C_2^3 and multiplicative group D_8. Furthermore, $\mathcal{G}(X, r)$ is not right nilpotent. Hence it is impossible to decompose the skew brace $\mathcal{G}(X, r)$ as in Theorem 15.2.7.

Example 15.2.9 Let $A = \mathbb{Z}/2 \times \mathbb{Z}/2 \times \mathbb{Z}/2$ with the skew brace structure given by

$$(x_1, y_1, z_1) \circ (x_2, y_2, z_2) = (x_1 + x_2 + z_1 y_2, y_1 + y_2, z_1 + z_2).$$

Then $(A, \circ)$ is isomorphic to the dihedral group of eight elements. Thus the skew brace A has the same additive and multiplicative groups as the skew brace $\mathcal{G}(X, r)$ of Example 15.2.8 but it has a factorization as in Theorem 15.2.7. This shows A is right nilpotent.

15.3 Exercises

15.3.1 Let G be a metabelian group. Prove that the following statements hold.

(1) If H is a subgroup of G, then H is metabelian.
(2) If $f \colon G \to H$ is a group homomorphism, then $f(H)$ is metabelian.

15.3.2 Prove that $\mathbf{SL}_2(3)$ is metabelian.

15.3.3 Let A be a skew brace. If there exists a proper strong left ideal I, then (A, r_A) is decomposable as $A = I \cup (A \setminus I)$.

15.3.4 Prove that the assumptions of Theorem 15.2.5 cannot be relaxed.

15.3.5 Let A be a nonzero skew brace that has a factorization $A = B + C$ through left ideals B and C, where both are trivial as skew braces. If B is a strong left ideal of A, then B or C contains a nonzero ideal I of A that acts trivially on A.

15.3.6 Let A be a skew brace with abelian multiplicative group. Prove that the additive group of A is meta-abelian.

15.3.7 Let A be a finite skew brace with cyclic multiplicative group. Prove that the additive group of A is supersolvable.

15.4 Open Problems

Open problem 15.4.1 Let A be a skew brace with solvable additive group. Is the multiplicative group of A solvable?

15.5 Notes

Theorem 15.2.5 was proved by Jespers, Kubat, Van Antwerpen, and Vendramin in [74]. Exercises 15.3.4 and 15.3.5 also appear in there. In [13], several simple skew braces (of abelian type) were constructed. As there are simple skew braces. In particular, there are simple skew braces of size $p^a q^b$ for primes p and q and $a, b \geq 1$. Therefore one cannot expect a naive result similar to that of Kegel–Wielandt in the context of skew braces.

Theorem 15.1.3 was proved by Tsang and Qin in [111]. Exercises 15.3.6 and 15.3.7 also appear in [111].

Problem 15.4.1 was formulated by Byott in [29]; see [30, 63, 108, 111] for partial results.

Chapter 16
Transitive Groups

In this chapter, we recall the classic classification of finite solvable primitive groups. We study indecomposable solutions to the YBE and show that homomorphic images of indecomposable solutions are also indecomposable. We introduce simple solutions to the YBE and show that for cardinalities greater than 2, they are indecomposable. Furthermore, the simple solutions of non-prime cardinalities are irretractable. We continue with the classification of the finite primitive involutive solutions. Finally, we provide some methods to construct simple involutive solutions.

16.1 Solvable Primitive Groups

Let X be a non-empty set. A subgroup G of $\mathbb{S}_X$ is *transitive* if for every $x, y \in X$ there exists $\sigma \in G$ such that $\sigma(x) = y$. A partition $\mathcal{P} = \{X_i : i \in I\}$ of X is a *system of imprimitivity* for G if for every $X_i \in \mathcal{P}$, $\sigma(X_i) \in \mathcal{P}$ for all $\sigma \in G$. Note that $\mathcal{P}_0 = \{\{x\} : x \in X\}$ and $\mathcal{P}_\infty = \{X\}$ are two systems of imprimitivity for G. $\mathcal{P}_0$ and $\mathcal{P}_\infty$ are called the trivial imprimitivity systems for G.

A transitive subgroup G of $\mathbb{S}_X$ is said to be *primitive* if there is no non-trivial system of imprimitivity for G.

Example 16.1.1 Let G be group. Then G acts on G by left multiplication, that is, the map $f : G \to \mathbb{S}_G$ defined by $f(x)(y) = xy$, for all $x, y \in G$, is a homomorphism of groups. Note that $f(G)$ is a transitive subgroup of $\mathbb{S}_G$ and $G \cong f(G)$.

Let G be a subgroup of $\mathbb{S}_X$. Note that for every $g \in G$, the stabilizers

$$G_x = \{h \in G : h(x) = x\}$$

of $x \in X$ and $G_{g(x)}$ of $g(x)$ are conjugate, in fact $G_{g(x)} = gG_xg^{-1}$. We write the orbit of $x \in X$ by

F. Cedó and L. Vendramin, *Groups, Radical Rings, and the Yang-Baxter Equation*, Progress in Mathematics 361, https://doi.org/10.1007/978-3-032-01524-2_16

$$G(x) = \{g(x) : g \in G\}.$$

Hence if G is transitive, then all stabilizers of elements of X are conjugate. Note that if G is finite, then $G = |G_x| \cdot |G(x)|$ for all $x \in X$.

Example 16.1.2 Let G be an abelian transitive subgroup of $\mathbb{S}_n$. Then the stabilizer G_i of i is equal to the stabilizer G_1 of 1 for all $i \in \{1, \ldots, n\}$. Hence

$$\{\mathrm{id}\} = \bigcup_{i=1}^{n} G_i = G_1.$$

Therefore $|G| = n$. In particular, if G is cyclic and $G = \langle \sigma \rangle$, then σ is a cycle of length n.

Lemma 16.1.3 *Let G be an abelian non-trivial primitive subgroup of $\mathbb{S}_n$. Then n is prime and $G \cong \mathbb{Z}/(n)$.*

Proof Since G is abelian non-trivial and transitive, we have that $|G| = n > 1$. Let p be a prime divisor of n. Let $\sigma \in G$ be an element of order p. Let $H = \langle \sigma \rangle$. Let $H(i) = \{h(i) : h \in H\}$. Then $\mathcal{S} = \{H(i) : 1 \leq i \leq n\}$ is a system of imprimitivity for G, because $g(H(i)) = H(g(i))$ for all $1 \leq i \leq n$. Since G is primitive, $H(i) = \{1, \ldots, n\}$ for all i. Hence $G = H$ and the result follows. □

Theorem 16.1.4 (Galois) *Let G be a non-trivial, primitive, and solvable subgroup of $\mathbb{S}_n$. Then*

(1) There exist a prime number p and a positive integer m such that $n = p^m$.
(2) There exists a unique minimal normal subgroup N of G and $N \cong (\mathbb{Z}/(p))^m$. Furthermore $C_G(N) = N$.
(3) N is transitive and if A is the stabilizer of 1, then $G = NA$ is the inner semidirect product of N and A.

Proof (1). Let N be a minimal normal subgroup of G. By Lemma 6.1.1, N is an elementary abelian p-group for some prime number p. Hence $N \cong (\mathbb{Z}/(p))^m$ for some positive integer m. Let $N(i) = \{h(i) : h \in N\}$ for all $i \in \{1, \ldots, n\}$. Note that for every $g \in G$,

$$N(g(i)) = gNg^{-1}(g(i)) = g(N(i))$$

for all $i \in \{1, \ldots, n\}$. Hence $\{N(i) : i \in \{1, \ldots, n\}\}$ is a system of imprimitivity for G. Since G is primitive and N is non-trivial, we have that

$$N(i) = \{1, \ldots, n\}$$

for all $i \in \{1, \ldots, n\}$. Hence N is transitive. By Example 16.1.2, $n = |N| = p^m$.

(2) and (3). Let A be the stabilizer of 1. Let H be a subgroup of G such that $A \subseteq H$. Let $\mathcal{S} = \{g(H(1)) : g \in G\}$. If $i \in g_1(H(1)) \cap g_2(H(1))$, then there exist $h_1, h_2 \in H$ such that $i = g_1h_1(1) = g_2h_2(1)$. Hence $h_1^{-1}g_1^{-1}g_2h_2 \in A$ and thus, since

$A \subseteq H$, we have that $g_1^{-1}g_2 \in H$. Therefore $g_1(H(1)) = g_2(H(1))$. This proves that $\mathcal{S}$ is a system of imprimitivity for G. Since G is primitive, either $H(1) = \{1\}$ or $g(H(1)) = H(1) = \{1, \dots, n\}$ for all $g \in G$. Hence either $A = H$ or $H = G$. Thus A is a maximal subgroup of G. Let N be a minimal normal subgroup of G. We have seen that $N \cong (\mathbb{Z}/(p))^m$, for some prime number p and some positive integer m. We also know that N is transitive. Since A is a maximal subgroup, $G = NA$. Note that $N \cap A$ is a normal subgroup of G. Since N is a minimal normal subgroup of G and $N \not\subseteq A$, we have that $N \cap A = \{\mathrm{id}\}$. Hence G is the inner semidirect product of N and A. Note that $N \subseteq C_G(N)$. Let $h \in C_G(N)$, $x \in N$ and $g \in G$. We have that

$$gh(g^{-1}xg)h^{-1}g^{-1} = gg^{-1}xgg^{-1} = x.$$

Hence $ghg^{-1} \in C_G(N)$. Thus $C_G(N)$ is a normal subgroup of G. Since $G = NA$, we have that $C_G(N) \cap A$ is a normal subgroup of G. Let $i \in \{1, \dots, n\}$. There exists $g \in G$ such that $i = g(1)$. Hence

$$(C_G(N) \cap A)(i) = (C_G(N) \cap A)(g(1)) = g(C_G(N) \cap A)(1) = g(1) = i$$

for all $i \in \{1, \dots, n\}$. Therefore $C_G(N) \cap A = \{\mathrm{id}\}$. Since $G = NA = C_G(N)A$ and

$$|C_G(N)| = \frac{|G|}{|A|} = |N|,$$

we have that $N = C_G(N)$. Let M be a minimal normal subgroup of G. Suppose that $N \neq M$. Then $N \cap M = \{\mathrm{id}\}$ and $M \subseteq C_G(N) = N$, a contradiction. Therefore N is the unique minimal normal subgroup of G and the result follows. □

16.2 Indecomposable Solutions

Definition 16.2.1 A solution (X, r) to the YBE is said to be *decomposable* if there is a decomposition $X = X_1 \cup X_2$ of X into a disjoint union of non-empty subsets X_1 and X_2 such that $r(X_1 \times X_1) \subseteq X_1 \times X_1$ and $r(X_2 \times X_2) \subseteq X_2 \times X_2$ and the restrictions $(X_1, r|_{X_1 \times X_1})$ and $(X_2, r|_{X_2 \times X_2})$ are solutions to the YBE. A solution (X, r) to the YBE is then *indecomposable* if it is not decomposable.

If (X, r) is a decomposable solution to the YBE and $X = X_1 \cup X_2$ is a decomposition, then $r(X_1 \times X_2) \subseteq X_2 \times X_1$ and $r(X_2 \times X_1) \subseteq X_1 \times X_2$, see Exercise 16.5.1.

Proposition 16.2.2 *A solution (X, r) to the YBE is indecomposable if and only if the group*

$$\langle \sigma_x, \tau_x : x \in X \rangle$$

acts transitively on X.

Proof Let Y be an orbit of X under the action of the group $\langle \sigma_x, \tau_x : x \in X \rangle$. Let $x, y \in Y$. Then $r(x, y) = (\sigma_x(y), \tau_y(x)) \in Y \times Y$. Also it is clear that

$$r((X \setminus Y) \times (X \setminus Y)) \subseteq (X \setminus Y) \times (X \setminus Y).$$

It is clear that $r|_{Y\times Y} : Y \times Y \to Y \times Y$ is injective. Let $x, y \in Y$. Let $r^{-1}(x, y) = (z, t)$. Then $\sigma_z(t) = x$ and $\tau_t(z) = y$. Hence $z = \tau_t^{-1}(y) \in Y$ and $t = \sigma_z^{-1}(x) \in Y$ and thus $r|_{Y\times Y}$ is bijective. Also it is easy to see that $\sigma_x|_Y : Y \to Y$ and $\tau_x|_Y : Y \to Y$ are bijective and that $(Y, r|_{Y\times Y})$ is a solution to the YBE. Similarly, one can check that if $X \setminus Y$ is not empty, then $(X \setminus Y, r|_{(X\setminus Y)\times(X\setminus Y)})$ is a solution to the YBE. Therefore, the result follows. □

Note that this group is in general not isomorphic to the permutation group of the solution. For involutive solutions (X, r) to the YBE, we identify $\mathcal{G}(X, r)$ with the subgroup $\langle \sigma_x : x \in X \rangle$ of $\mathbb{S}_X$ (see Proposition 9.4.3 and the comments below its proof).

Proposition 16.2.3 *An involutive solution (X, r) to the YBE is indecomposable if and only if $\mathcal{G}(X, r)$ acts transitively on X.*

Proof We know that in this case $\mathcal{G}(X, r) = \langle \sigma_x : x \in X \rangle$. Let Y be an orbit of X under the action of $\mathcal{G}(X, r)$. Let $x, y \in Y$. Then $\tau_y(x) = \sigma^{-1}_{\sigma_x(y)}(x) \in Y$. Hence Y also is an orbit of X under the action of $\langle \sigma_x, \tau_x : x \in X \rangle$. Therefore the result follows from Proposition 16.2.2. □

Proposition 16.2.4 *Let (X, r) be an indecomposable solution to the YBE. Let $f : (X, r) \to (Y, s)$ be a surjective homomorphism of solutions. Then (Y, s) is indecomposable and $|f^{-1}(a)| = |f^{-1}(b)|$ for all $a, b \in Y$. Furthermore $\{f^{-1}(a) : a \in Y\}$ is a system of imprimitivity for $\langle \sigma_x, \tau_x : x \in X \rangle$.*

Proof We write $r(x, y) = (\sigma_x(y), \tau_y(x))$ and $s(a, b) = (\sigma'_a(b), \tau'_b(a))$ for all $x, y \in X$ and all $a, b \in Y$. We have that

$$f(\sigma_x(y)) = \sigma'_{f(x)}(f(y)) \text{ and } f(\tau_y(x)) = \tau'_{f(y)}(f(x))$$

for all $x, y \in X$. Hence

$$f(y) = f(\sigma_x\sigma_x^{-1}(y)) = \sigma'_{f(x)}(f(\sigma_x^{-1}(y))),$$

and thus

$$f(\sigma_x^{-1}(y)) = (\sigma'_{f(x)})^{-1}(f(y)).$$

Similarly we have that

$$f(\tau_y^{-1}(x)) = (\tau'_{f(y)})^{-1}(f(x)).$$

Let $a, b \in Y$. There exist $x, y \in X$ such that $f(x) = a$ and $f(y) = b$. Since (X, r) is indecomposable, by Proposition 16.2.2, there exist elements $x_1, \ldots, x_m \in X$ and $\varepsilon_1, \ldots, \varepsilon_m \in \{-1, 1\}$ such that

$$y = \varphi_{x_1}^{\varepsilon_1} \cdots \varphi_{x_m}^{\varepsilon_m}(x),$$

where $\varphi_{x_i} \in \{\sigma_{x_i}, \tau_{x_i}\}$ for all $i = 1, \ldots, m$. Hence

$$b = f(y) = (\varphi'_{f(x_1)})^{\varepsilon_1} \cdots (\varphi'_{f(x_m)})^{\varepsilon_m}(f(x)) = (\varphi'_{f(x_1)})^{\varepsilon_1} \cdots (\varphi'_{f(x_m)})^{\varepsilon_m}(a),$$

where

$$\varphi'_{f(x_i)} = \begin{cases} \sigma'_{f(x_i)} & \text{if } \varphi_{x_i} = \sigma_{x_i} \\ \tau'_{f(x_i)} & \text{if } \varphi_{x_i} = \tau_{x_i} \end{cases}$$

Hence $\langle \sigma'_c, \tau'_c : c \in Y \rangle$ acts transitively on Y. By Proposition 16.2.2, (Y, s) is indecomposable.

Let $H = \langle \sigma_x, \tau_x : x \in X \rangle$. Let $a, b \subset Y$, $x \in f^{-1}(a)$ and $y \in f^{-1}(b)$. There exists $h \in H$ such that $h(x) = y$. We shall prove that the restriction of h to $f^{-1}(a)$ gives a bijection from $f^{-1}(a)$ to $f^{-1}(b)$. We know that

$$h = \varphi_{x_1}^{\varepsilon_1} \cdots \varphi_{x_m}^{\varepsilon_m},$$

for some $x_1, \ldots, x_m \in X$ and $\varepsilon_1, \ldots, \varepsilon_m \in \{-1, 1\}$, where $\varphi_{x_i} \in \{\sigma_{x_i}, \tau_{x_i}\}$ for all $i = 1, \ldots, m$. Let

$$\varphi'_{f(x_i)} = \begin{cases} \sigma'_{f(x_i)} & \text{if } \varphi_{x_i} = \sigma_{x_i} \\ \tau'_{f(x_i)} & \text{if } \varphi_{x_i} = \tau_{x_i} \end{cases}$$

and

$$h' = (\varphi'_{f(x_1)})^{\varepsilon_1} \cdots (\varphi'_{f(x_m)})^{\varepsilon_m}.$$

Let $x' \in f^{-1}(a)$. Then

$$f(h(x')) = h'(f(x')) = h'(a) = h'(f(x)) = f(h(x)) = f(y) = b.$$

Hence the map $f^{-1}(a) \to f^{-1}(b) \colon x' \mapsto h(x')$ is well defined and it is easy to see that it is bijective. Therefore the result follows. □

Definition 16.2.5 A solution (X, r) to the YBE is said to be *simple* if $|X| > 1$ and for every surjective homomorphism $f \colon (X, r) \to (Y, s)$ of solutions either f is an isomorphism or $|Y| = 1$.

Example 16.2.6 Let p be a prime number. The solution $(\mathbb{Z}/(p), r)$ to the YBE, where $r(a, b) = (b + 1, a - 1)$ for all $a, b \in \mathbb{Z}/(p)$ is a simple solution. Note that it is indecomposable and a permutation solution. Thus it is a multipermutation solution of level 1.

Theorem 16.2.7 *Let (X, r) be a finite indecomposable involutive solution to the YBE. Assume that $|X| = p$ is a prime number. Then (X, r) is isomorphic to $(\mathbb{Z}/(p), r)$, where $r(x, y) = (y + 1, x - 1)$ for all $x, y \in \mathbb{Z}/(p)$.*

Proof By Proposition 16.2.3, $\mathcal{G} = \mathcal{G}(X, r)$ is transitive. Hence p is a divisor of $\mathcal{G}$. Since $\mathcal{G}$ is a subgroup of $\mathbb{S}_X$ and $|\mathbb{S}_X| = p!$, we have that $|\mathcal{G}| = pn$ for some positive integer n coprime with p. Let $x \in X$. Let H_x be the stabilizer of x in $\mathcal{G}$. By Theorem 9.5.2, $\mathcal{G}$ is solvable. Let A_p and $A_{p'}$ be the Sylow p-subgroup and the Hall p'-subgroup of the additive group of the skew brace $\mathcal{G}$. Let $t \in H_x$. Since $\mathcal{G} = A_p \oplus A_{p'}$, there exist unique elements $t_1 \in A_p$ and $t_2 \in A_{p'}$ such that $\sigma_x = t_1 + t_2$. Then

$$\sigma_x = \sigma_{t(x)} = \lambda_t(\sigma_x) = \lambda_t(t_1) + \lambda_t(t_2).$$

Hence $\lambda_t(t_i) = t_i$ for $i = 1, 2$. Suppose that $t_1 = 0$. For every $y \in X$, there exists $h \in \mathcal{G}$ such that $h(x) = y$. Now we have

$$\sigma_y = \sigma_{h(x)} = \lambda_h(\sigma_x) = \lambda_h(t_2) \in A_{p'}.$$

Hence $\mathcal{G} = \langle \sigma_y : y \in X \rangle \subseteq A_{p'}$, a contradiction. Thus $t_1 \neq 0$ and therefore $\lambda_t(a) = a$ for all $a \in A_p$ and all $t \in H_x$. Since $|H_x| = n = |A_{p'}|$, there exists $g \in \mathcal{G}$ such that $A_{p'} = gH_xg^{-1}$. Let $a \in A_p$ and $t \in H_x$. We have

$$\lambda_{gtg^{-1}}(a) = \lambda_g\lambda_t\lambda_{g^{-1}}(a) = \lambda_g\lambda_{g^{-1}}(a) = a.$$

Hence $\lambda_b(a) = a$ for all $a \in A_p$ and $b \in A_{p'}$. Now

$$ab = \lambda_a(b)\lambda^{-1}_{\lambda_a(b)}(a) = \lambda_a(b)a,$$

and thus

$$aba^{-1} = \lambda_a(b) \in A_{p'}$$

for all $a \in A_p$ and all $b \in A_{p'}$. Since $\mathcal{G} = A_pA_{p'}$, we have that $A_{p'}$ is a normal subgroup of $\mathcal{G}$. Hence $H_x = A_{p'}$ for all $x \in X$. Therefore $A_{p'} = \bigcap_{x \in X} H_x = \{\mathrm{id}\}$. Hence $\mathcal{G}$ is the trivial skew brace of order p. Since for $x, y \in X$ there exists $h \in \mathcal{G}$ such that $y = h(x)$, we have that

$$\sigma_y = \sigma_{h(x)} = \lambda_h(\sigma_x) = \sigma_x.$$

Therefore σ_x should be a cycle of length p and the result follows. □

In [57], Etingof, Soloviev, and Guralnick classify indecomposable solutions to the YBE of prime cardinality.

Theorem 16.2.8 (Etingof–Soloviev–Guralnick) *Let (X, r) be a finite indecomposable solution to the YBE. Assume that $|X| = p$ is a prime number. Then (X, r) is isomorphic to one of the following solutions:*

(1) $(\mathbb{Z}/(p), r_{a,b})$, *where* $r_{a,b}(x, y) = (y + a, x + b)$ *and* $(0, 0) \neq (a, b) \in (\mathbb{Z}/(p))^2$.

(2) $(\mathbb{Z}/(p), r_{q_1,q_2,h})$, *where*

$$r_{q_1,q_2,h}(x, y) = (q_1^{-1}y + (1 - q_2q_1^{-1})x + h, q_2x - q_1h),$$

$h \in \mathbb{Z}/(p)$, $q_1, q_2 \in \mathbb{Z}/(p) \setminus \{0\}$ *and* $q_1 \neq q_2$.

(3) $(\mathbb{Z}/(p), r'_{q_1,q_2,h})$, *where*

$$r'_{q_1,q_2,h}(x, y) = (q_1^{-1}y + h, q_2x + (1 - q_2q_1^{-1})y - q_2h),$$

$h \in \mathbb{Z}/(p)$, $q_1, q_2 \in \mathbb{Z}/(p) \setminus \{0\}$ *and* $q_1 \neq q_2$. □

Proposition 16.2.9 *Let* (X, r) *be a simple solution to the YBE. If* $|X| > 2$, *then* (X, r) *is indecomposable.*

Proof Let us assume that (X, r) is decomposable. Decompose $X = X_1 \cup X_2$ for non-empty disjoint subsets X_1 and X_2 of X such that $r(X_i \times X_i) \subseteq X_i \times X_i$ for $i \in \{1, 2\}$ and $(X_1, r|_{X_1\times X_1})$ and $(X_2, r|_{X_2\times X_2})$ are solutions to the YBE.

Let $Y = \{1, 2\}$ and $s\colon Y \times Y \to Y \times Y$, $s(x, y) = (y, x)$. Since $X = X_1 \cup X_2$ is a decomposition, it follows that $r(X_i \times X_j) \subseteq X_j \times X_i$ for all $i, j \in \{1, 2\}$. Let $f\colon X \to Y$, $f(x) = i$ if $x \in X_i$. Since f is then a surjective homomorphism of solutions and f is not an isomorphism (because $|X| > 2$), the simplicity of (X, r) implies that $|Y| = 1$, a contradiction. □

Proposition 16.2.10 *Let* (X, r) *be a simple solution to the YBE. If* $|X|$ *is not a prime number, then* (X, r) *is irretractable.*

Proof Let us assume that (X, r) is retractable. Let $(X, r) \to \operatorname{Ret}(X, r)$, $x \mapsto [x]$, be the canonical map. Since it is a surjective homomorphism of solutions and (X, r) is retractable, the simplicity of (X, r) implies that $|\operatorname{Ret}(X, r)| = 1$. Therefore (X, r) is a permutation solution, say $r(x, y) = (\sigma(y), \tau(x))$ for some permutations $\sigma, \tau\colon X \to X$ such that $\sigma\tau = \tau\sigma$. Since $|X| > 2$, the solution (X, r) is indecomposable by Proposition 16.2.9. This implies that $G = \langle\sigma, \tau\rangle$ is an abelian transitive subgroup of $\mathbb{S}_X$. Since G is finitely generated, there exists a subgroup H of prime index in G. Let $Y = \{H(x) : x \in X\}$. Let $s\colon Y \times Y \to Y \times Y$ be the map defined by $s(H(x), H(y)) = (H(\sigma(y)), H(\tau(x)))$ for all $x, y \in X$. It is easy to check that (Y, s) is a solution to the YBE. Then the map $f\colon (X, r) \to (Y, s)$, defined by $f(x) = H(x)$ for all $x \in X$, is a surjective homomorphism of solutions. Note that for every $x \in X$ the stabilizer of x in G is $\{\mathrm{id}\}$. Now we have $|X| > |Y| = (G : H) > 1$, a contradiction because (X, r) is simple. Therefore (X, r) is irretractable and the result follows. □

16.3 Primitive Involutive Solutions

Definition 16.3.1 An involutive solution (X, r) to the YBE is called *primitive* if $\mathcal{G}(X, r)$ is a primitive subgroup of $\mathbb{S}_X$.

Theorem 16.3.2 *Let (X, r) be a finite primitive involutive solution to the YBE with $|X| > 1$. Then $|X|$ is prime. Furthermore, $\sigma_x = \sigma_y$, for all $x, y \in X$, and σ_x is a cycle of length $|X|$.*

Proof Let (X, r) be a finite primitive involutive solution to the YBE. Suppose that $|X| > 1$ and that $|X|$ is not prime. Let $\mathcal{G} = \mathcal{G}(X, r)$. Then, by Proposition 16.2.3, (X, r) is indecomposable and by Theorem 16.1.4 $|X| = p^m$, for some prime p and integer $m > 1$. Furthermore, by Theorem 9.5.2, $\mathcal{G}$ is solvable and, by Theorem 16.1.4, $\mathcal{G} = NA$ is the inner semidirect product of N by A, where $N \cong (\mathbb{Z}/(p))^m$ is the unique minimal normal subgroup of $\mathcal{G}$ and A is the stabilizer of some $x \in X$ and a maximal subgroup of $\mathcal{G}$. Moreover, N is self-centralizing in $\mathcal{G}(X, r)$.

Recall that $\mathcal{G}$ has a natural structure of skew brace of abelian type. By Proposition 16.2.4, (X, r) is simple. By Proposition 16.2.10, (X, r) is irretractable. Hence the multiplicative group of the skew brace $\mathcal{G}$ acts primitively on $\widehat{X} = \{\sigma_x : x \in X\}$, by the lambda map. Since A is the stabilizer of some $x \in X$, we have that $\lambda_g(\sigma_x) = \sigma_{g(x)} = \sigma_x$, for all $g \in A$. Thus A also is the stabilizer of σ_x, for some $x \in X$.

Note that

$$\mathcal{G} = B_1 \oplus \cdots \oplus B_k,$$

where B_i is the Sylow p_i-subgroup of $(\mathcal{G}, +)$ (here $p_1, \ldots, p_k$ are the prime divisors of $|\mathcal{G}|$). Furthermore, $\mathcal{G} = B_1 \cdots B_k$ and $B_i B_j = B_j B_i$, for all i, j.

Without loss of generality, we may assume $p = p_1$. Since the multiplicative group of $\mathcal{G}$ is solvable, and $B_2 \cdots B_k = B_2 \oplus \cdots \oplus B_k$ is a Hall p_1'-subgroup of the multiplicative group of $\mathcal{G}$, there exists $g \in \mathcal{G}$ such that $B_2 \cdots B_k \subseteq gAg^{-1}$. Since gAg^{-1} is the stabilizer of $\sigma_{g(x)}$, we may assume that $B_2 \oplus \cdots \oplus B_k \subseteq A$. Since N is a normal subgroup of $\mathcal{G}$, $N \subseteq B_1$.

Let $2 \leq i \leq k$ and $b \in B_i$. Since N is a normal subgroup of $\mathcal{G}$, for every $n \in N$, there exists $n' \in N$ such that $nb = bn'$. Hence, because $N \subseteq B_1$, we have that $\lambda_n(b) = b$. So $(\lambda_n)_{|B_i} = \mathrm{id}$, for every $i = 2, \ldots, k$ and every $n \in N$. Since $\lambda_n \in \mathrm{Aut}(\mathcal{G}, +)$, it follows that

$$(\lambda_n)_{|B_2 \oplus \cdots \oplus B_k} = \mathrm{id}\,. \tag{16.1}$$

Since $\mathcal{G}$ is the inner direct sum $\mathcal{G} = B_1 \oplus \cdots \oplus B_k$, there exist unique $x_i \in B_i$, for $i = 1, \ldots, k$, such that $\sigma_x = x_1 + \cdots + x_k$. As $\lambda_a(\sigma_x) = \sigma_x$, for all $a \in A$ and because each B_i is a left ideal, we get that $\lambda_a(x_i) = x_i$ for any $a \in A$. Hence, since $\mathcal{G} = NA$, using (16.1), we get that

$$\lambda_g(x_i) = x_i, \quad \text{for any } i \geq 2 \text{ and any } g \in \mathcal{G}. \tag{16.2}$$

Note that, because of (16.1), for every $n \in N$,

$$\lambda_n(\sigma_x) = \lambda_n(x_1) + \cdots + \lambda_n(x_k) = \lambda_n(x_1) + x_2 + \cdots + x_k.$$

Since $\mathcal{G}$ acts transitively on $\widehat{X}$, $\mathcal{G} = NA$ and A stabilizes σ_x,

$$\begin{aligned}\widehat{X} = \{\lambda_g(\sigma_x) : g \in \mathcal{G}\} &= \{\lambda_n(\sigma_x) : n \in N\} \\ &= \{\lambda_n(x_1) + x_2 + \cdots + x_k : n \in N\}.\end{aligned}$$

For every subset S of $\mathcal{G}$, we denote by $\langle S\rangle_{\cdot}$ the subgroup of $(\mathcal{G}, \cdot)$ generated by S, and $\langle S\rangle_+$ will denote the subgroup of $(\mathcal{G}, +)$ generated by S. Note that

$$\mathcal{G} = \langle\widehat{X}\rangle_{\cdot} = \langle\widehat{X}\rangle_+ = \langle X_1\rangle_+ + \cdots + \langle X_k\rangle_+,$$

where $X_1 = \{\lambda_n(x_1) : n \in N\}$, and $X_i = \{x_i\}$, for $i = 2, \ldots, k$. Thus, $B_j = \langle X_j\rangle_+$, for every j, and

$$\mathcal{G} = \langle X_1\rangle_+ \oplus \langle x_2\rangle_+ \oplus \cdots \oplus \langle x_k\rangle_+ = B_1 \oplus B_2 \cdots \oplus B_k. \tag{16.3}$$

Furthermore, from (16.2) and (16.3) we get that

$$(\lambda_g)_{|B_2\oplus\cdots\oplus B_k} = \mathrm{id} \quad \text{for any} g \in \mathcal{G}. \tag{16.4}$$

Note that, for $b_1 \in B_1$ and $b \in B_i$ with $i \geq 2$, there exists $b' \in B_1$ such that

$$b^{-1}b_1b = b^{-1}\lambda_{b_1}(b)b' = b^{-1}bb' = b' \in B_1.$$

Hence, we have shown that B_1 is a normal subgroup of $(\mathcal{G}, \cdot)$. Let $Z(B_1)$ be the center of $(B_1, \cdot)$. As $B_1 \supseteq N$ is a non-trivial p-group, we have that $Z(B_1)$ is a normal non-trivial subgroup of $(\mathcal{G}, \cdot)$. Since N is the unique minimal normal subgroup of $(\mathcal{G}, \cdot)$, We have that $N \subseteq Z(B_1)$. Since N is self-centralizing in $(\mathcal{G}, \cdot)$, we have that $N = Z(B_1) = B_1$.

Thus, $N = B_1$ is an ideal and $A = B_2 + \cdots + B_k$ is a left ideal of the left brace $\mathcal{G}$. Note that by (16.4), $A = B_2 \oplus \cdots \oplus B_k$ is a trivial brace. Thus, using (16.4) again, we have

$$\sigma_x = x_1 + \cdots + x_k = x_1 + \lambda_{x_1}(x_2 + \cdots + x_k) = x_1(x_2 + \cdots + x_k) = x_1x_2\cdots x_k,$$

and $x_1 \in N$ and $x_2 \cdots x_k \in A$. Since A is the stabilizer of σ_x in $\mathcal{G}$, for every $a \in A$, we have

$$\begin{aligned}\sigma_x = \lambda_a(\sigma_x) &= a\sigma_x - a = ax_1x_2\cdots x_k - a \\ &= ax_1a^{-1}ax_2\cdots x_k - a = ax_1a^{-1} + \lambda_{ax_1a^{-1}}(ax_2\cdots x_k) - a \\ &= ax_1a^{-1} + ax_2\cdots x_k - a \quad \text{(by (16.4))}\end{aligned}$$

$$\begin{aligned}
&= ax_1a^{-1} + x_2 \cdots x_k \quad \text{(since } A \text{ is a trivial brace)}\\
&= ax_1a^{-1} + \lambda_{ax_1a^{-1}}(x_2 \cdots x_k) \quad \text{(by (16.4))}\\
&= ax_1a^{-1}x_2 \cdots x_n.
\end{aligned}$$

Hence $ax_1a^{-1} = x_1$, for all $a \in A$. Since $x_1 \in N$, we get that $\langle x_1 \rangle$ is a normal subgroup of $\mathcal{G}$, and thus $N = \langle x_1 \rangle$. But N is not cyclic, a contradiction. Therefore $|X|$ is prime.

The second part of the result follows by Theorem 16.2.7. □

16.4 Simple Involutive Solutions

In this section, we give some methods to construct finite simple involutive solutions to the YBE. First we give some properties of the permutation group of such a solution.

Proposition 16.4.1 *Let (X, r) be a finite simple involutive solution to the YBE such that $|X| > 1$ is not prime. If I is a minimal nonzero ideal of $\mathcal{G} = \mathcal{G}(X, r)$, then $\mathcal{G}/I$ is a trivial cyclic skew brace, and $I = \mathcal{G}^2 = \mathcal{G} * \mathcal{G} = \langle \sigma_x - \sigma_y : x, y \in X \rangle_+$. Furthermore* $\operatorname{Soc}(\mathcal{G}) = \{0\}$.

Proof By Propositions 16.2.9 and 16.2.10, (X, r) is indecomposable and irretractable.

Let I be a minimal nonzero ideal of $\mathcal{G}$. Let $\pi \colon \mathcal{G} \to \mathcal{G}/I$ be the natural map. Let $g \in I$ be a nonzero element. We have that $\lambda_g(\sigma_x) = \sigma_{g(x)}$, for all $x \in X$. Hence

$$\pi(\sigma_{g(x)}) = \pi(\lambda_g(\sigma_x)) = \lambda_{\pi(g)}(\pi(\sigma_x)) = \lambda_0(\pi(\sigma_x)) = \pi(\sigma_x).$$

Thus the restriction $\pi|_{\sigma(X)}$ of π induces an epimorphism $x \mapsto \pi(\sigma_x)$ of solutions from (X, r) to $(\pi(\sigma(X)), s)$, where

$$s(\pi(\sigma_x), \pi(\sigma_y)) = (\pi(\sigma_{\sigma_x(y)}), \pi(\sigma_{\sigma^{-1}_{\sigma_x(y)}(x)})).$$

Since $g \neq 0$, there exists $x \in X$ such that $x \neq g(x)$. Since $\pi(\sigma_x) = \pi(\sigma_{g(x)})$ and (X, r) is simple, we have that $\pi(\sigma_x) = \pi(\sigma_y)$, for all $x, y \in X$. Hence

$$\langle \sigma_x - \sigma_y : x, y \in X \rangle_+ \subseteq I.$$

Note that

$$\begin{aligned}
\mathcal{G}^2 &= \langle g * h : g, h \in \mathcal{G} \rangle_+\\
&= \langle \lambda_g(h) - h : g, h \in \mathcal{G} \rangle_+\\
&= \langle \lambda_g(\sigma_x) - \sigma_x : g \in \mathcal{G},\ x \in X \rangle_+ \quad (\text{since } \mathcal{G} = \langle \sigma_x : x \in X \rangle_+)\\
&= \langle \sigma_{g(x)} - \sigma_x : g \in \mathcal{G},\ x \in X \rangle_+
\end{aligned}$$

$$= \langle \sigma_y - \sigma_x : x, y \in X \rangle_+, \quad \text{(since } (X, r) \text{ is indecomposable).}$$

By the minimality of I we get $I = \mathcal{G}^2 = \langle \sigma_x - \sigma_y : x, y \in X \rangle_+$. Now it is easy to see that if $x \in X$, $\mathcal{G}/I = \langle \pi(\sigma_x) \rangle_+ = \langle \pi(\sigma_x) \rangle$ is the trivial skew brace of the cyclic group $\langle \pi(\sigma_x) \rangle$. Finally let $g \in \mathrm{Soc}(\mathcal{G})$. We have that

$$\sigma_{g(x)} = \lambda_g(\sigma_x) = \sigma_x,$$

for all $x \in X$. Since (X, r) is irretractable, we get that $g(x) = x$ for all $x \in X$. This proves that $\mathrm{Soc}(\mathcal{G}) = \{0\}$, and the result follows. □

Proposition 16.4.2 *Let A be an abelian additive group. Let $(j_a)_{a \in A}$ be a family of elements of A such that $j_a = j_{-a}$ for all $a \in A$. For every $(a_1, a_2) \in A^2$, let $\sigma_{(a_1,a_2)} \colon A^2 \to A^2$ be the map defined by*

$$\sigma_{(a_1,a_2)}(c_1, c_2) = (c_1 + a_2, c_2 - j_{c_1+a_2-a_1})$$

for all $(c_1, c_2) \in A^2$. Let $r \colon A^2 \times A^2 \to A^2 \times A^2$ be the map defined by

$$r((a_1, a_2), (c_1, c_2)) = (\sigma_{(a_1,a_2)}(c_1, c_2), \sigma^{-1}_{\sigma_{(a_1,a_2)}(c_1,c_2)}(a_1, a_2))$$

for all $(a_1, a_2)(c_1, c_2) \in A^2$. Then (A^2, r) is an involutive solution to the YBE.

Proof Note that $\sigma_{(a_1,a_2)}$ is bijective and

$$\sigma^{-1}_{(a_1,a_2)}(c_1, c_2) = (c_1 - a_2, c_2 + j_{c_1-a_1})$$

for all $(a_1, a_2), (c_1, c_2) \in A^2$. By Theorem 4.2.9, (A^2, r) is an involutive solution if and only if

$$\sigma_{(a_1,a_2)}\sigma_{\sigma^{-1}_{(a_1,a_2)}(c_1,c_2)} = \sigma_{(c_1,c_2)}\sigma_{\sigma^{-1}_{(c_1,c_2)}(a_1,a_2)}$$

for all $(a_1, a_2), (c_1, c_2) \in A^2$ and the map

$$T \colon A^2 \to A^2, \quad T(a_1, a_2) = \sigma^{-1}_{(a_1,a_2)}(a_1, a_2)$$

is bijective. Note that

$$T(a_1, a_2) = (a_1 - a_2, a_2 + j_0)$$

for all $(a_1, a_2) \in A^2$. Hence T is bijective and

$$T^{-1}(a_1, a_2) = (a_1 + a_2 - j_0, a_2 - j_0)$$

for all $(a_1, a_2) \in A^2$.

Let $(a_1, a_2), (c_1, c_2), (x, y) \in A^2$. Then

$$
\begin{aligned}
&\sigma_{(a_1,a_2)}\sigma_{\sigma^{-1}_{(a_1,a_2)}(c_1,c_2)}(x,y)\\
&= \sigma_{(a_1,a_2)}\sigma_{(c_1-a_2,c_2+j_{c_1-a_1})}(x,y)\\
&= \sigma_{(a_1,a_2)}(x+c_2+j_{c_1-a_1}, y-j_{x+c_2+j_{c_1-a_1}-c_1+a_2})\\
&= (x+c_2+j_{c_1-a_1}+a_2, y-j_{x+c_2+j_{c_1-a_1}-c_1+a_2}-j_{x+c_2+j_{c_1-a_1}+a_2-a_1})\\
&= \sigma_{(c_1,c_2)}(x+a_2+j_{c_1-a_1}, y-j_{x+c_2+j_{c_1-a_1}+a_2-a_1})\\
&= \sigma_{(c_1,c_2)}\sigma_{(a_1-c_2,a_2+j_{a_1-c_1})}(x,y)\\
&= \sigma_{(c_1,c_2)}\sigma_{\sigma^{-1}_{(c_1,c_2)}(a_1,a_2)}(x,y).
\end{aligned}
$$

Hence the result follows. □

Proposition 16.4.3 *Let (A^2, r) be the involutive solution to the YBE constructed in Proposition 16.4.2. Then (A^2, r) is indecomposable if and only if $A = \langle j_a : a \in A\rangle$.*

Proof Let $W = \langle j_a : a \in A\rangle$. We shall prove that the orbit of $(0, 0)$ under the action of $\mathcal{G}(A^2, r)$ on A^2 is $A \times W$. First note that $\sigma_{(a_1,a_2)}(a, w) = (a + a_2, w - j_{a+a_2-a_1}) \in A \times W$ for all $a, a_1, a_2 \in A$ and all $w \in W$. Hence $\mathcal{G}(A^2, r)(A \times W) = A \times W$. Let $a \in A$ and $w \in W$. There exist $a_1, \dots, a_n \in A$ and $\varepsilon_1, \dots, \varepsilon_n \in \{-1, 1\}$ such that $w = \sum_{i=1}^{n} \varepsilon_i j_{a-a_i}$. Now we have that

$$
\begin{aligned}
\sigma^{\varepsilon_n}_{(a_n,0)}\cdots\sigma^{\varepsilon_1}_{(a_1,0)}\sigma^{-1}_{(0,0)}\sigma_{(0,a)}(0,0) &= \sigma^{\varepsilon_n}_{(a_n,0)}\cdots\sigma^{\varepsilon_1}_{(a_1,0)}\sigma^{-1}_{(0,0)}(a,-j_a)\\
&= \sigma^{\varepsilon_n}_{(a_n,0)}\cdots\sigma^{\varepsilon_1}_{(a_1,0)}(a,0)\\
&= \sigma^{\varepsilon_n}_{(a_n,0)}\cdots\sigma^{\varepsilon_2}_{(a_2,0)}(a,\varepsilon_1 j_{a-a_1})\\
&= \sigma^{\varepsilon_n}_{(a_m,0)}(a,\varepsilon_1 j_{a-a_1}+\cdots+\varepsilon_{n-1}j_{a-a_{n-1}})\\
&= (a,\varepsilon_1 j_{a-a_1}+\cdots+\varepsilon_n j_{a-a_n}) = (a,w).
\end{aligned}
$$

Hence the orbit of $(0, 0)$ under the action of $\mathcal{G}(A^2, r)$ is $A \times W$, and the result follows. □

Proposition 16.4.4 *Let (A^2, r) be the involutive solution to the YBE constructed in Proposition 16.4.2. Then (A^2, r) is irretractable if and only if for every nonzero $a \in A$ there exits $c \in A$ such that $j_c \neq j_{c+a}$.*

Proof Let $(a_1, a_2), (c_1, c_2) \in A^2$ be such that $\sigma_{(a_1,a_1)} = \sigma_{(c_1,c_2)}$. Hence

$$
(x+a_2, y-j_{x+a_2-a_1}) = \sigma_{(a_1,a_2)}(x,y) = \sigma_{(c_1,c_2)}(x,y) = (x+c_2, y-j_{x+c_2-c_1})
$$

for all $x, y \in A$. Thus $a_2 = c_2$ and $j_{x+a_2-a_1} = j_{x+a_2-c_1}$ for all $x \in A$. Therefore the result follows. □

Consider the solution (A^2, r) to the YBE constructed in Proposition 16.4.2. For a nonzero element $a \in A$, let

$$
V_{a,1} = \langle j_c - j_{c+a} : c \in A\rangle,
$$

$$V_{a,i} = V_{a,i-1} + \langle j_c - j_{c+v} : c \in A,\ v \in V_{a,i-1} \rangle \qquad \text{for } i > 1.$$

Let $V_a = \sum_{i=1}^{\infty} V_{a,i}$. Then $V_a = \bigcup_{i=1}^{\infty} V_{a,i}$.

Theorem 16.4.5 *Let (A^2, r) be the involutive solution to the YBE constructed in Proposition 16.4.2. Suppose that A is finite and nonzero. Then (A^2, r) is simple if and only if $V_a = A$ for every nonzero element $a \in A$.*

Proof Suppose that (A^2, r) is simple. Let $a \in A \setminus \{0\}$ and $\pi\colon A \to A/V_a$ be the natural map. Note that

$$\begin{aligned}
&\sigma_{(a_1+v_1,a_2+v_2)}(c_1 + v_3, c_2 + v_4)\\
&= (c_1 + a_2 + v_3 + v_2, c_2 + v_4 - j_{c_1+a_2+v_3+v_2-a_1-v_1})\\
&= (c_1 + a_2 + v_3 + v_2, c_2 - j_{c_1+a_2-a_1} + v_4 + j_{c_1+a_2-a_1} - j_{c_1+a_2+v_3+v_2-a_1-v_1}),
\end{aligned}$$

for all $a_1, a_2, c_1, c_2 \in A$ and $v_1, v_2, v_3, v_4 \in V_a$. Since

$$j_{c_1+a_2-a_1} - j_{c_1+a_2+v_3+v_2-a_1-v_1} \in V_a,$$

the map

$$\begin{aligned}
&\sigma'_{(\pi(a_1),\pi(a_2))}\colon (A/V_a)^2 \to (A/V_a)^2,\\
&\sigma'_{(\pi(a_1),\pi(a_2))}(\pi(c_1), \pi(c_2)) = (\pi(c_1) + \pi(a_2), \pi(c_2) - \pi(j_{c_1+a_2-a_1}))
\end{aligned}$$

is well defined. Furthermore, if

$$r'\colon (A/V_a)^2 \times (A/V_a)^2 \to (A/V_a)^2 \times (A/V_a)^2$$

is given by

$$\begin{aligned}
&r'((\pi(a_1), \pi(a_2)), (\pi(c_1), \pi(c_2)))\\
&= \left(\sigma'_{(\pi(a_1),\pi(a_2))}(\pi(c_1), \pi(c_2)), \left(\sigma'_{\sigma'_{(\pi(a_1),\pi(a_2))}(\pi(c_1),\pi(c_2))}\right)^{-1}(\pi(a_1), \pi(a_2))\right),
\end{aligned}$$

then $((A/V_a)^2, r')$ is a solution to the YBE and the map $f\colon A^2 \to (A/V_a)^2$, defined by $f(a_1, a_2) = (\pi(a_1), \pi(a_2))$, is an epimorphism of solutions. Since (A^2, r) is simple, by Proposition 16.2.10 (A^2, r) is irretractable. By Proposition 16.4.4, $V_{a,1} \neq \{0\}$. Hence, since (A^2, r) is simple, we have that $V_a = A$.

Conversely, assume that $V_a = A$ for every nonzero element $a \in A$. In particular, $\langle j_a : a \in A \rangle = A$, and thus (A^2, r) is indecomposable by Proposition 16.4.3. Furthermore $V_{a,1} = \mathrm{gr}(j_c - j_{c+a} : c \in A) \neq \{0\}$ for every nonzero element $a \in A$, and thus (A^2, r) is irretractable by Proposition 16.4.4. Let

$$f\colon (A^2, r) \to (Y, s)$$

be an epimorphism of solutions that is not an isomorphism. Since (A^2, r) is indecomposable, by Proposition 16.2.4, we have that (Y, s) also is indecomposable and $|f^{-1}(y)| = |f^{-1}(y')|$ for all $y, y' \in Y$. We write

$$s(y_1, y_2) = (\sigma'_{y_1}(y_2), (\sigma'_{\sigma'_{y_1}(y_2)})^{-1}(y_1))$$

for all $y_1, y_2 \in Y$.

We will use the following consequence of the definition of a homomorphism of solutions:

$$f(\sigma_{(u,v)}(x, y)) = f(\sigma_{(u',v')}(x', y'))$$

whenever $f(u, v) = f(u', v')$ and $f(x, y) = f(x', y')$.

Note also that if $f(u, v) = f(u', v')$ and $f(x, y) = f(x', y')$, then

$$\begin{aligned}
\sigma'_{f(u,v)} f(\sigma^{-1}_{(u,v)}(x, y)) &= f(\sigma_{(u,v)}(\sigma^{-1}_{(u,v)}(x, y)) \\
&= f(x, y) \\
&= f(x', y') \\
&= f(\sigma_{(u',v')}(\sigma^{-1}_{(u',v')}(x', y')) \\
&= \sigma'_{f(u',v')} f(\sigma^{-1}_{(u',v')}(x', y')),
\end{aligned}$$

and since $\sigma'_{f(u,v)} = \sigma'_{f(u',v')}$ is bijective, we get that

$$f(\sigma^{-1}_{(u,v)}(x, y)) = f(\sigma^{-1}_{(u',v')}(x', y')). \tag{16.5}$$

Suppose first that there exist $a_1, a_2, c \in A$ such that $f(a_1, c) = f(a_2, c)$ and $a_1 \neq a_2$. Let $a = a_2 - a_1$. We have that

$$f(u + c, v - j_{u+c-a_1}) = f(\sigma_{(a_1,c)}(u, v)) = f(\sigma_{(a_2,c)}(u, v)) = f(u + c, v - j_{u+c-a_2}),$$

for all $u, v \in A$. Thus, it also follows that

$$f(u, v) = f(u, v + j_{u-a_1} - j_{u-a_2}) = f(u, v + z(j_{u-a_1} - j_{u-a_2})),$$

for all $u, v \in A$ and $z \in \mathbb{Z}$.

Hence, we also get

$$\begin{aligned}
f(x + v, y - j_{x+v-u}) &= f(\sigma_{(u,v)}(x, y)) \\
&= f(\sigma_{(u,v+z(j_{u-a_1}-j_{u-a_2}))}(x, y - t(j_{x-a_1} - j_{x-a_2}))) \\
&= f(x + v + z(j_{u-a_1} - j_{u-a_2}), y - t(j_{x-a_1} - j_{x-a_2}) - j_{x+v+z(j_{u-a_1}-j_{u-a_2})-u})
\end{aligned}$$

for all $u, v, x, y \in A$ and $z, t \in \mathbb{Z}$. In particular,

$$f(x, y) = f(x + z(j_{u-a_1} - j_{u-a_2}), \\ y + j_{x-u} - t(j_{x-v-a_1} - j_{x-v-a_2}) - j_{x+z(j_{u-a_1}-j_{u-a_2})-u})$$

for all $x, y, u, v \in A$ and $z, t \in \mathbb{Z}$. Hence, for $z = 0$ and $v = x - w$,

$$f(x, y) = f(x, y - t(j_{w-a_1} - j_{w-a_2})),$$

for all $x, y, w \in A$ and $t \in \mathbb{Z}$. Thus

$$f(x, y) = f(x, y + v_1)$$

for all $x, y \in A$ and $v_1 \in V_{a,1}$. Suppose that $f(x, y) = f(x, y + v_n)$, for all $x, y \in A$ and all $v_n \in V_{a,n}$. Then we have that

$$\begin{aligned} f(u + y, v - j_{u+y-x}) &= f(\sigma_{(x,y)}(u, v)) \\ &= f(\sigma_{(x,y+v_n)}(u, v + v'_n)) \\ &= f(u + y + v_n, v + v'_n - j_{u+y+v_n-x}) \end{aligned}$$

for all $x, y, u, v \in A$ and $v_n, v'_n \in V_{a,n}$. This implies that

$$f(u, v) = f(u + v_n, v + v'_n + j_w - j_{v_n+w}),$$

for all $u, v, w \in A$ and $v_n, v'_n \in V_{a,n}$. Hence we also have

$$\begin{aligned} &f(x + v, y - j_{x+v-u}) \\ &\quad = f(\sigma_{(u,v)}(x, y)) \\ &\quad = f(\sigma_{(u+v_n, v+v'_n+j_{w_1}-j_{v_n+w_1})}(x + z_n, y + z'_n + j_{w_2} - j_{z_n+w_2})) \\ &\quad = f(x + z_n + v + v'_n + j_{w_1} - j_{v_n+w_1}, \\ &\qquad y + z'_n + j_{w_2} - j_{z_n+w_2} - j_{x+z_n+v+v'_n+j_{w_1}-j_{v_n+w_1}-u-v_n}) \end{aligned}$$

for all $u, v, x, y \in A$ and $v_n, v'_n z_n, z'_n \in V_{a,n}$. For $v_n = 0$ and $z_n = -v'_n$, this leads to

$$f(x + v, y - j_{x+v-u}) = f(x + v, y + z'_n + j_{w_2} - j_{z_n+w_2} - j_{x+v-u})$$

for all $x, y, u, v, w_2 \in A$ and $z_n, z'_n \in V_{a,n}$. Thus

$$f(x, y) = f(x, y + z'_n + j_{w_2} - j_{z_n+w_2})$$

for all $x, y, w_2 \in A$ and $z_n, z'_n \in V_{a,n}$. Hence

$$f(x, y) = f(x, y + v_{n+1})$$

for all $x, y \in A$ and $v_{n+1} \in V_{a,n+1}$. By induction, we get that

$$f(x, y) = f(x, y + v)$$

for all $x, y \in A$ and $v \in V_a = A$. Therefore

$$f(x, 0) = f(x, y)$$

for all $x, y \in A$. This implies that

$$f(0, 0) = f(0, -j_0) = f(\sigma_{(0,0)}(0, 0)) = f(\sigma_{(0,x)}(0, y)) = f(x, y - j_x) = f(x, y)$$

for all $x, y \in A$. Therefore $|Y| = 1$ in this case, as desired. Hence we may assume that $f(x, y) = f(u, v)$ and $y = v$ imply $x = u$. We will show that this case leads to a contradiction.

Let $f(0, 0) = y_0$. Let $f^{-1}(y_0) = \{(a_i, c_i) : i = 1, \ldots, m\}$, where $m > 1$ and $c_i \neq c_j$ for all $i \neq j$. We may assume that $(0, 0) = (a_1, c_1)$. Using (16.5), we have that

$$\begin{aligned} y_0 &= f(0, 0) \\ &= f(\sigma_{(a_i,c_i)}\sigma^{-1}_{(a_u,c_u)}(0, 0)) \\ &= f(\sigma_{(a_i,c_i)}(-c_u, j_{a_u})) \\ &= f(c_i - c_u, j_{a_u} - j_{c_i-c_u-a_i}) \end{aligned}$$

for all $i, u \in \{1, \ldots, m\}$. Thus, for every $u = 1, \ldots, m$,

$$f^{-1}(y_0) = \{(a_i, c_i) : i = 1, \ldots, m\} = \{(c_i - c_u, j_{a_u} - j_{c_i-c_u-a_i}) : i = 1, \ldots, m\}.$$

Let $H = \{a_i : i = 1, \ldots, m\}$. Hence,

$$H = \{a_i : i = 1, \ldots, m\} = \{c_i : i = 1, \ldots, m\} = \{j_0 - j_{c_i-a_i} : i = 1, \ldots, m\} \tag{16.6}$$

is a subgroup of A. Therefore, for every $a, c \in H$,

$$j_a = j_c \text{ implies that } a = c.$$

In particular, $a = -a$ for all $a \in H$. Now there exists $l \in \{1, \ldots, m\}$ such that $c_l = a_2$, and using (16.5), we have

$$\begin{aligned} y_0 &= f(0, 0) \\ &= f(\sigma^{-1}_{(0,0)}\sigma_{(a_l,c_l)}(a_2, c_2)) \\ &= f(\sigma^{-1}_{(0,0)}(0, c_2 - j_{a_l})) \\ &= f(0, c_2 + j_0 - j_{a_l}) \end{aligned}$$

and

$$\begin{aligned} y_0 &= f(0,0) \\ &= f(\sigma_{(0,0)}\sigma^{-1}_{(a_l,c_l)}(a_2,c_2)) \\ &= f(\sigma_{(0,0)}(0,c_2+j_{a_2-a_l})) \\ &= f(0,c_2+j_{a_2-a_l}-j_0). \end{aligned}$$

Hence $c_2 = j_0 - j_{a_2-a_l} = j_0 - j_{a_l}$, and this implies that $a_2 - a_l = a_l$, thus $a_2 = 0 = a_1$, a contradiction. Therefore the result follows. □

16.5 Exercises

16.5.1 Let (X, r) be a finite solution to the YBE. Let X_1, X_2 be disjoint non-empty subsets of X such that $X = X_1 \cup X_2$. Suppose that

$$r(X_1 \times X_1) \subseteq X_1 \times X_1 \quad \text{and} \quad r(X_2 \times X_2) \subseteq X_2 \times X_2.$$

Prove that the restrictions $(X_1, r|_{X_1\times X_1})$ and $(X_2, r|_{X_2\times X_2})$ are solutions to the YBE and $r(X_1 \times X_2) \subseteq X_2 \times X_1$ and $r(X_2 \times X_1) \subseteq X_1 \times X_2$.

16.5.2 Let (X, r) be a finite involutive permutation solution. Thus there exists $\sigma \in \mathbf{S}_X$ such that

$$r(x, y) = (\sigma(y), \sigma^{-1}(x))$$

for all $x, y \in X$. Prove that (X, r) is indecomposable if and only if σ is a cycle of length $|X|$.

16.5.3 Let A be an additive group. Prove that the group $\mathrm{Hol}(A)$ acts transitively on A and the stabilizer $a \in A$ is isomorphic to $\mathrm{Aut}(A)$.

16.6 Open Problems

Open problem 16.6.1 Construct indecomposable (involutive) solutions of small size.

Open problem 16.6.2 Let $n, m > 1$ be integers. Does it exist a simple involutive solution of the YBE of cardinality m^2n?

16.7 Notes

The results of Sect. 16.1 are classical.

Indecomposable solutions to the YBE were introduced by Etingof, Schedler, and Soloviev in [56]. They proved Theorem 16.2.7. The classification of indecomposable solutions of prime cardinality (Theorem 16.2.8) is due to Etingof, Soloviev, and Guralnick; see [57].

Proposition 16.2.4 for involutive solutions was proved by Castelli, Catino, and Pinto in [32].

Simple solutions were first considered by Andruskiewitsch and Graña in [5]. Propositions 16.2.9 and 16.2.10 were proved for involutive solutions in [43].

Theorem 16.3.2 was proved by Cedó, Jespers, and Okniński in [42].

The first examples of simple involutive solutions of non-prime cardinality were constructed in [43], where Proposition 16.4.1 is proved. Propositions 16.4.2, 16.4.3, and 16.4.4 and Theorem 16.4.5 were proved in [44].

In [43, Theorem 4.12], it is claimed that for integers $n, m > 1$ there exists a simple involutive solution of the YBE of cardinality m^2n, but the proof is incorrect [45] (see Open Problem 16.6.2).

Chapter 17
Involutive Solutions

Let (X, r) be an involutive solution to the YBE. We know that its permutation group $\mathcal{G}(X, r)$ has a natural structure of skew brace of abelian type. If (Y, r') is a solution isomorphic to (X, r), then $\mathcal{G}(Y, r') \cong \mathcal{G}(X, r)$ as skew braces. Thus the classification of the involutive solutions can be done in two steps:

(1) Classify all the skew braces of abelian type.
(2) For each skew brace B of abelian type, classify all the involutive solutions (X, r) to the YBE such that $\mathcal{G}(X, r) \cong B$.

In this chapter, we focus on the second step.

17.1 Extensions of Skew Braces of Abelian Type

In this section, we study some extensions of skew braces of abelian type and how to construct involutive solutions to the YBE using these extensions.

Proposition 17.1.1 *Let H be an additive abelian group and B be a skew brace of abelian type. Suppose that $\varphi\colon (B, \circ) \to \operatorname{Aut}(H, +)$ is an injective homomorphism and $\eta\colon (H, +) \to (B, +)$ is a surjective homomorphism such that they satisfy*

$$\eta(\varphi(g)(h)) = \lambda_g(\eta(h))$$

for all $g \in B$ and $h \in H$. Then, the multiplication in H defined by

$$h_1 \cdot h_2 := h_1 + \varphi(\eta(h_1))(h_2),$$

for $h_1, h_2 \in H$, defines a structure of skew brace on H such that η is a homomorphism of skew braces with $Soc(H) = \ker(\eta)$ and $H/Soc(H) \cong B$ as skew braces.

F. Cedó and L. Vendramin, *Groups, Radical Rings, and the Yang-Baxter Equation*, Progress in Mathematics 361, https://doi.org/10.1007/978-3-032-01524-2_17

Two of these structures, determined by φ, η and φ', η', respectively, are isomorphic if and only if there exists an $F \in \mathrm{Aut}(H, +)$ such that

$$\varphi'(\eta'(h)) = F^{-1}\varphi(\eta(F(h)))F$$

for all $h \in H$.

Conversely, suppose that G is a skew brace of abelian type. Then, the map $\varphi\colon (G/Soc(G), \circ) \to \mathrm{Aut}(G, +)$, induced by the map $\lambda\colon (G, \circ) \to \mathrm{Aut}(G, +)$, and the natural map $\eta\colon G \to G/Soc(G)$ satisfy the above properties.

Proof We shall see that $(H, \cdot)$ is a group. Let $a, b, c \in H$. We have

$$\begin{aligned} a \cdot (b \cdot c) &= a \cdot (b + \varphi(\eta(b))(c)) = a + \varphi(\eta(a))(b + \varphi(\eta(b))(c)) \\ &= a + \varphi(\eta(a))(b) + \varphi(\eta(a))(\varphi(\eta(b))(c)) \\ &= a + \varphi(\eta(a))(b) + \varphi(\eta(a) \circ \eta(b))(c) \\ &= a + \varphi(\eta(a))(b) + \varphi(\eta(a) + \lambda_{\eta(a)}(\eta(b)))(c) \\ &= a + \varphi(\eta(a))(b) + \varphi(\eta(a) + \eta(\varphi(\eta(a))(b)))(c) \\ &= (a + \varphi(\eta(a))(b)) + \varphi(\eta(a + \varphi(\eta(a))(b)))(c) \\ &= (a \cdot b) \cdot c. \end{aligned}$$

Note that,

$$\begin{aligned} 0 \cdot a &= 0 + \varphi(\eta(0))(a) = \varphi(0)(a) = \mathrm{id}(a) = a, \\ a \cdot 0 &= a + \varphi(\eta(a))(0) = a + 0 = a. \end{aligned}$$

Furthermore,

$$a \cdot (-\varphi(\eta(a)')(a)) = a + \varphi(\eta(a))(-\varphi(\eta(a)')(a)) = a - \varphi(0)(a) = a - a = 0$$

and

$$\begin{aligned} (-\varphi(\eta(a)')(a)) \cdot a &= -\varphi(\eta(a)')(a) + \varphi(\eta(-\varphi(\eta(a)')(a)))(a) \\ &= -\varphi(\eta(a)')(a) + \varphi(-\eta(\varphi(\eta(a)')(a)))(a) \\ &= -\varphi(\eta(a)')(a) + \varphi(-\lambda_{\eta(a)'}(\eta(a)))(a) \\ &= -\varphi(\eta(a)')(a) + \varphi(\eta(a)')(a) = 0. \end{aligned}$$

Hence $(H, \cdot)$ is a group. Now we have

$$\begin{aligned} a \cdot (b + c) &= a + \varphi(\eta(a))(b + c) = a + \varphi(\eta(a))(b) + \varphi(\eta(a))(c) \\ &= a + \varphi(\eta(a))(b) - a + a + \varphi(\eta(a))(b + c) = a \cdot b - a + a \cdot c. \end{aligned}$$

Thus $(H, +, \cdot)$ is a skew brace of abelian type. Note that

$$\begin{aligned}\eta(a \cdot b) &= \eta(a + \varphi(\eta(a))(b)) \\ &= \eta(a) + \eta(\varphi(\eta(a))(b)) \\ &= \eta(a) + \lambda_{\eta(a)}(\eta(b)) \\ &= \eta(a) \circ \eta(b).\end{aligned}$$

Hence η is a homomorphism of skew braces. Since φ is injective,

$$\begin{aligned}\mathrm{Soc}(H) &= \{a \in H : a \cdot b = a + b \text{ for all } b \in H\} \\ &= \{a \in H : \varphi(\eta(a))(b) = b \text{ for all } b \in H\} \\ &= \{a \in H : \eta(a) = 0\} = \ker(\eta).\end{aligned}$$

Therefore, since η is surjective, $H/\mathrm{Soc}(H) \cong B$ as skew braces.

Let

$$\varphi' \colon (B, \circ) \to \mathrm{Aut}(H, +)$$

be an injective homomorphism and

$$\eta' \colon (H, +) \to (B, +)$$

be a surjective homomorphism such that they satisfy $\eta'(\varphi'(g)(h)) = \lambda_g(\eta'(h))$ for all $g \in B$ and $h \in H$. Define the multiplication $\circ$ in H by

$$h_1 \circ h_2 := h_1 + \varphi'(\eta'(h_1))(h_2)$$

for $h_1, h_2 \in H$. Note that $F \in \mathrm{Aut}(H, +)$ is an isomorphism from $(H, +, \circ)$ to $(H, +, \cdot)$ if and only if

$$F(h_1 \circ h_2) = F(h_1) \cdot F(h_2),$$

which is equivalent to

$$F(h_1 + \varphi'(\eta'(h_1))(h_2)) = F(h_1) + \varphi(\eta(F(h_1)))(F(h_2)).$$

Hence F is an isomorphism if and only if

$$F^{-1}\varphi(\eta(F(h)))F = \varphi'(\eta'(h))$$

for all $h \in H$. This proves the first part of the result. The second part follows easily. □

The next result gives a first way to construct involutive solutions to the YBE.

Proposition 17.1.2 *Let B be a skew brace of abelian type, and let X be a non-empty set. Suppose that $\eta \colon \mathbb{Z}^{(X)} \to (B, +)$ is a surjective homomorphism, and that $\varphi \colon (B, \circ) \to \mathrm{Aut}(\mathbb{Z}^{(X)})$ is an injective homomorphism such that $\varphi(a)|_X$ is a bijection of X for all $a \in B$, and $\eta(\varphi(a)(m)) = \lambda_a(\eta(m))$ for all $a \in B$ and $m \in \mathbb{Z}^{(X)}$.*

Let r be the map

$$r\colon X \times X \to X \times X, \quad (x, y) \mapsto (\sigma_x(y), \sigma^{-1}_{\sigma_x(y)}(x)),$$

where $\sigma_x(y) = \varphi(\eta(x))(y)$. *Then* (X, r) *is an involutive solution to the YBE and* $\mathcal{G}(X, r) \cong B$ *as skew braces. Moreover, any involutive solution* (Z, t) *to the YBE with* $\mathcal{G}(Z, t) \cong B$ *is of this form.*

Proof By Proposition 17.1.1, the abelian group $\mathbb{Z}^{(X)}$ with multiplication

$$a \cdot b := a + \varphi(\eta(a))(b),$$

for $a, b \in \mathbb{Z}^X$, is a skew brace of abelian type and η becomes a homomorphism of skew braces. Note that for $x, y \in X \subseteq \mathbb{Z}^{(X)}$, we have that

$$\lambda_x(y) = -x + x \cdot y = -x + x + \varphi(\eta(x))(y) = \varphi(\eta(x))(y) = \sigma_x(y).$$

Therefore (X, r) is an involutive solution to the YBE because it is the restriction of the solution to the YBE associated with the skew brace $\mathbb{Z}^{(X)}$ (cf. Theorem 8.2.1 and Lemma 8.2.2). The skew brace $\mathbb{Z}^{(X)}$ is equal to the skew brace $G(X, r)$. Recall that the addition of the skew brace

$$\mathcal{G}(X, r) = \left\{\varphi(\eta(m))|_X : m \in \mathbb{Z}^{(X)}\right\}$$

is defined by $\varphi(\eta(m_1))|_X + \varphi(\eta(m_2))|_X = \varphi(\eta(m_1 + m_1))|_X$ for all $m_1, m_2 \in \mathbb{Z}^{(X)}$. Therefore, the map $B \to \mathcal{G}(X, r)$ defined by $a \mapsto \varphi(a)|_X$ is an isomorphism of skew braces.

On the other hand, observe that, if (Z, t) is an involutive solution to the YBE and $\xi\colon B \to \mathcal{G}(Z, t)$ is an isomorphism of skew braces, then the unique homomorphism $\eta\colon (\mathbb{Z}^{(Z)}, +) \to (B, +)$ such that $\eta(z) = \xi^{-1}(\phi(z))$ for all $z \in Z$, where

$$\phi\colon G(Z, t) \to \mathcal{G}(Z, t)$$

is the natural map, is surjective, and the map $\varphi : (B, \circ) \to \operatorname{Aut}(\mathbb{Z}^{(Z)})$, where $\varphi(a)$ is the unique automorphism of $\mathbb{Z}^{(Z)}$ such that $\varphi(a)(z) = \xi(a)(z)$, for all $z \in Z$, is an injective homomorphism. Furthermore, for $a \in B$, there exists $m \in \mathbb{Z}^{(Z)}$ such that $a = \eta(m)$ and, since ϕ is a homomorphism of skew braces,

$$\begin{aligned}\eta(\varphi(a)(z)) &= \eta(\xi(a)(z)) = \xi^{-1}(\phi(\xi(\eta(m))(z)))\\ &= \xi^{-1}(\phi(\phi(m)(z))) = \xi^{-1}(\phi(-m + mz)) = \xi^{-1}(-\phi(m) + \phi(m)\phi(z))\\ &= -\eta(m) + \eta(m)\eta(z) = -a + a\eta(z)\\ &= \lambda_a(\eta(z))\end{aligned}$$

for all $z \in Z$. Therefore $\eta(\varphi(a)(n)) = \lambda_a(\eta(n))$ for all $a \in B$ and all $n \in \mathbb{Z}^{(Z)}$. Now we have that

$$\varphi(\eta(x))(y) = \xi(\eta(x))(y) = \phi(x)(y)$$

for all $x, y \in Z$. Hence (Z, t) is exactly the same solution to the YBE as the solution obtained by the given construction using the maps η and φ. □

This proposition reduces the problem of finding all the involutive solutions to the YBE to the problem of finding maps η and φ with the required properties. Note that the map η is determined by its restriction to X and, since η is surjective, $\eta(X)$ should generate $(B, +)$. Note also that the monomorphism φ factors through the symmetric group $\mathbb{S}_X$, that is, $\varphi = \varphi_1\varphi_2$ where $\varphi_2 \colon (B, \circ) \to \mathbb{S}_X$ is a monomorphism and $\varphi_1 : \mathbb{S}_X \to \operatorname{Aut}(\mathbb{Z}^{(X)})$ is the natural map. Therefore the problem of finding all the solutions to the YBE is reduced to the problem of finding all the faithful $(B, \circ)$-sets X and all the homomorphisms of $(B, \circ)$-sets $X \to (B, +)$ such that the image of X generates $(B, +)$. In the next sections, we give a method that describes how to solve this problem using only the structure of the skew brace B.

17.2 Construction of Solutions

As we have seen in the previous section, given a skew brace B of abelian type, to construct an involutive solution (X, r) to the YBE such that $\mathcal{G}(X, r) \cong B$ as skew braces, it is enough to find two maps $\eta \colon X \to B$ and $\varphi \colon B \to \mathbb{S}_X$ satisfying some properties. In this section, we show how to construct the sets X and the maps η and φ using only information about the structure of the skew brace B.

Recall that $\lambda : (B, \circ) \to \operatorname{Aut}(B, +)$ is an action. The stabilizer of $a \in B$ by the action λ is denoted $\operatorname{St}(a)$. For $a \in B$, let $B_a = \{\lambda_b(a) \mid b \in B\}$ be the orbit of a, and let $\mathcal{O} = \{B_a \mid a \in B\}$ be the set of orbits of the action λ. For each $i \in \mathcal{O}$, choose an element $a_i \in i$. Let I be a subset of $\mathcal{O}$, such that $Y = \bigcup_{i \in I} i$ (and thus $Y = \{\lambda_b(a_i) \mid i \in I,\ b \in B\}$) satisfies $B = \langle Y \rangle_+$, the additive subgroup generated by Y. For each $i \in I$, let J_i be a non-empty set and let $\{K_{i,j}\}_{j \in J_i}$ be a family of subgroups of $\operatorname{St}(a_i)$ such that

$$\bigcap_{i \in I} \bigcap_{j \in J_i} \bigcap_{b \in B} b \circ K_{i,j} \circ b' = \{0\}.$$

Note that if one of the subgroups $K_{i,j}$ is trivial, then this last condition is satisfied.

The next result is the main result of this section.

Theorem 17.2.1 *With the above notation, let* $X := \bigsqcup_{i \in I} \bigsqcup_{j \in J_i} B/K_{i,j}$ *be the disjoint union of the sets of left cosets* $B/K_{i,j}$ *of the group* $(B, \circ)$*. Let* $\eta \colon X \to B$ *be the map defined by* $\eta(b \circ K_{i,j}) = \lambda_b(a_i)$*, and define* $\varphi \colon B \to \mathbb{S}_X$ *to be the natural action of* $(B, \circ)$ *on* X *given by left multiplications on the cosets in* $B/K_{i,j}$*, that is:*

$$\varphi(c)(b \circ K_{i,j}) := c \circ b \circ K_{i,j}.$$

Then φ is injective and $\eta(\varphi(a)(x)) = \lambda_a(\eta(x))$ for all $x \in X$ and $a \in B$. Moreover (X, r), where r is the map

$$r\colon X \times X \to X \times X, \quad (x, y) \mapsto (\sigma_x(y),\ \sigma^{-1}_{\sigma_x(y)}(x)),$$

with $\sigma_x(y) = \varphi(\eta(x))(y)$, is an involutive solution to the YBE such that $\mathcal{G}(X, r) \cong B$ as skew braces. Furthermore, any involutive solution (Z, t), with $\mathcal{G}(Z, t) \cong B$ as skew braces, is isomorphic to such a solution.

Proof First we shall prove that φ is injective. Let $c \in B$ be an element such that $\varphi(c) = \mathrm{id}$. Hence $c \circ b \circ K_{i,j} = b \circ K_{i,j}$, for all $b \in B$, $i \in I$ and $j \in J_i$. Thus

$$c \in \bigcap_{i\in I} \bigcap_{j\in J_i} \bigcap_{b\in B} b \circ K_{i,j} \circ b' = \{0\}.$$

Therefore φ is injective. Let $a, b \in B$. We have that

$$\begin{aligned}\eta(\varphi(a)(b \circ K_{i,j})) &= \eta(a \circ b \circ K_{i,j}) = \lambda_{a\circ b}(a_i)\\ &= \lambda_a(\lambda_b(a_i)) = \lambda_a(\eta(b \circ K_{i,j})).\end{aligned}$$

Hence $\eta(\varphi(a)(x)) = \lambda_a(\eta(x))$ for all $x \in X$ and $a \in B$. Hence, by Proposition 17.1.2, (X, r) is an involutive solution to the YBE and $\mathcal{G}(X, r) \cong B$ as skew braces.

Let (Z, t) be an involutive solution to the YBE such that $\mathcal{G}(Z, t) \cong B$ as skew braces. Let $\xi : \mathcal{G}(Z, t) \to B$ be an isomorphism of skew braces. Let $\eta = \xi\phi$, where $\phi\colon G(Z, t) \to \mathcal{G}(Z, t)$ is the natural projection. Then $\eta(Z)$ is a subset of B, invariant by λ which generates B additively. Let $Y = \eta(Z)$.

We also have an injective homomorphism $\varphi : (B, \circ) \to \mathrm{Aut}(\mathbb{Z}^{(Z)})$, such that $\varphi(b)(z) = \xi^{-1}(b)(z)$ for all $b \in B$ and $z \in Z$. Therefore Z is a left B-set with the action induced by φ. Let $a \in B$ and $z \in Z$. Let $g \in G(Z, t)$ such that $\phi(g) = \xi^{-1}(a)$. We have

$$\begin{aligned}\eta(\varphi(a)(z)) &= \xi(\phi(\xi^{-1}(a)(z))) = \xi(\phi(\phi(g)(z)))\\ &= \xi(\phi(-g + gz)) = -\xi(\phi(g)) + \xi(\phi(g)) \circ \xi(\phi(z))\\ &= -a + a \circ \eta(z) = \lambda_a(\eta(z)).\end{aligned}$$

Therefore the restriction $\eta|_Z\colon Z \longrightarrow Y$ of η is a B-map.

Let $Y = \bigcup_{i\in I} i$ be the decomposition of Y as a disjoint union of orbits under the action λ. Since $\eta|_Z$ is a surjective B-map, for all $i \in I$, the action φ splits $(\eta|_Z)^{-1}(i)$ into orbits: $(\eta|_Z)^{-1}(i) = \bigcup_{j\in J_i} Z_{i,j}$ and $\eta(Z_{i,j}) = i$. So we have $Z = \bigcup_{i\in I} \bigcup_{j\in J_i} Z_{i,j}$, where $\eta(Z_{i,j}) = i$ for all i, j.

For each $i \in I$, we choose an element $a_i \in i$, and for each $j \in J_i$, we choose $z_{i,j} \in Z_{i,j}$ such that $\eta(z_{i,j}) = a_i$. Note that $\mathrm{St}(z_{i,j}) \leq \mathrm{St}(a_i) \leq B$, since any $b \in \mathrm{St}(z_{i,j})$ satisfies $\varphi(b)(z_{i,j}) = z_{i,j}$ and, applying η, we obtain

$$a_i = \eta(z_{i,j}) = \eta(\varphi(b)(z_{i,j})) = \lambda_b(\eta(z_{i,j})) = \lambda_b(a_i),$$

and thus $b \in \mathrm{St}(a_i)$.

Recall that the maps $f_i : i \to B/\mathrm{St}(a_i)$ and $f_{i,j} : Z_{i,j} \to B/\mathrm{St}(z_{i,j})$ defined by $f_i(\lambda_b(a_i)) = b \circ \mathrm{St}(a_i)$ and $f_{i,j}(\varphi(b)(z_{i,j})) = b \circ \mathrm{St}(z_{i,j})$ are isomorphisms of B-sets. Then, $\eta\mid_{Z_{i,j}}$ is determined by the canonical projection

$$\begin{aligned} \pi : B/\mathrm{St}(z_{i,j}) &\to B/\mathrm{St}(a_i) \\ t \circ \mathrm{St}(z_{i,j}) &\mapsto t \circ \mathrm{St}(a_i), \end{aligned}$$

that is,

$$\begin{aligned} \eta(f_{i,j}^{-1}(b \circ \mathrm{St}(z_{i,j}))) &= \eta(\varphi(b)(z_{i,j})) = \lambda_b(\eta(z_{i,j})) \\ &= \lambda_b(a_i) = f_i^{-1}(b \circ \mathrm{St}(a_i)) \\ &= f_i^{-1}(\pi(b \circ \mathrm{St}(z_{i,j}))). \end{aligned}$$

Note that $\varphi(c)(z) = z$ for all $z \in Z$, if and only if $c \circ b \circ \mathrm{St}(z_{i,j}) = b \circ \mathrm{St}(z_{i,j})$ for all $b \in B$ and all i, j, equivalently, $c \in \bigcap_{i,j} \bigcap_{b \in B} b \circ \mathrm{St}(z_{i,j}) \circ b'$. Since φ is injective, we have $\bigcap_{i,j} \bigcap_{b \in B} b \circ \mathrm{St}(z_{i,j}) \circ b' = \{0\}$. Put $H_i := \mathrm{St}(a_i)$ and $K_{i,j} := \mathrm{St}(z_{i,j})$. Let (X, r) be the solution defined as in the statement of the theorem. We shall prove that $(X, r) \cong (Z, t)$. Let $f : Z \to X$ be the map defined by $f(z) = f_{i,j}(z)$ for all $z \in Z_{i,j}$. Clearly f is bijective. Let $x, z \in Z$. We may assume that $z \in Z_{i,j}$ and $x \in Z_{p,q}$. Therefore there exist $b_1, b_2 \in B$ such that $z = \varphi(b_1)(z_{i,j})$ and $x = \varphi(b_2)(z_{p,q})$. We have

$$\begin{aligned} f(\phi(x)(z)) &= f(\varphi(\eta(x))(z)) = f_{i,j}(\varphi(\eta(x))(\varphi(b_1)(z_{i,j}))) \\ &= f_{i,j}(\varphi(\eta(x)b_1)(z_{i,j})) = \eta(x) \circ b_1 \circ \mathrm{St}(z_{i,j}) \\ &= \eta(x) \circ b_1 \circ K_{i,j} = \eta(\varphi(b_2)(z_{p,q})) \circ b_1 \circ K_{i,j} \\ &= \lambda_{b_2}(\eta(z_{p,q})) \circ b_1 \circ K_{i,j} = \lambda_{b_2}(a_p) \circ b_1 \circ K_{i,j} \\ &= f_{b_2 \circ K_{p,q}}(b_1 \circ K_{i,j}) \\ &= f_{f(x)}(f(z)). \end{aligned}$$

Therefore f is an isomorphism of solutions to the YBE. □

As a consequence of Theorem 17.2.1, given a skew brace B of abelian type, to construct all the involutive solutions (X, r) to the YBE such that $\mathcal{G}(X, r) \cong B$ as skew braces one can proceed as follows:

(1) Find the decomposition of B as disjoint union of orbits, $B = \bigcup_{i \in K} B_i$, by the action $\lambda : (B, \circ) \to \mathrm{Aut}(B, +)$. Then choose one element a_i in each orbit B_i for all $i \in K$.

(2) Find all the subsets I of K such that the subset $Y = \bigcup_{i \in I} B_i$ generates the additive group of B.

(3) Given such an Y, find for each $i \in I$ a non-empty family $\{K_{i,j}\}_{j\in J_i}$ of subgroups of $\mathrm{St}(a_i)$ such that $\bigcap_{i,j}\bigcap_{b\in B} b \circ K_{i,j} \circ b' = \{0\}$. Note that the $K_{i,j}$ could be equal for different (i, j).

(4) Construct a solution as in the statement of Theorem 17.2.1 using the families $\{K_{i,j}\}_{j\in J_i}$, for $i \in I$.

Note that by Theorem 17.2.1, any involutive solution (X, r) to the YBE such that $\mathcal{G}(X, r) \cong B$ (as skew braces) is isomorphic to one constructed in this way. Different solutions to the YBE constructed in this way from a skew brace B of abelian type could be isomorphic. In the next section, we characterize when two of these solutions are isomorphic.

17.3 Isomorphism of Solutions

Let B be a skew brace of abelian type and let $\mathcal{O}, I, a_i, J_i, K_{i,j}$ be as in the beginning of Sect. 17.2. Let (X, r) be the solution to the YBE of the statement of Theorem 17.2.1.

Let $I' \subseteq \mathcal{O}$ such that $Y' = \bigcup_{i'\in I'} i'$ satisfies $B = \langle Y'\rangle_+$. For each $i' \in I'$, let $\{L_{i',j'}\}_{j'\in J'_{i'}}$ be a non-empty family of subgroups of $\mathrm{St}(a_{i'})$ such that

$$\bigcap_{i'\in I'}\bigcap_{j'\in J'_{i'}}\bigcap_{b\in B} b \circ L_{i',j'} \circ b' = \{0\}.$$

Let (X', r') be the corresponding solution to the YBE defined as in the statement of Theorem 17.2.1, that is:

$$r'(b_1 \circ L_{i'_1,j'_1}, b_2 \circ L_{i'_2,j'_2}) = (\lambda_{b_1}(a_{i'_1}) \circ b_2 \circ L_{i'_2,j'_2}, \lambda_{\lambda_{b_1}(a_{i'_1})\circ b_2}(a_{i'_2})' \circ b_1 \circ L_{i'_1,j'_1}).$$

We shall characterize when (X, r) and (X', r') are isomorphic in the following result.

Theorem 17.3.1 *The solutions (X, r) and (X', r') are isomorphic if and only if there exist an automorphism ψ of the skew brace B, a bijective map $\alpha\colon I \to I'$, a bijective map $\beta_i\colon J_i \to J'_{\alpha(i)}$ and $z_{i,j} \in B$, for each $i \in I$ and $j \in J_i$, such that*

$$\psi(a_i) = \lambda_{z_{i,j}}(a_{\alpha(i)}) \quad \text{and} \quad \psi(K_{i,j}) = z_{i,j} \circ L_{\alpha(i),\beta_i(j)} \circ z'_{i,j},$$

for all $i \in I_1$ and $j \in J_i$.

Proof Suppose that there exist an automorphism ψ of the skew brace B, a bijective map $\alpha\colon I \to I'$, a bijective map $\beta_i\colon J_i \to J'_{\alpha(i)}$ and $z_{i,j} \in B$, for $i \in I$, and for $j \in J_i$, such that

$$\psi(a_i) = \lambda_{z_{i,j}}(a_{\alpha(i)}) \quad \text{and} \quad \psi(K_{i,j}) = z_{i,j} \circ L_{\alpha(i),\beta_i(j)} \circ z'_{i,j}.$$

Observe that we also have $\psi(\lambda_b(a_i)) = \lambda_{\psi(b)\circ z_{i,j}}(a_{\alpha(i)})$ for every $b \in B$, because ψ is a homomorphism of skew braces. We define

$$F\colon X \to X', \quad F(b \circ K_{i,j}) = \psi(b) \circ z_{i,j} \circ L_{\alpha(i),\beta_i(j)}$$

for all $i \in I$, $j \in J_i$ and $b \in B$. Since $\psi(K_{i,j}) = z_{i,j} \circ L_{\alpha(i),\beta_i(j)} \circ z'_{i,j}$, we get that F is well defined. It is easy to check that F is an isomorphism of the solutions (X, r) and (X', r').

Conversely, suppose that there exists an isomorphism $F\colon X \to X'$ of the solutions (X, r) and (X', r'). We can write $F(b \circ K_{i,j}) = \varphi(b \circ K_{i,j}) \circ L_{\alpha(b,i,j),\beta(b,i,j)}$, for some maps $\varphi\colon X \to B$, $\alpha\colon X \to I'$ and $\beta\colon X \to \bigcup_{i'\in I'} J'_{i'}$. We shall prove that $\alpha(b, i, j) = \alpha(0, i, k)$ and $\beta(b, i, j) = \beta(0, i, j)$ for all $b \in B$, $i \in I$ and $j, k \in J_i$. Since F is a homomorphism of solutions to the YBE, we have

$$F(\lambda_{b_1}(a_{i_1}) \circ b_2 \circ K_{i_2,j_2}) = \lambda_{\varphi(b_1\circ K_{i_1,j_1})}(a_{\alpha(b_1,i_1,j_1)})F(b_2 \circ K_{i_2,j_2}) \qquad (17.1)$$

for all $b_1, b_2 \in B$, $i_1, i_2 \in I$, $j_1 \in J_{i_1}$, and $j_2 \in J_{i_2}$. Hence

$$\begin{aligned}&\varphi(\lambda_{b_1}(a_{i_1}) \circ b_2 \circ K_{i_2,j_2}) \circ L_{\alpha(\lambda_{b_1}(a_{i_1})\circ b_2,i_2,j_2),\beta(\lambda_{b_1}(a_{i_1})\circ b_2,i_2,j_2)}\\ &= \lambda_{\varphi(b_1\circ K_{i_1,j_1})}(a_{\alpha(b_1,i_1,j_1)})\varphi(b_2 \circ K_{i_2,j_2}) \circ L_{\alpha(b_2,i_2,j_2),\beta(b_2,i_2,j_2)}\end{aligned}$$

for all $b_1, b_2 \in B$, $i_1, i_2 \in I$, $j_1 \in J_{i_1}$, and $j_2 \in J_{i_2}$. Thus

$$\begin{aligned}\alpha(\lambda_{b_1}(a_{i_1}) \circ b_2, i_2, j_2) &= \alpha(b_2, i_2, j_2),\\ \beta(\lambda_{b_1}(a_{i_1}) \circ b_2, i_2, j_2) &= \beta(b_2, i_2, j_2).\end{aligned}$$

Since $B = \langle Y\rangle_+$ and Y is B-invariant (by the action λ), we know that Y also generates the multiplicative group of B. Therefore

$$\alpha(b_2, i_2, j_2) = \alpha(0, i_2, j_2), \quad \beta(b_2, i_2, j_2) = \beta(0, i_2, j_2).$$

Note also that

$$\lambda_{\varphi(b_1\circ K_{i_1,j_1})}(a_{\alpha(b_1,i_1,j_1)}) \circ F(b_2 \circ K_{i_2,j_2}) = \lambda_{\varphi(b_1\circ K_{i_1,j})}(a_{\alpha(b_1,i_1,j)}) \circ F(b_2 \circ K_{i_2,j_2})$$

for all $b_1, b_2 \in B$, $i_1, i_2 \in I$, $j_1, j \in J_{i_1}$ and $j_2 \in J_{i_2}$. Since

$$\bigcap_{i',j'}\bigcap_{b\in B} b \circ L_{i',j'} \circ b' = \{0\},$$

we have that

$$\lambda_{\varphi(b_1\circ K_{i_1,j_1})}(a_{\alpha(b_1,i_1,j_1)}) = \lambda_{\varphi(b_1\circ K_{i_1,j})}(a_{\alpha(b_1,i_1,j)})$$

for all $b_1 \in B, i_1 \in I$ and $j_1, j \in J_{i_1}$. Therefore $a_{\alpha(b_1,i_1,j_1)}, a_{\alpha(b_1,i_1,j)} \in \alpha(1, i_1, k)$ for all $b_1 \in B, i_1 \in I$ and $j_1, j \in J_{i_1}$ and thus $\alpha(b, i, j) = \alpha(0, i, k)$ for all $b \in B, i \in I_1$ and $j, k \in J_i$. For each $i \in I$ we choose an element $j_i \in J_i$. Since F is bijective, the map $I \to I'$ defined by $i \mapsto \alpha(0, i, j_i)$ is bijective and for each $i \in I$ the map $J_i \to J'_{\alpha(0,i,j_i)}$ defined by $j \mapsto \beta(0, i, j)$ is bijective. We shall see that there exists an automorphism ψ of the left brace B such that

$$\psi(\lambda_b(a_i)) = \lambda_{\psi(b)\varphi(K_{i,j_i})}(a_{\alpha(0,i,j_i)})$$

and

$$\psi(K_{i,j}) = \varphi(K_{i,j_i}) \circ L_{\alpha(0,i,j_i),\beta(0,i,j)} \circ \varphi(K_{i,j_i})'$$

for all $b \in B$, $i \in I$ and $j \in J_i$. Let $0 = \lambda_{b_1}(a_{i_1})^{\varepsilon_1} \circ \cdots \circ \lambda_{b_m}(a_{i_m})^{\varepsilon_m}$, for some $b_1, \ldots, b_m \in B, i_1, \ldots, i_m \in I$ and $\varepsilon_1, \ldots, \varepsilon_m \in \{1, -1\}$ (here, $a^{-1} = a'$ for $a \in B$). By (17.1), we have

$$\begin{aligned} F(b \circ K_{i,j}) &= F(\lambda_{b_1}(a_{i_1})^{\varepsilon_1} \circ \cdots \circ \lambda_{b_m}(a_{i_m})^{\varepsilon_m} \circ b \circ K_{i,j}) \\ &= \lambda_{\varphi(b_1 \circ K_{i_1,j_{i_1}})}(a_{\alpha(0,i_1,j_{i_1})})^{\varepsilon_1} \\ &\quad \circ F(\lambda_{b_2}(a_{i_2})^{\varepsilon_2} \circ \cdots \circ \lambda_{b_m}(a_{i_m})^{\varepsilon_m} \circ b \circ K_{i,j}) \\ &= \lambda_{\varphi(b_1 \circ K_{i_1,j_{i_1}})}(a_{\alpha(0,i_1,j_{i_1})})^{\varepsilon_1} \circ \cdots \circ \lambda_{\varphi(b_m \circ K_{i_m,j_{i_m}})}(a_{\alpha(0,i_m,j_{i_m})})^{\varepsilon_m} \\ &\quad \circ F(b \circ K_{i,j}) \end{aligned}$$

for all $b \in B$, $i \in I$ and $j \in J_i$. Since $\bigcap_{i',j'} \bigcap_{b \in B} b \circ L_{j,l} \circ b' = \{0\}$, we have that $\lambda_{\varphi(b_1 \circ K_{i_1,j_{i_1}})}(a_{\alpha(0,i_1,j_{i_1})})^{\varepsilon_1} \circ \cdots \circ \lambda_{\varphi(b_m \circ K_{i_m,j_{i_m}})}(a_{\alpha(0,i_m,j_{i_m})})^{\varepsilon_m} = 0$. Therefore there exists a unique homomorphism $\psi\colon B \to B$ of multiplicative groups such that $\psi(\lambda_b(a_i)) = \lambda_{\varphi(b \circ K_{i,j_i})}(a_{\alpha(0,i,j_i)})$. Since Y generates the multiplicative group of B, by (17.1) one can see that

$$\varphi(b \circ K_{i,j}) \circ L_{\alpha(0,i,j_i),\beta(0,i,j)} = F(b \circ K_{i,j}) = \psi(b) \circ \varphi(K_{i,j}) \circ L_{\alpha(0,i,j_i),\beta(0,i,j)}.$$

Therefore, since $L_{\alpha(0,i,j_i),\beta(0,i,j)} \subseteq \mathrm{St}(a_{\alpha(0,i,j_i)})$, we have

$$\lambda_{\varphi(b \circ K_{i,j})}(a_{\alpha(0,i,j_i)}) = \lambda_{\psi(b) \circ \varphi(K_{i,j})}(a_{\alpha(0,i,j_i)}).$$

Hence $\psi(\lambda_b(a_i)) = \lambda_{\psi(b) \circ \varphi(K_{i,j_i})}(a_{\alpha(0,i,j_i)})$. Now we have that

$$\begin{aligned} \psi(b + a_i) &= \psi(b \circ \lambda_{b'}(a_i)) = \psi(b) \circ \psi(\lambda_{b'}(a_i)) \\ &= \psi(b)\lambda_{\psi(b)' \circ \varphi(K_{i,j_i})}(a_{\alpha(0,i,j_i)}) \\ &= \psi(b) \circ \lambda_{\psi(b)'}(\lambda_{\varphi(K_{i,j_i})}(a_{\alpha(0,i,j_i)})) \\ &= \psi(b) + \lambda_{\varphi(K_{i,j_i})}(a_{\alpha(0,i,j_i)}) \\ &= \psi(b) + \psi(\lambda_0(a_i)) \\ &= \psi(b) + \psi(a_i). \end{aligned}$$

Now it is easy to see that ψ is a homomorphism of skew braces. Since F is bijective and $F(b \circ c \circ K_{i,j}) = \psi(b) \circ F(c \circ K_{i,j})$, it follows that ψ is bijective. Furthermore $b \in K_{i,j}$ if and only if

$$\begin{aligned} \varphi(K_{i,j}) \circ L_{\alpha(0,i,j_i),\beta(0,i,j)} &= F(K_{i,j}) \\ &= F(b \circ K_{i,j}) \\ &= \psi(b) \circ F(K_{i,j}) \\ &= \psi(b) \circ \varphi(K_{i,j}) \circ L_{\alpha(0,i,j_i),\beta(0,i,j)}. \end{aligned}$$

Therefore the result follows. □

Summarizing, the last theorem says that two solutions constructed as in Theorem 17.2.1 are isomorphic if we can find an automorphism of the brace B that brings each $K_{i,j}$ to one $L_{i',j'}$, taking into account that maybe the $L_{i',j'}$'s are permuted (that is the reason for the α and β_i maps), and that maybe we have chosen another element of the orbit in the process (that is the reason why the image a_i is $\lambda_{z_{i,j}}(a_{\alpha(i)})$ and not just $a_{\alpha(i)}$, and it is also the reason why the $L_{\alpha(i),\beta_i(j)}$ is conjugated by $z_{i,j}$).

The following is an example of how to use Theorem 17.2.1 and Theorem 17.3.1 to compute all the finite solutions associated to a given finite left brace up to isomorphism. We use the easiest examples of skew braces of abelian type: trivial skew braces of order p, where p is a prime number.

Example 17.3.2 Let p be a prime number. Consider the trivial skew brace over $G = \mathbb{Z}/(p)$. Then, the orbits are $\{\alpha\}$ for every $\alpha \in \mathbb{Z}/(p)$. Since any orbit has only one element, we have $\mathrm{St}(\alpha) = G$, and the possible $K_{i,j}$'s in this case are 0 and G.

Let Y be a subset of $\mathbb{Z}/(p)$ with at least a nonzero element. Let $K_{\alpha,j} = G$ for $\alpha \in Y$ and $j \in \{1, \dots, k_\alpha\}$, and let $K'_{\alpha,k} = 0$ for $\alpha \in Y$ and $k \in \{1, \dots, m_\alpha\}$, where k_α and m_α are non-negative integers such that $k_\alpha + m_\alpha > 0$. Write $G/K_{\alpha,j} = \{y_{\alpha,j}\}$, and $G/K'_{\alpha,k} = \{y^1_{\alpha,k}, \dots, y^p_{\alpha,k}\}$, where $y^l_{\alpha,k} = l + K'_{\alpha,k}$. Assume that at least one m_α is positive. Then the corresponding solution of the YBE is (X, r), where

$$X = \bigcup_{\alpha \in Y}\left(\left(\bigcup_{1\le j\le k_\alpha}\{y_{\alpha,j}\}\right) \cup \left(\bigcup_{1\le k\le m_\alpha}\{y^1_{\alpha,k}, \dots, y^p_{\alpha,k}\}\right)\right)$$

and $r(x, y) = (\sigma_x(y), \sigma^{-1}_{\sigma_x(y)}(x))$, with the sigma maps given by

$$\sigma_{y_{\alpha,j}} = \sigma_{y^l_{\alpha,k}} = \tau^\alpha, \text{ for all } \alpha \in Y, \text{ for all } j, k \text{ and for all } l \in \{1, \dots, p\},$$

where $\tau \in \mathbb{S}_X$ is the product of all the cycles $(y^1_{\alpha,k}, y^2_{\alpha,k}, \dots, y^p_{\alpha,k})$ for any $\alpha \in Y$ and $k \in \{1, \dots, m_\alpha\}$.

Finally observe that, in this case, $\mathrm{Aut}(G, +, \circ) = \mathrm{Aut}(G, +) \cong (\mathbb{Z}/(p))^*$, and the effect of an automorphism of G over a solution is to change $\sigma_{y_{\alpha,j}} = \sigma_{y^l_{\alpha,k}} = \tau^\alpha$ to the isomorphic solution $\sigma_{y_{\alpha,j}} = \sigma_{y^l_{\alpha,k}} = \tau^{A\alpha}$, where $A \in (\mathbb{Z}/(p))^*$.

17.4 Exercises

17.4.1 Let p and q be two distinct prime numbers. Describe all the finite involutive solutions (X, r) to the YBE such that $\mathcal{G}(X, r)$ is the trivial skew brace over the cyclic group $\mathbb{Z}/(pq)$.

17.4.2 Let p be a prime number. Give an example of a finite non-involutive solution (X, r) to the YBE such that $\mathcal{G}(X, r)$ is the trivial skew brace over the cyclic group $\mathbb{Z}/(p)$.

17.5 Notes

Proposition 17.1.1 is due to Bachiller (see (7, Theorem 2.1)).

A cohomological version of Proposition 17.1.2 appears in (19, Theorem 4.6) and a first version without a formal statement in (56, pages 182–183).

Sections 17.2 and 17.3 are based on (11).

In (10), Bachiller generalized the results of this chapter to general solutions to the YBE. Hence the construction of all (finite) solutions to the YBE is reduced to the construction of all (finite) skew braces.

Chapter 18
Simple Skew Braces

As we have seen in Chap. 17, the classification of finite involutive solutions to the YBE is reduced to the classification of the finite skew braces of abelian type. A first step in the classification of the finite skew braces of abelian type is the classification of the finite simple skew braces of abelian type.

In this chapter, we study some constructions of finite simple skew braces of abelian type.

18.1 Finite Simple Skew Braces

Definition 18.1.1 A nonzero skew brace B is *simple* if $\{0\}$ and B are the only ideals of B.

Example 18.1.2 Let G be a simple additive group. Then the trivial skew brace $(G, +, +)$ is a simple skew brace. These are the simple trivial skew braces.

Thus the classification of the finite simple trivial skew braces is equivalent to the classification of the finite simple groups. In particular, the finite simple trivial skew braces of abelian type are the trivial skew braces of prime cardinality.

By Theorem 9.5.2, the multiplicative group of a finite skew brace of nilpotent type is solvable. Let B be a nonzero finite skew brace of nilpotent type. Let $p_1, \dots, p_k$ be the distinct prime divisors of $|B|$. Let B_{p_i} be the Sylow p_i-subgroup of the additive group of B, for $i = 1, \dots, k$. Note that B_{p_i} is a left ideal of B, and the multiplicative groups of these left ideals B_{p_i} form a Sylow system of the multiplicative group of B. Thus

$$B = \prod_{i=1}^{k} B_{p_i}.$$

F. Cedó and L. Vendramin, *Groups, Radical Rings, and the Yang-Baxter Equation*, Progress in Mathematics 361, https://doi.org/10.1007/978-3-032-01524-2_18

Note that for $a \in B_{p_i}, b \in B_{p_j}$, where $i \neq j$, there exist unique $c \in B_{p_j}$ and $d \in B_{p_i}$ such that $a \circ b = c \circ d$. On the other hand,

$$a \circ b = \lambda_a(b) \circ \lambda_a(b)' \circ a \circ b = \lambda_a(b) \circ (b' + \lambda_{b'}(a') - b')'.$$

Hence $c = \lambda_a(b)$ and $d = (b' + \lambda_{b'}(a') - b')'$.

Lemma 18.1.3 *Let B be a finite simple skew brace of nilpotent type with nilpotent multiplicative group. Then B is a trivial skew brace of prime order.*

Proof Let p be a prime divisor of $|B|$. Let B_p be the Sylow p-subgroup of the additive group of B. We know that B_p is a left ideal of B. In particular, B_p also is a Sylow p-subgroup of the multiplicative group of B. Since the multiplicative group of B is nilpotent, we have that B_p is an ideal of B. Since B is simple, $B = B_p$. Let G be the multiplicative group of B and let A be the additive group of B. Then the semidirect product $H = A \rtimes G$ via the lambda map is a p-group, and thus it is nilpotent. Let $a \in A$ and $g \in G$. We have that

$$(0, g)(a, 0)(0, g')(-a, 0) = (\lambda_g(a), g)(-\lambda_g^{-1}(a), g') = (\lambda_g(a) - a, 0) = (g * a, 0).$$

Therefore $[\{0\} \times G, A \times \{0\}] = B * B \times \{0\}$. Since H is nilpotent, it is clear that $B * B \neq B$. Since $B * B$ is an ideal of B and B is simple, it follows that $B * B = \{0\}$. Hence B is a trivial skew brace. Since B is simple, $|B| = p$. □

The above result shows that if B is a finite non-trivial simple skew brace of abelian type, then there exist two distinct prime divisors of $|B|$. To understand the structure of B we need to study the structure of the left ideals B_p, where B_p is the Sylow p-subgroup of $(B, +)$, and the restriction of the lambda map to B_p acting on B_q, where p, q are distinct prime divisors of $|B|$.

Let B be a finite simple skew brace of abelian type of order $p^n q^m$ for distinct primes p, q and positive integers n, m. Let B_p and B_q be the Sylow p-subgroup and the Sylow q-subgroup of the additive group of B, respectively. Let

$$\alpha\colon B_q \to \operatorname{Aut}(B_p, +) \text{ and } \beta\colon B_p \to \operatorname{Aut}(B_q, +)$$

be restrictions of the lambda map of B. Let $a \in B_p$ and $b \in B_q$ and suppose that $(a, b) \neq (0, 0)$. Since B is simple, we have that the ideal generated by $a + b$ is B. In particular, if $a \neq 0$, then the ideal generated by a is B. Let I_0 be a minimal nonzero ideal of B_p. Clearly, I_0 is not an ideal of B. This means that there exist $d \in B_q$ and $c \in I_0$ such that $c * d = (\lambda_c - \mathrm{id})(d) = (\beta_c - id)(d) \neq 0$. Let J_0 be the ideal of B_q generated by $\{(\beta_c - id)(d) : c \in I_0,\ d \in B_q\}$. Let I_1 be the ideal of B_p generated by

$$I_0 \cup \{(\alpha_d - id)(c) : d \in J_0,\ c \in B_p\}.$$

Let J_1 be the ideal of B_q generated by

$$J_0 \cup \{(\beta_c - id)(d) : c \in I_1,\ d \in B_q\}.$$

For $k > 1$ let I_k be the ideal of B_p generated by

$$I_{k-1} \cup \{(\alpha_d - id)(c) : d \in J_{k-1},\ c \in B_p\},$$

and let J_k be the ideal of B_q generated by

$$J_{k-1} \cup \{(\beta_c - id)(d) : c \in I_k,\ d \in B_q\}.$$

Then there exists a positive integer k such that $I_k = B_p$ and $J_k = B_q$.

We shall study the main constructions of the finite simple braces of abelian type of order $p^n q^m$ for distinct primes p, q and positive integers n, m.

18.2 Matched Product of Skew Braces

Let B be a skew brace. Suppose that $(B, +)$ is the direct product of two left ideals B_1 and B_2 of B. Note that for $a, x \in B_1$ and $b, y \in B_2$, we have

$$\begin{aligned}\lambda_{a+b}(x + y) &= \lambda_{a+b}(x) + \lambda_{a+b}(y)\\ &= \lambda_a\lambda_{\lambda_a^{-1}(b)}(x) + \lambda_a\lambda_{\lambda_a^{-1}(b)}(y)\\ &= \lambda_b\lambda_{\lambda_b^{-1}(a)}(x) + \lambda_b\lambda_{\lambda_b^{-1}(a)}(y),\end{aligned}$$

where the last equality follows because $a + b = b + a$.

Let $\alpha\colon (B_2, \circ) \to \mathrm{Aut}(B_1, +)$ and $\beta\colon (B_1, \circ) \to \mathrm{Aut}(B_2, +)$ be the restrictions of thc lambda map, i.e.:

$$\alpha_b(x) = \lambda_b(x) \text{ and } \beta_a(y) = \lambda_a(y),$$

where $\alpha_b = \alpha(b)$ and $\beta_a = \beta(a)$, for all $a, x \in B_1$ and $b, y \in B_2$. Then

$$\alpha_b\lambda^{(1)}_{\alpha_b^{-1}(a)} = \lambda^{(1)}_a\alpha_{\beta_a^{-1}(b)}$$

and

$$\beta_a\lambda^{(2)}_{\beta_a^{-1}(b)} = \lambda^{(2)}_b\beta_{\alpha_b^{-1}(a)},$$

for all $a \in B_1$ and $b \in B_2$, where $\lambda^{(1)}$ is the lambda map of B_1 and $\lambda^{(2)}$ is the lambda map of B_2.

Definition 18.2.1 Let B_1 and B_2 be two skew braces. Let $\alpha\colon (B_2, \circ) \to \mathrm{Aut}(B_1, +)$ and $\beta\colon (B_1, \circ) \to \mathrm{Aut}(B_2, +)$ be group homomorphisms. We say that $(B_1, B_2, \alpha, \beta)$ is a *matched pair* of skew braces if

$$\alpha_b \lambda^{(1)}_{\alpha_b^{-1}(a)} = \lambda^{(1)}_a \alpha_{\beta_a^{-1}(b)}$$

and

$$\beta_a \lambda^{(2)}_{\beta_a^{-1}(b)} = \lambda^{(2)}_b \beta_{\alpha_b^{-1}(a)},$$

where $\alpha_b = \alpha(b)$ and $\beta_a = \beta(a)$, for all $a \in B_1$ and $b \in B_2$, and where $\lambda^{(1)}$ is the lambda map of B_1 and $\lambda^{(2)}$ is the lambda map of B_2.

Lemma 18.2.2 *Let $(B, +)$ be a group (not necessarily abelian). Let*

$$\lambda \colon B \to \operatorname{Aut}(B, +), \quad a \mapsto \lambda_a$$

be a map such that $\lambda_a \lambda_b = \lambda_{a+\lambda_a(b)}$, for all $a, b \in B$. Then B with the sum $+$ and the product defined by $a \circ b := a + \lambda_a(b)$, for all $a, b \in B$, is a skew brace.

Proof Note that, if 0 is the neutral element of $(B, +)$, then $\lambda_0 \lambda_0 = \lambda_{\lambda_0(0)} = \lambda_0$. Hence $\lambda_0 = \mathrm{id}$. Since $a \circ 0 = a + \lambda_a(0) = a + 0 = a$ and $0 \circ a = 0 + \lambda_0(a) = a$, we have that 0 is the neutral element of $(B, \circ)$. Since $\lambda_a \lambda_{\lambda_a^{-1}(-a)} = \lambda_{a-a} = \lambda_0 = \mathrm{id}$, we get that $\lambda_a^{-1} = \lambda_{\lambda_a^{-1}(-a)}$. Now $a \circ \lambda_a^{-1}(-a) = a - a = 0$ and

$$\lambda_a^{-1}(-a) \circ a = \lambda_a^{-1}(-a) + \lambda_{\lambda_a^{-1}(-a)}(a) = -\lambda_a^{-1}(a) + \lambda_a^{-1}(a) = 0.$$

Hence $\lambda_a^{-1}(-a)$ is the symmetric element of a in $(B, \circ)$. We shall check the associativity property:

$$\begin{aligned} a \circ (b \circ c) &= a + \lambda_a(b + \lambda_b(c)) = a + \lambda_a(b) + \lambda_a(\lambda_b(c)) \\ &= a + \lambda_a(b) + \lambda_{a+\lambda_a(b)}(c) = (a \circ b) + \lambda_{a\circ b}(c) \\ &= (a \circ b) \circ c. \end{aligned}$$

Hence $(B, \circ)$ is a group. Note that

$$\begin{aligned} a \circ (b + c) &= a + \lambda_a(b + c) = a + \lambda_a(b) + \lambda_a(c) \\ &= a + \lambda_a(b) - a + a + \lambda_a(c) = a \circ b - a + a \circ c. \end{aligned}$$

Therefore $(B, +, \circ)$ is a skew brace, and the result follows. □

Theorem 18.2.3 *Let $(B_1, B_2, \alpha, \beta)$ be a matched pair of skew braces. Then $B_1 \times B_2$ is a skew brace with sum given by*

$$(a, b) + (c, d) = (a + c, b + d)$$

and lambda map given by

$$\lambda_{(a,b)}(c, d) := \left(\lambda^{(1)}_a \alpha_{\beta_a^{-1}(b)}(c), \lambda^{(2)}_b \beta_{\alpha_b^{-1}(a)}(d) \right).$$

Moreover $B_1 \times \{0\}$ and $\{0\} \times B_2$ are left ideals isomorphic to B_1 and B_2 as skew braces, respectively. This skew brace is the matched product *of B_1 and B_2, and it is denoted by $B_1 \bowtie B_2$.*

Conversely, if B is a skew brace such that $(B, +)$ is the direct product of two left ideals L_1 and L_2 of B, then $(L_1, L_2, \alpha, \beta)$ is a matched pair of left braces, where α and β are the restrictions of the lambda map of B, and B is isomorphic to the corresponding matched product $L_1 \bowtie L_2$.

Proof To prove the first part, by Lemma 18.2.2, it is enough to check that

$$\lambda_{(a,b)}\lambda_{(c,d)} = \lambda_{(a,b)+\lambda_{(a,b)}(c,d)}$$

for all $a, c \in B_1$ and $b, d \in B_2$. Let $x \in B_1$ and $y \in B_2$. We have that

$$\begin{aligned}\lambda_{(a,b)}\lambda_{(c,d)}(x, y) &= \lambda_{(a,b)}\left(\lambda_c^{(1)}\alpha_{\beta_c^{-1}(d)}(x), \lambda_d^{(2)}\beta_{\alpha_d^{-1}(c)}(y)\right)\\ &= \left(\lambda_a^{(1)}\alpha_{\beta_a^{-1}(b)}\lambda_c^{(1)}\alpha_{\beta_c^{-1}(d)}(x), \lambda_b^{(2)}\beta_{\alpha_b^{-1}(a)}\lambda_d^{(2)}\beta_{\alpha_d^{-1}(c)}(y)\right)\end{aligned}$$

and

$$\begin{aligned}\lambda_{(a,b)+\lambda_{(a,b)}(c,d)}(x, y) &= \lambda_{(a+\lambda_a^{(1)}\alpha_{\beta_a^{-1}(b)}(c),b+\lambda_b^{(2)}\beta_{\alpha_b^{-1}(a)}(d))}(x, y)\\ &= \left(\lambda^{(1)}_{a+\lambda_a^{(1)}\alpha_{\beta_a^{-1}(b)}(c)}\alpha_{\beta^{-1}_{a+\lambda_a^{(1)}\alpha_{\beta_a^{-1}(b)}(c)}(b+\lambda_b^{(2)}\beta_{\alpha_b^{-1}(a)}(d))}(x),\right.\\ &\qquad \left.\lambda^{(2)}_{b+\lambda_b^{(2)}\beta_{\alpha_b^{-1}(a)}(d)}\beta_{\alpha^{-1}_{b+\lambda_b^{(2)}\beta_{\alpha_b^{-1}(a)}(d)}(a+\lambda_a^{(1)}\alpha_{\beta_a^{-1}(b)}(c))}(y)\right).\end{aligned}$$

Thus, it is enough to check the following equalities:

$$\lambda_a^{(1)}\alpha_{\beta_a^{-1}(b)}\lambda_c^{(1)}\alpha_{\beta_c^{-1}(d)} = \lambda^{(1)}_{a+\lambda_a^{(1)}\alpha_{\beta_a^{-1}(b)}(c)}\alpha_{\beta^{-1}_{a+\lambda_a^{(1)}\alpha_{\beta_a^{-1}(b)}(c)}(b+\lambda_b^{(2)}\beta_{\alpha_b^{-1}(a)}(d))} \tag{18.1}$$

and

$$\lambda_b^{(2)}\alpha_{\alpha_b^{-1}(a)}\lambda_d^{(2)}\beta_{\alpha_d^{-1}(c)} = \lambda^{(2)}_{b+\lambda_b^{(2)}\beta_{\alpha_b^{-1}(a)}(d)}\beta_{\alpha^{-1}_{b+\lambda_b^{(2)}\beta_{\alpha_b^{-1}(a)}(d)}(a+\lambda_a^{(1)}\alpha_{\beta_a^{-1}(b)}(c))}. \tag{18.2}$$

We shall prove (18.1). The equality (18.2) is proved similarly. We have

$$\begin{aligned}&\lambda^{(1)}_{a+\lambda_a^{(1)}\alpha_{\beta_a^{-1}(b)}(c)}\alpha_{\beta^{-1}_{a+\lambda_a^{(1)}\alpha_{\beta_a^{-1}(b)}(c)}(b+\lambda_b^{(2)}\beta_{\alpha_b^{-1}(a)}(d))}\\ &= \lambda^{(1)}_{a\circ\alpha_{\beta_a^{-1}(b)}(c)}\alpha_{\beta^{-1}_{a\circ\alpha_{\beta_a^{-1}(b)}(c)}(b+\lambda_b^{(2)}\beta_{\alpha_b^{-1}(a)}(d))}\end{aligned}$$

$$
\begin{aligned}
&= \lambda^{(1)}_{a\circ\alpha_{\beta_a^{-1}(b)}(c)}\alpha_{\beta^{-1}_{a\circ\alpha_{\beta_a^{-1}(b)}(c)}}(b)+\beta^{-1}_{a\circ\alpha_{\beta_a^{-1}(b)}(c)}(\lambda^{(2)}_b\beta_{\alpha_b^{-1}(a)}(d))\\
&= \lambda^{(1)}_{a\circ\alpha_{\beta_a^{-1}(b)}(c)}\alpha_{\beta^{-1}_{a\circ\alpha_{\beta_a^{-1}(b)}(c)}}(b)+\lambda^{(2)}_{\beta^{-1}_{a\circ\alpha_{\beta_a^{-1}(b)}(c)}(b)}\beta^{-1}_{\alpha_b^{-1}(a\circ\alpha_{\beta_a^{-1}(b)}(c))}\beta_{\alpha_b^{-1}(a)}(d)\\
&= \lambda^{(1)}_{a\circ\alpha_{\beta_a^{-1}(b)}(c)}\alpha_{\beta^{-1}_{a\circ\alpha_{\beta_a^{-1}(b)}(c)}}(b)\circ\beta^{-1}_{\alpha_b^{-1}(a\circ\alpha_{\beta_a^{-1}(b)}(c))}\beta_{\alpha_b^{-1}(a)}(d)\\
&= \lambda^{(1)}_{a\circ\alpha_{\beta_a^{-1}(b)}(c)}\alpha_{\beta^{-1}_{a\circ\alpha_{\beta_a^{-1}(b)}(c)}}(b)\circ\beta^{-1}_{\alpha_b^{-1}(a+\lambda^{(1)}_a\alpha_{\beta_a^{-1}(b)}(c))}\beta_{\alpha_b^{-1}(a)}(d)\\
&= \lambda^{(1)}_{a\circ\alpha_{\beta_a^{-1}(b)}(c)}\alpha_{\beta^{-1}_{a\circ\alpha_{\beta_a^{-1}(b)}(c)}}(b)\circ\beta^{-1}_{\alpha_b^{-1}(a+\alpha_b\lambda^{(1)}_{\alpha_b^{-1}(a)}(c))}\beta_{\alpha_b^{-1}(a)}(d)\\
&= \lambda^{(1)}_{a\circ\alpha_{\beta_a^{-1}(b)}(c)}\alpha_{\beta^{-1}_{a\circ\alpha_{\beta_a^{-1}(b)}(c)}}(b)\circ\beta^{-1}_{\alpha_b^{-1}(a)+\lambda^{(1)}_{\alpha_b^{-1}(a)}(c)}\beta_{\alpha_b^{-1}(a)}(d)\\
&= \lambda^{(1)}_{a\circ\alpha_{\beta_a^{-1}(b)}(c)}\alpha_{\beta^{-1}_{a\circ\alpha_{\beta_a^{-1}(b)}(c)}}(b)\circ\beta^{-1}_{\alpha_b^{-1}(a)\circ c}\beta_{\alpha_b^{-1}(a)}(d)\\
&= \lambda^{(1)}_{a\circ\alpha_{\beta_a^{-1}(b)}(c)}\alpha_{\beta^{-1}_{a\circ\alpha_{\beta_a^{-1}(b)}(c)}}(b)\circ\beta_c^{-1}(d)\\
&= \lambda^{(1)}_a\lambda^{(1)}_{\alpha_{\beta_a^{-1}(b)}(c)}\alpha_{\beta^{-1}_{\alpha_{\beta_a^{-1}(b)}(c)}}\beta_a^{-1}(b)\alpha_{\beta_c^{-1}}(d)\\
&= \lambda^{(1)}_a\alpha_{\beta_a^{-1}(b)}\lambda^{(1)}_c\alpha_{\beta_c^{-1}(d)}.
\end{aligned}
$$

This finishes the proof of the first part.

The second part follows easily by the argument above Definition 18.2.1. □

The matched product of two skew braces can be generalized to the matched product of more than two skew braces following the case of skew braces of abelian type (see [12] for the matched product of several skew braces of abelian type).

Let B be a finite skew brace of nilpotent type. Let $B_1, \dots, B_n$ be the Sylow subgroups of $(B, +)$. Then $(B, +)$ is the direct product of these Sylow subgroups and, moreover, $B_1, \dots, B_n$ are left ideals of B. Thus B is the matched product of $B_1, \dots, B_n$. In particular, every finite simple non-trivial skew brace of nilpotent type is the matched product of two or more skew braces of nilpotent type of order a power of a prime. But, in fact all the known finite simple non-trivial skew braces of abelian type are constructed using another kind of product, the so-called asymmetric product.

18.3 Asymmetric Product of Skew Braces of Abelian Type

Let S and T be two (additive) abelian groups. Recall that a (normalized) *symmetric 2-cocycle* on T with values in S is a map $b\colon T \times T \to S$ such that

(1) $b(0, 0) = 0$;
(2) $b(t_1, t_2) = b(t_2, t_1)$;
(3) $b(t_1 + t_2, t_3) + b(t_1, t_2) = b(t_1, t_2 + t_3) + b(t_2, t_3)$, for all $t_1, t_2, t_3 \in T$.

As a consequence, we get that $b(t, 0) = b(0, t) = 0$, for all $t \in T$.

Theorem 18.3.1 *Let T and S be two skew braces of abelian type. Let $b\colon T \times T \to S$ be a symmetric 2-cocycle on $(T, +)$ with values in $(S, +)$, and let $\alpha\colon (S, \circ) \to \mathrm{Aut}(T, +, \circ)$ be a homomorphism of groups such that*

$$s \circ b(t_2, t_3) + b(t_1 \circ \alpha_s(t_2 + t_3), t_1) = b(t_1 \circ \alpha_s(t_2), t_1 \circ \alpha_s(t_3)) + s, \quad (18.3)$$

where $\alpha_s = \alpha(s)$ for all $s \in S$ and $t_1, t_2, t_3 \in T$. Then the addition and multiplication over $T \times S$ given by

$$(t_1, s_1) + (t_2, s_2) = (t_1 + t_2,\ s_1 + s_2 + b(t_1, t_2)),$$

$$(t_1, s_1) \circ (t_2, s_2) = (t_1 \circ \alpha_{s_1}(t_2),\ s_1 \circ s_2)$$

define a structure of skew brace of abelian type on $T \times S$. We call this skew brace the asymmetric product *of T by S (via b and α) and denote it by $T \rtimes_\circ S$.*

Proof It is easy to check that $(T \times S, +)$ is an abelian group and that $(T \times S, \circ)$ is a group. In fact, $(T \times S, +)$ is the extension of the abelian groups S and T via the 2-cocycle b, and $(T \times S, \circ)$ is the semidirect product $T \rtimes_\alpha S$ of T by S via α. Thus it is enough to check that

$$(t_1, s_1) \circ ((t_2, s_2) + (t_3, s_3)) = (t_1, s_1) \circ (t_2, s_2) - (t_1, s_1) + (t_1, s_1) \circ (t_3, s_3).$$

We have that

$$\begin{aligned}
&(t_1, s_1) \circ (t_2, s_2) - (t_1, s_1) + (t_1, s_1) \circ (t_3, s_3)\\
&= (t_1 \circ \alpha_{s_1}(t_2), s_1 \circ s_2) - (t_1, s_1) + (t_1 \circ \alpha_{s_1}(t_3), s_1 \circ s_3)\\
&= (t_1 \circ \alpha_{s_1}(t_2) - t_1, s_1 \circ s_2 - s_1 - b(t_1 \circ \alpha_{s_1}(t_2) - t_1, t_1)) + (t_1 \circ \alpha_{s_1}(t_3), s_1 \circ s_3)\\
&= (t_1 \circ \alpha_{s_1}(t_2) - t_1 + t_1 \circ \alpha_{s_1}(t_3), s_1 \circ s_2 - s_1 - b(t_1 \circ \alpha_{s_1}(t_2) - t_1, t_1) + s_1 \circ s_3\\
&\qquad + b(t_1 \circ \alpha_{s_1}(t_2) - t_1, t_1 \circ \alpha_{s_1}(t_3)))\\
&= (t_1 \circ (\alpha_{s_1}(t_2) + \alpha_{s_1}(t_3)), s_1 \circ (s_2 + s_3) - b(t_1 \circ \alpha_{s_1}(t_2) - t_1, t_1)\\
&\qquad + b(t_1 \circ \alpha_{s_1}(t_2) - t_1, t_1 \circ \alpha_{s_1}(t_3)))\\
&= (t_1 \circ (\alpha_{s_1}(t_2 + t_3)), s_1 \circ (s_2 + s_3) + b(t_1 \circ \alpha_{s_1}(t_2), t_1 \circ \alpha_{s_1}(t_3))\\
&\qquad - b(t_1 \circ \alpha_{s_1}(t_2) - t_1 + t_1 \circ \alpha_{s_1}(t_3), t_1))\\
&= (t_1 \circ (\alpha_{s_1}(t_2 + t_3)), s_1 \circ (s_2 + s_3) + b(t_1 \circ \alpha_{s_1}(t_2), t_1 \circ \alpha_{s_1}(t_3))\\
&\qquad - b(t_1 \circ (\alpha_{s_1}(t_2) + \alpha_{s_1}(t_3)), t_1))\\
&= (t_1 \circ (\alpha_{s_1}(t_2 + t_3)), s_1 \circ (s_2 + s_3) + b(t_1 \circ \alpha_{s_1}(t_2), t_1 \circ \alpha_{s_1}(t_3))\\
&\qquad - b(t_1 \circ \alpha_{s_1}(t_2 + t_3)), t_1))\\
&= (t_1 \circ (\alpha_{s_1}(t_2 + t_3)), s_1 \circ (s_2 + s_3) - s_1 + s_1 \circ b(t_2, t_3)) \qquad \text{(by (18.3))}\\
&= (t_1 \circ (\alpha_{s_1}(t_2 + t_3)), s_1 \circ (s_2 + s_3 + b(t_2, t_3))
\end{aligned}$$

$$= (t_1, s_1) \circ (t_2 + t_3, s_2 + s_3 + b(t_2, t_3))$$
$$= (t_1, s_1) \circ ((t_2, s_2) + (t_3, s_3)).$$

Therefore the result follows. □

Note that the lambda map of $T \rtimes_\circ S$ is defined by

$$\lambda_{(t_1,s_1)}(t_2, s_2) = \left(\lambda_{t_1}\alpha_{s_1}(t_2),\ \lambda_{s_1}(s_2) - b(\lambda_{t_1}\alpha_{s_1}(t_2), t_1)\right), \tag{18.4}$$

and its socle is

$$\mathrm{Soc}(T \rtimes_\circ S) = \{(t, s) : \lambda_s = \mathrm{id}_S,\ \lambda_t \circ \alpha_s = \mathrm{id}_T,\ b(t, t_1) = 0 \text{ for all } t_1 \in T\}.$$

Moreover, the subset $T \times \{0\}$ is a normal subgroup of $(T \rtimes_\circ S, \circ)$, and $\{0\} \times S$ is a left ideal of $T \rtimes_\circ S$.

A particular case of this theorem is when we assume that b is a symmetric bilinear form. In this case, b is a 2-cocycle and condition (18.3) becomes

$$\lambda_s(b(t_2,\ t_3)) = b(\lambda_{t_1}\alpha_s(t_2),\ \lambda_{t_1}\alpha_s(t_3)),$$

which is equivalent to the two conditions:

$$\lambda_s(b(t_2,\ t_3)) = b(\alpha_s(t_2),\ \alpha_s(t_3)), \tag{18.5}$$
$$b(t_2,\ t_3) = b(\lambda_{t_1}(t_2),\ \lambda_{t_1}(t_3)), \tag{18.6}$$

for all $s \in S$ and $t_1, t_2, t_3 \in T$. Furthermore, if T and S are trivial skew braces of abelian type, then these two conditions (18.5) and (18.6) reduce to

$$b(t_2,\ t_3) = b(\alpha_s(t_2),\ \alpha_s(t_3)),$$

for all $s \in S$ and $t_2, t_3 \in T$, i.e., α_s is in the orthogonal group of the bilinear form b, for all $s \in S$.

18.4 Construction of Simple Braces of Abelian Type

Let p, q be two distinct primes and n_1, n_2 be positive integers. Consider the trivial skew braces over the abelian groups

$$T = (\mathbb{Z}/(p))^{n_1} \times (\mathbb{Z}/(q))^{n_2} \quad \text{and} \quad S = \mathbb{Z}/(p) \times \mathbb{Z}/(q)$$

and let

$$b_1\colon (\mathbb{Z}/(p))^{n_1} \times (\mathbb{Z}/(p))^{n_1} \to \mathbb{Z}/(p), \quad b_2\colon (\mathbb{Z}/(q))^{n_2} \times (\mathbb{Z}/(q))^{n_2} \to \mathbb{Z}/(q)$$

be two non-degenerate symmetric bilinear forms. Let c_1 be an element of order q in the orthogonal group of b_1 and let c_2 be an element of order p in the orthogonal group of b_2. Let

$$\alpha\colon S \to \operatorname{Aut}(T), \quad (i,j) \mapsto \alpha_{(i,j)}$$

be given by

$$\alpha_{(i,j)}(u,v) = (c_1^j(u), c_2^i(v))$$

for all $(i,j) \in S$ and $(u,v) \in T$. Let $b\colon T \times T \to S$ be the map defined by $b((u_1,v_1),(u_2,v_2)) = (b_1(u_1,u_2), b_2(v_1,v_2))$ for all $(u_1,v_1),(u_2,v_2) \in T$. Then

$$\begin{aligned} b(\alpha_{(i,j)}(u_1,v_1), \alpha_{(i,j)}(u_2,v_2)) &= (b_1(c_1^j(u_1), c_1^j(u_2)), b_2(c_2^i(v_1), c_2^i(v_2))) \\ &= (b_1(u_1,u_2), b_2(v_1,v_2)). \end{aligned}$$

Hence we can consider the asymmetric product $T \rtimes_\circ S$ of T by S (via b and α).

Theorem 18.4.1 *With the above notation, the skew brace $T \rtimes_\circ S$ is simple if and only if $c_1 - \mathrm{id}$ and $c_2 - \mathrm{id}$ are invertible.*

Proof Let $J = \{((u,v),(i,j)) : u \in \operatorname{im}(c_1 - \mathrm{id}),\ v \in \operatorname{im}(c_2 - \mathrm{id}),\ (i,j) \in S\}$. We shall prove that J is an ideal of $T \rtimes_\circ S$. Clearly J is a subgroup of $(T \rtimes_\circ S, +)$. Let $((u_1,v_1),(i_1,j_1)) \in J$ and $((u_2,v_2),(i_2,j_2)) \in T \rtimes_\circ S$. We have that

$$\begin{aligned} &\lambda_{((u_2,v_2),(i_2,j_2))}((u_1,v_1),(i_1,j_1)) \\ &= -((u_2,v_2),(i_2,j_2)) + ((u_2,v_2),(i_2,j_2)) \circ ((u_1,v_1),(i_1,j_1)) \\ &= ((c_1^{j_2}(u_1), c_2^{i_2}(v_1)), (i_1 - b_1(c_1^{j_2}(u_1),u_2), j_1 - b_2(c_2^{i_2}(v_1),v_2))) \in J. \end{aligned}$$

Hence J is a left ideal of $T \rtimes_\circ S$. Furthermore,

$$\begin{aligned} &((u_1,v_1),(i_1,j_1)) * ((u_2,v_2),(i_2,j_2)) \\ &\quad = -((u_1,v_1),(i_1,j_1)) + ((u_1,v_1),(i_1,j_1)) \circ ((u_2,v_2),(i_2,j_2)) \\ &\qquad - ((u_2,v_2),(i_2,j_2)) \\ &\quad = ((c_1^{j_1}(u_2), c_2^{i_1}(v_2)), (i_2 - b_1(c_1^{j_1}(u_2),u_1), j_2 - b_2(c_2^{i_1 i_2}(v_2),v_1))) \\ &\qquad - ((u_2,v_2),(i_2,j_2)) \\ &\quad = ((c_1^{j_1}(u_2) - u_2, c_2^{i_1}(v_2) - v_2), (-b_1(c_1^{j_1}(u_2),u_1) - b_1(c_1^{j_1}(u_2) - u_2, u_2), \\ &\qquad - b_2(c_2^{i_1 i_2}(v_2),v_1) - b_2(c_2^{i_1}(v_2) - v_2, v_2))) \in J. \end{aligned}$$

Thus J is an ideal of $T \rtimes_\circ S$. Therefore if $T \rtimes_\circ S$ is simple, then $c_1 - \mathrm{id}$ and $c_2 - \mathrm{id}$ are invertible.

Conversely, suppose that $c_1 - \mathrm{id}$ and $c_2 - \mathrm{id}$ are invertible. Let I be a nonzero ideal of $T \rtimes_\circ S$. Let $((u_1,v_1),(i_1,j_1)) \in I$ be a nonzero element. Suppose first that $(u_1,v_1) = (0,0)$. In this case, $(i_1,j_1) \neq (0,0)$. We may assume that $i_1 \neq 0$. Note that $q((0,0),(i_1,j_1)) = ((0,0),(qi_1,0)) \in I$. Hence we may assume that $j_1 = 0$.

Let m be a positive integer such that $mi_1 = 1 \in \mathbb{Z}/(p)$. Then $m((0,0),(i_1,0)) = ((0,0),(1,0)) \in I$. Now,

$$\begin{aligned}((0,0),(1,0)) &* ((u_2,v_2),(i_2,j_2))\\ &= ((0,(c_2-\mathrm{id})(v_2)),(0,-b_2((c_2-\mathrm{id})(v_2),v_2))) \in I\end{aligned}$$

for all $((u_2,v_2),(i_2,j_2)) \in T \rtimes_\circ S$. Since $c_2 - \mathrm{id}$ is invertible, for every $w \in (\mathbb{Z}/(q))^{n_2}$ there exists $l_w \in \mathbb{Z}/(q)$ such that $((0,w),(0,l_w)) \in I$. Let $l \in \mathbb{Z}/(q)$. Note that if $w \neq 0$, since b_2 is non-degenerate, there exists $t \in (\mathbb{Z}/(q))^{n_2}$ such that $b_2(w,t) = l_w - l$. Hence

$$\lambda_{((0,t),(0,0))}((0,w),(0,l_w)) = ((0,w),(0,l_w - b_2(w,t))) = ((0,w),(0,l)) \in I$$

for all $w \in (\mathbb{Z}/(q))^{n_2} \setminus \{0\}$ and $l \in \mathbb{Z}/(q)$. Thus, if $w \neq 0$, then

$$((0,w),(0,l)) - ((0,w),(0,0)) = ((0,0),(0,l)) \in I$$

for all $l \in \mathbb{Z}/(q)$. In particular, $((0,0),(0,1)) \in I$. Now by a similar argument one can see that $((u,0),(k,0)) \in I$, for all $u \in (\mathbb{Z}/(p))^{n_1}$ and $k \in \mathbb{Z}/(p)$. Now

$$((u,0),(k,0)) + ((0,w),(0,l)) = ((u,w),(k,l)) \in I$$

for all $u \in (\mathbb{Z}/(p))^{n_1}$, $w \in (\mathbb{Z}/(q))^{n_2}$, $k \in \mathbb{Z}/(p)$, and $l \in \mathbb{Z}/(q)$. Thus $I = T \rtimes_\circ S$ in this case.

Suppose now that $(u_1,v_1) \neq (0,0)$. In this case, we may assume that $u_1 \neq 0$. Then $q((u_1,v_1),(i_1,j_1)) = ((qu_1,0),(i_2,j_2)) \in I$, for some $(i_2,j_2) \in S$. Furthermore

$$q((qu_1,0),(i_2,j_2)) = ((q^2u_1,0),(i_3,0)) \in I$$

for some $i_3 \in \mathbb{Z}/(p)$. Since $q^2u_1 \neq 0$ and the bilinear form b_1 is non-degenerate, there exists $u \in (\mathbb{Z}/(p))^{n_1}$ such that $b_1(q^2u_1,u) = i_3 - k$. Hence

$$\begin{aligned}\lambda_{((u,0),(0,0))}((q^2u_1,0),(i_3,0)) &= ((q^2u_1,0),(i_3 - b_1(q^2u_1,u)))\\ &= ((q^2u_1,0),(k,0)) \in I\end{aligned}$$

for all $k \in \mathbb{Z}/(p)$, and then

$$((q^2u_1,0),(1,0)) - ((q^1u_1,0),(0,0)) = ((0,0),(1,0)) \in I.$$

Now, as above, one can prove that $I = T \rtimes_\circ S$, and the result follows. □

We shall give now some concrete constructions of simple skew braces of abelian type using Theorem 18.4.1. Let p, q be two distinct primes. Let n, m be two positive integers. Let

$$D_1 = \begin{pmatrix} 0 & 0 \dots 0 & -1 \\ 1 & 0 \dots 0 & -1 \\ 0 & \ddots \quad \ddots \quad \vdots & \vdots \\ \vdots & \ddots \quad \ddots \quad 0 & -1 \\ 0 \dots & 0 \; 1 & -1 \end{pmatrix}, \quad E_1 = \begin{pmatrix} 1-q & 1 & \dots & 1 \\ 1 & \ddots & \ddots & \vdots \\ \vdots & \ddots & \ddots & 1 \\ 1 & \dots & 1 & 1-q \end{pmatrix} \in M_{q-1}(\mathbb{Z}/(p))$$

and

$$D_2 = \begin{pmatrix} 0 & 0 \dots 0 & -1 \\ 1 & 0 \dots 0 & -1 \\ 0 & \ddots \quad \ddots \quad \vdots & \vdots \\ \vdots & \ddots \quad \ddots \quad 0 & -1 \\ 0 \dots & 0 \; 1 & -1 \end{pmatrix}, \quad E_2 = \begin{pmatrix} 1-p & 1 & \dots & 1 \\ 1 & \ddots & \ddots & \vdots \\ \vdots & \ddots & \ddots & 1 \\ 1 & \dots & 1 & 1-p \end{pmatrix} \in M_{p-1}(\mathbb{Z}/(q)).$$

Note that D_1 is the companion matrix of the polynomial $1 + x + \cdots + x^{q-1}$ and thus it has order q. Similarly D_2 has order p. It is easy to check that $D_1^t E_1 D_1 = E_1$ and $D_2^t E_2 D_2 = E_2$. Consider in $M_{n_1(q-1)}(\mathbb{Z}/(p))$ the block diagonal matrices with n_1 blocks of degree $q - 1$:

$$C_1 = \begin{pmatrix} D_1 & 0 & \dots & 0 \\ 0 & \ddots & \ddots & \vdots \\ \vdots & \ddots & \ddots & 0 \\ 0 & \dots & 0 & D_1 \end{pmatrix} \text{ and } B_1 = \begin{pmatrix} E_1 & 0 & \dots & 0 \\ 0 & \ddots & \ddots & \vdots \\ \vdots & \ddots & \ddots & 0 \\ 0 & \dots & 0 & E_1 \end{pmatrix}.$$

Similarly consider in $M_{n_2(p-1)}(\mathbb{Z}/(q))$ the block diagonal matrices with n_2 blocks of degree $p - 1$:

$$C_2 = \begin{pmatrix} D_2 & 0 & \dots & 0 \\ 0 & \ddots & \ddots & \vdots \\ \vdots & \ddots & \ddots & 0 \\ 0 & \dots & 0 & D_2 \end{pmatrix} \text{ and } B_2 = \begin{pmatrix} E_2 & 0 & \dots & 0 \\ 0 & \ddots & \ddots & \vdots \\ \vdots & \ddots & \ddots & 0 \\ 0 & \dots & 0 & E_2 \end{pmatrix}.$$

Let $T = (\mathbb{Z}/(p))^{n_1(q-1)} \times (\mathbb{Z}/(q))^{n_2(p-1)}$ and $S = \mathbb{Z}/(p) \times \mathbb{Z}/(q)$ be the trivial skew braces over these abelian groups. Let $b\colon T \times T \to S$ be the symmetric bilinear map defined by

$$b((u_1, v_1), (u_2, v_2)) = (u_1 B_1 u_2^t, v_1 B_2 v_2^t),$$

for all $u_1, u_2 \in (\mathbb{Z}/(p))^{n_1(q-1)}$ and $v_1, v_2 \in (\mathbb{Z}/(q))^{n_2(p-1)}$.

Note that $\det(B_1) = (-q)^{q-2} \neq 0$ and $\det(B_2) = (-p)^{p-2} \neq 0$, thus the symmetric bilinear forms associated with these matrices are non-degenerate. Let $\alpha\colon S \to \operatorname{Aut}(T)$ be the map defined by $\alpha(i, j) = \alpha_{(i,j)}$ and $\alpha_{(i,j)}(u, v) = (u(C_1^t)^j, v(C_2^t)^i)$, for all $(i, j) \in S$ and $(u, v) \in T$. Since $C_k^t B_k C_k = B_k$, for $k = 1, 2$, we can consider the asymmetric product $T \rtimes_\circ S$ of T by S (via b and α). Since $C_k - I$, where I is the identity matrix, is invertible for $k \in \{1, 2\}$, by Theorem 18.4.1, $T \rtimes_\circ S$ is a simple skew brace of abelian type.

Another construction is based on the following fact. Let D be an invertible matrix in $M_n(K)$, where K is a field. Consider the block matrices

$$C = \begin{pmatrix} D^t & 0 \\ 0 & D^{-1} \end{pmatrix} \in M_{2n}(K) \quad \text{and} \quad B = \begin{pmatrix} 0 & I_n \\ I_n & 0 \end{pmatrix} \in M_{2n}(K).$$

Then one can check that $C^t BC = B$. Take any invertible matrix $D_1 \in M_{m_1}(\mathbb{Z}/(p))$ of order q and any matrix $D_2 \in M_{m_2}(\mathbb{Z}/(q))$ of order p, such that $D_k - I$ also is invertible for $k = 1, 2$. Consider in $M_{2n_1m_1}(\mathbb{Z}/(p))$ the block square matrices with $2n_1$ blocks of degree m_1 in the diagonal:

$$C_1 = \begin{pmatrix} D_1^t & 0 & 0 & \dots & 0 & 0 \\ 0 & D_1^{-1} & 0 & \dots & 0 & 0 \\ 0 & 0 & \ddots & \ddots & \vdots & \vdots \\ \vdots & \vdots & \ddots & \ddots & 0 & 0 \\ 0 & 0 & \dots & 0 & D_1^t & 0 \\ 0 & 0 & \dots & 0 & 0 & D_1^{-1} \end{pmatrix} \text{ and } B_1 = \begin{pmatrix} 0 & I_{m_1} & 0 & \dots & 0 & 0 \\ I_{m_1} & 0 & 0 & \dots & 0 & 0 \\ 0 & 0 & \ddots & \ddots & \vdots & \vdots \\ \vdots & \vdots & \ddots & \ddots & 0 & 0 \\ 0 & 0 & \dots & 0 & 0 & I_{m_1} \\ 0 & 0 & \dots & 0 & I_{m_1} & 0 \end{pmatrix}.$$

Similarly consider in $M_{2n_2m_2}(\mathbb{Z}/(q))$ the block square matrices with $2n_2$ blocks of degree m_2 in the diagonal:

$$C_2 = \begin{pmatrix} D_2^t & 0 & 0 & \dots & 0 & 0 \\ 0 & D_2^{-1} & 0 & \dots & 0 & 0 \\ 0 & 0 & \ddots & \ddots & \vdots & \vdots \\ \vdots & \vdots & \ddots & \ddots & 0 & 0 \\ 0 & 0 & \dots & 0 & D_2^t & 0 \\ 0 & 0 & \dots & 0 & 0 & D_2^{-1} \end{pmatrix} \text{ and } B_1 = \begin{pmatrix} 0 & I_{m_2} & 0 & \dots & 0 & 0 \\ I_{m_2} & 0 & 0 & \dots & 0 & 0 \\ 0 & 0 & \ddots & \ddots & \vdots & \vdots \\ \vdots & \vdots & \ddots & \ddots & 0 & 0 \\ 0 & 0 & \dots & 0 & 0 & I_{m_2} \\ 0 & 0 & \dots & 0 & I_{m_2} & 0 \end{pmatrix}.$$

Let $T = (\mathbb{Z}/(p))^{2n_1m_1} \times (\mathbb{Z}/(q))^{2n_2m_2}$ and $S = \mathbb{Z}/(p) \times \mathbb{Z}/(q)$ be the trivial skew braces over these abelian groups. Let $b\colon T \times T \to S$ be the symmetric bilinear map defined by

$$b((u_1, v_1), (u_2, v_2)) = (u_1 B_1 u_2^t, v_1 B_2 v_2^t),$$

for all $u_1, u_2 \in (\mathbb{Z}/(p))^{2n_1m_1}$ and $v_1, v_2 \in (\mathbb{Z}/(q))^{2n_2m_2}$. Let $\alpha\colon S \to \operatorname{Aut}(T)$ be the map defined by $\alpha(i, j) = \alpha_{(i,j)}$ and $\alpha_{(i,j)}(u, v) = (u(C_1^t)^j, v(C_2^t)^i)$, for all $(i, j) \in S$ and $(u, v) \in T$. Since $C_k^t B_k C_k = B_k$ for $k = 1, 2$, we can consider the asymmetric product $T \rtimes_\circ S$ of T by S (via b and α). Since $C_k - I$, where I is the identity matrix, is invertible for $k = 1, 2$, by Theorem 18.4.1, $T \rtimes_\circ S$ is a simple skew brace of abelian type.

For example, for $p = 2$ and $q = 31$, with the first construction, we obtain simple skew braces of abelian type of order $2^{30n_1}31^{n_2}$ for all positive integers n_1, n_2. With the second construction taking

$$D_1 = \begin{pmatrix} 0\,0\,0\,0\,1 \\ 1\,0\,0\,0\,0 \\ 0\,1\,0\,0\,1 \\ 0\,0\,1\,0\,0 \\ 0\,0\,0\,1\,0 \end{pmatrix} \in M_5(\mathbb{Z}/(2)),$$

we obtain simple skew braces of abelian type of order $2^{10n_1}31^{2n_2}$ for all positive integers n_1, n_2. Using the two methods, one can easily construct simple skew braces of abelian type of order $2^{10n_1}31^{n_2}$ for all positive integers n_1, n_2.

18.5 Exercises

18.5.1 Construct a concrete example of a simple skew brace of abelian type of order $2^{10}31$.

18.6 Open Problems

Open problems 18.6.1 Is every finite simple skew brace of abelian type the asymmetric product of two nonzero skew braces of abelian type?

18.7 Notes

The first example of a simple non-trivial skew brace of abelian type is due to Bachiller [9]. In the same paper, Bachiller introduced the matched product of two skew braces of abelian type. One can see that the definition of matched product of two skew braces is the same that in the case of skew braces of abelian type. In [12], Bachiller, Cedó, Jespers, and Okniński introduced the matched product of more than two skew braces of abelian type (this can be generalized to arbitrary skew braces without changes). In the same paper, the authors construct several families of simple non-trivial skew braces of abelian type using matched products.

The asymmetric product of skew braces of abelian type was introduced by Catino, Colazzo, and Stefanelli in [33], and Theorem 18.3.1 is due to them ([33, Theorem 3]). In [13], Bachiller, Cedó, Jespers, and Okniński used the asymmetric product to construct new families of simple non-trivial skew braces of abelian type. In fact, every known simple non-trivial skew brace of abelian type is an asymmetric product of two nonzero skew braces of abelian type (see [13, 40, 41]). Theorem 18.4.1 is based on [12, Theorem 3.6] and [13, Theorem 6.2]. The two concrete constructions of simple non-trivial simple skew braces of abelian type appear in [12, 40].

References

1. E. Acri, R. Lutowski, L. Vendramin, Retractability of solutions to the Yang-Baxter equation and p-nilpotency of skew braces. Internat. J. Algebra Comput. **30**(1), 91–115 (2020)
2. O. Akgün, M. Mereb, L. Vendramin, Enumeration of set-theoretic solutions to the Yang-Baxter equation. Math. Comp. **91**(335), 1469–1481 (2022)
3. B. Amberg, S. Franciosi, F. de Giovanni, *Products of groups*. Oxford Mathematical Monographs (The Clarendon Press, Oxford University Press, New York, 1992). Oxford Science Publications
4. S.A. Amitsur, Nil radicals. Historical notes and some new results, in *Rings, modules and radicals (Proceedings of the International Colloquium, Keszthely, 1971)*, vol. 6 (Colloquium Mathematical Society János Bolyai, 1973), pp. 47–65
5. N. Andruskiewitsch, M. Graña, From racks to pointed Hopf algebras. Adv. Math. **178**(2), 177–243 (2003)
6. M. Ashford and O. Riordan. Counting racks of order n. *Electron. J. Combin.*, 24(2):Paper No. 2.32, 20, 2017
7. D. Bachiller, Classification of braces of order p^3. J. Pure Appl. Algebra **219**(8), 3568–3603 (2015)
8. D. Bachiller, Counterexample to a conjecture about braces. J. Algebra **453**, 160–176 (2016)
9. D. Bachiller, Extensions, matched products, and simple braces. J. Pure Appl. Algebra **222**(7), 1670–1691 (2018)
10. D. Bachiller, Solutions of the Yang Baxter equation associated to skew left braces, with applications to racks. J. Knot Theory Ramifications **27**(8), 1850055 (2018)
11. D. Bachiller, F. Cedó, E. Jespers, Solutions of the Yang-Baxter equation associated with a left brace. J. Algebra **463**, 80–102 (2016)
12. D. Bachiller, F. Cedó, E. Jespers, J. Okniński, Iterated matched products of finite braces and simplicity; new solutions of the Yang-Baxter equation. Trans. Amer. Math. Soc. **370**(7), 4881–4907 (2018)
13. D. Bachiller, F. Cedó, E. Jespers, J. Okniński, Asymmetric product of left braces and simplicity; new solutions of the Yang-Baxter equation. Commun. Contemp. Math. **21**(8), 1850042, 30 pp. (2019)
14. D. Bachiller, F. Cedó, L. Vendramin, A characterization of finite multipermutation solutions of the Yang-Baxter equation. Publ. Mat. **62**(2), 641–649 (2018)
15. A. Ballester-Bolinches, R. Esteban-Romero, V. Pérez-Calabuig, Enumeration of left braces with additive group $C_4 \times C_4 \times C_4$. Math. Comp. **93**(346), 911–919 (2024)
16. A. Ballester-Bolinches, M. Ferrara, V. Pérez-Calabuig, M. Trombetti, A note on right-nil and strong-nil skew braces. arXiv:2310.11123v2
17. V.G. Bardakov, M.V. Neshchadim, M.K. Yadav, Computing skew left braces of small orders. Internat. J. Algebra Comput. **30**(4), 839–851 (2020)
18. R.J. Baxter, Partition function of the eight-vertex lattice model. Ann. Physics **70**, 193–228 (1972)

F. Cedó and L. Vendramin, *Groups, Radical Rings, and the Yang-Baxter Equation*, Progress in Mathematics 361, https://doi.org/10.1007/978-3-032-01524-2_1

19. N. Ben David, Y. Ginosar, On groups of I-type and involutive Yang-Baxter groups. *J. Algebra* 458, 197–206 (2016)
20. G.M. Bergman, Errata: A ring primitive on the right but not on the left. Proc. Amer. Math. Soc. **15**, 1000 (1964)
21. G.M. Bergman, A ring primitive on the right but not on the left. Proc. Amer. Math. Soc. **15**, 473–475 (1964)
22. G.M. Bergman, Right orderable groups that are not locally indicable. Pacific J. Math. **147**(2), 243–248 (1991)
23. S.R. Blackburn, Enumerating finite racks, quandles and kei. Electron. J. Combin. **20**(3):Paper 43, 9 (2013)
24. B.H. Bowditch, A variation on the unique product property. J. London Math. Soc. (2) **62**(3), 813–826 (2000)
25. M. Brešar, *Introduction to noncommutative algebra* (Universitext. Springer, Cham, 2014)
26. E. Brieskorn, Automorphic sets and braids and singularities, in *Braids (Santa Cruz, CA, 1986)*. Contemporary Mathematics, vol. 78 (American Mathematical Society, Providence, RI, 1988), pp. 45–115
27. R.G. Burns, V.W.D. Hale, A note on group rings of certain torsion-free groups. Canad. Math. Bull. **15**, 441–445 (1972)
28. N.P. Byott, Nilpotent and abelian Hopf-Galois structures on field extensions. J. Algebra **381**, 131–139 (2013)
29. N.P. Byott, Solubility criteria for Hopf-Galois structures. New York J. Math. **21**, 883–903 (2015)
30. N.P. Byott, On insoluble transitive subgroups in the holomorph of a finite soluble group. J. Algebra **638**, 1–31 (2024)
31. A. Caranti, F. Dalla Volta, M. Sala, Abelian regular subgroups of the affine group and radical rings. Publ. Math. Debrecen **69**(3):297–308 (2006)
32. M. Castelli, F. Catino, G. Pinto, Indecomposable involutive set-theoretic solutions of the Yang-Baxter equation. J. Pure Appl. Algebra **223**(10), 4477–4493 (2019)
33. F. Catino, I. Colazzo, P. Stefanelli, Regular subgroups of the affine group and asymmetric product of radical braces. J. Algebra **455**, 164–182 (2016)
34. F. Catino, I. Colazzo, P. Stefanelli, Skew left braces with non-trivial annihilator. J. Algebra Appl. **18**(2), 1950033, 23 (2019)
35. F. Catino, R. Rizzo, Regular subgroups of the affine group and radical circle algebras. Bull. Aust. Math. Soc. **79**(1), 103–107 (2009)
36. F. Cedó, Left braces: solutions of the Yang-Baxter equation. Adv. Group Theory Appl. **5**, 33–90 (2018)
37. F. Cedó, T. Gateva-Ivanova, A. Smoktunowicz, Braces and symmetric groups with special conditions. J. Pure Appl. Algebra **222**(12), 3877–3890 (2018)
38. F. Cedó, E. Jespers, L. Kubat, A. Van Antwerpen, C. Verwimp, On various types of nilpotency of the structure monoid and group of a set-theoretic solution of the Yang-Baxter equation. J. Pure Appl. Algebra **227**(2), Paper No. 107194 (2023)
39. F. Cedó, E. Jespers, J. Okniński, Braces and the Yang-Baxter equation. Comm. Math. Phys. **327**(1), 101–116 (2014)
40. F. Cedó, E. Jespers, J. Okniński, An abundance of simple left braces with abelian multiplicative Sylow subgroups. Rev. Mat. Iberoam. **36**(5), 1309–1332 (2020)
41. F. Cedó, E. Jespers, J. Okniński, Every finite abelian group is a subgroup of the additive group of a finite simple left brace. J. Pure Appl. Algebra **225**(1), 106476, 10 (2021)
42. F. Cedó, E. Jespers, J. Okniński, Primitive set-theoretic solutions of the Yang-Baxter equation. Commun. Contemp. Math. **24**(9), Paper No. 2150105, 10 (2022)
43. F. Cedó, J. Okniński, Constructing finite simple solutions of the Yang-Baxter equation. Adv. Math. **391**, Paper No. 107968, 40 (2021)
44. F. Cedó, J. Okniński, New simple solutions of the Yang-Baxter equation and solutions associated to simple left braces. J. Algebra **600**, 125–151 (2022)

45. F. Cedó, J. Okniński, Corrigendum to "Constructing finite simple solutions of the Yang–Baxter equation. [Adv. Math. 391 (2021) 107968]. Adv. Math. **451**, Paper No. 109807, 3 (2024)
46. F. Cedó, A. Smoktunowicz, L. Vendramin, Skew left braces of nilpotent type. Proc. Lond. Math. Soc. (3) **118**(6), 1367–1392 (2019)
47. F. Chouraqui, Garside groups and Yang-Baxter equation. Comm. Algebra **38**(12), 4441–4460 (2010)
48. F. Chouraqui, Left orders in Garside groups. Internat. J. Algebra Comput. **26**(7), 1349–1359 (2016)
49. P.M. Cohn, *Algebra*, vol. 2. (Wiley, London-New York-Sydney, 1977). With errata to Vol. I
50. P.M. Cohn, *Algebra*, vol. 1, 2nd edn. (Wiley, New York, 1982)
51. M. Deakonesku, G.L. Uolls, On the orbits of automorphism groups. Sibirsk. Mat. Zh. **46**(3), 533–537 (2005)
52. P. Dehornoy, Set-theoretic solutions of the Yang-Baxter equation, RC-calculus, and Garside germs. Adv. Math. **282**, 93–127 (2015)
53. V.G. Drinfel'd, On some unsolved problems in quantum group theory, in *Quantum Groups (Leningrad, 1990)*. Lecture Notes in Mathematics, vol. 1510 (Springer, Berlin, 1992), pp. 1–8
54. M. Elhamdadi, S. Nelson, *Quandles–an Introduction to the Algebra of Knots*. Student Mathematical Library, vol. 74. (American Mathematical Society, Providence, RI, 2015)
55. P. Etingof, S. Gelaki, A method of construction of finite-dimensional triangular semisimple Hopf algebras. Math. Res. Lett. **5**(4), 551–561 (1998)
56. P. Etingof, T. Schedler, A. Soloviev, Set-theoretical solutions to the quantum Yang-Baxter equation. Duke Math. J. **100**(2), 169–209 (1999)
57. P. Etingof, A. Soloviev, R. Guralnick, Indecomposable set-theoretical solutions to the quantum Yang-Baxter equation on a set with a prime number of elements. J. Algebra **242**(2), 709–719 (2001)
58. W. Feit, J.G. Thompson, Solvability of groups of odd order. Pacific J. Math. **13**, 775–1029 (1963)
59. The GAP Group, *GAP – Groups, Algorithms, and Programming, Version 4.11.1* (2021)
60. G. Gardam, A counterexample to the unit conjecture for group rings. Ann. Math. (2) **194**(3), 967–979 (2021)
61. T. Gateva-Ivanova, Set-theoretic solutions of the Yang-Baxter equation, braces and symmetric groups. Adv. Math. **338**, 649–701 (2018)
62. T. Gateva-Ivanova, M. Van den Bergh, Semigroups of I-type. J. Algebra **206**(1), 97–112 (1998)
63. I. Gorshkov, T. Nasybullov, Finite skew braces with solvable additive group. J. Algebra **574**, 172–183 (2021)
64. L. Guarnieri, L. Vendramin, Skew braces and the Yang-Baxter equation. Math. Comp. **86**(307), 2519–2534 (2017)
65. I.N. Herstein, A counterexample in Noetherian rings. Proc. Nat. Acad. Sci. U.S.A. **54**, 1036–1037 (1965)
66. I.N. Herstein, Noncommutative rings, volume 15 of Carus Mathematical Monographs. Mathematical Association of America, Washington, DC. *Reprint of the 1968 original* (With an afterword by Lance W, Small, 1994)
67. J. Hoste, P.D. Shanahan, An enumeration process for racks. Math. Comp. **88**(317), 1427–1448 (2019)
68. I.M. Isaacs, *Finite Group Theory, Graduate Studies in Mathematics*, vol. 92 (American Mathematical Society, Providence, RI, 2008)
69. I.M. Isaacs, Group actions and orbits. Arch. Math. (Basel) **98**(5), 399–401 (2012)
70. N. Jacobson, The radical and semi-simplicity for arbitrary rings. Amer. J. Math. **67**, 300–320 (1945)
71. N. Jacobson, *Structure of Rings*. American Mathematical Society Colloquium Publications, vol. 37, Revised edition (American Mathematical Society, Providence, R.I., 1964)
72. A.V. Jategaonkar, A counter-example in ring theory and homological algebra. J. Algebra **12**, 418–440 (1969)

73. P. Jedlička, A. Pilitowska, A. Zamojska-Dzienio, The construction of multipermutation solutions of the Yang-Baxter equation of level 2. J. Combin. Theory Ser. A **176**, 105295, 35 (2020)
74. E. Jespers, L. Kubat, A. Van Antwerpen, L. Vendramin, Factorizations of skew braces. Math. Ann. **375**(3–4), 1649–1663 (2019)
75. E. Jespers, J. Okniński, Monoids and groups of I-type. Algebr. Represent. Theory **8**(5), 709–729 (2005)
76. E. Jespers, J. Okniński, *Noetherian Semigroup Algebras*. Algebra and Applications, vol. 7. (Springer, Dordrecht, 2007)
77. I. Kaplansky, *Problems in the theory of rings. Report of a conference on linear algebras, June, 1956* (National Academy of Sciences-National Research Council, Washington, D.C., 1957), pp. 1–3 Publ. 502
78. I. Kaplansky, “Problems in the theory of rings” revisited. Amer. Math. Monthly **77**, 445–454 (1970)
79. S. Kionke, J. Raimbault, On geometric aspects of diffuse groups. Doc. Math. **21**, 873–915 (2016). With an appendix by Nathan Dunfield
80. A. Konovalov, A. Smoktunowicz, L. Vendramin, On skew braces and their ideals. Exp. Math. **30**(1), 95–104 (2021)
81. G. Köthe, Die Struktur der Ringe, deren Restklassenring nach dem Radikal vollständig reduzibel ist. Math. Z. **32**(1), 161–186 (1930)
82. J. Krempa, Logical connections between some open problems concerning nil rings. Fund. Math. **76**(2), 121–130 (1972)
83. E. Landau, Über die Klassenzahl der binären quadratischen Formen von negativer Discriminante. Math. Ann. **56**(4), 671–676 (1903)
84. I. Lau, An associative left brace is a ring. J. Algebra Appl. **19**(9), 2050179, 6 (2020)
85. V. Lebed, L. Vendramin, On structure groups of set-theoretic solutions to the Yang-Baxter equation. Proc. Edinb. Math. Soc. (2) **62**(3), 683–717 (2019)
86. P. Linnell, D. Witte Morris, Amenable groups with a locally invariant order are locally indicable. Groups Geom. Dyn. **8**(2), 467–478 (2014)
87. J.-H. Lu, M. Yan, Y.-C. Zhu, On the set-theoretical Yang-Baxter equation. Duke Math. J. **104**(1), 1–18 (2000)
88. H. Meng, A. Ballester-Bolinches, R. Esteban-Romero, Left braces and the quantum Yang-Baxter equation. Proc. Edinb. Math. Soc. (2) **62**(2), 595–608 (2019)
89. M. Newman, A bound for the number of conjugacy classes in a group. J. London Math. Soc. **43**, 108–110 (1968)
90. P.P. Nielsen, Simplifying Smoktunowicz’s extraordinary example. Comm. Algebra **41**(11), 4339–4350 (2013)
91. J. Pakianathan, K. Shankar, Nilpotent numbers. Amer. Math. Monthly **107**(7), 631–634 (2000)
92. D.S. Passman, *The Algebraic Structure of Group Rings*. Pure and Applied Mathematics. (Wiley-Interscience [Wiley], New York-London-Sydney, 1977)
93. R. Plemmons, *Construction and analysis of non-equivalent finite semigroups, in Computational Problems in Abstract Algebra (Proceedings Conference, Oxford, 1967)* (Pergamon, Oxford, 1970), pp.223–228
94. S.D. Promislow, A simple example of a torsion-free, nonunique product group. Bull. London Math. Soc. **20**(4), 302–304 (1988)
95. J.H. Przytycki, Distributivity versus associativity in the homology theory of algebraic structures. Demonstratio Math. **44**(4), 823–869 (2011)
96. D.J.S. Robinson, *A course in the theory of groups*. Graduate Texts in Mathematics, vol. 80. (Springer, New York-Berlin, 1982)
97. D.J.S. Robinson, *A Course in the Theory of Groups*. Graduate Texts in Mathematics, vol. 80, 2nd ed. (Springer, New York, 1996)
98. W. Rump, A decomposition theorem for square-free unitary solutions of the quantum Yang-Baxter equation. Adv. Math. **193**(1), 40–55 (2005)

99. W. Rump, Braces, radical rings, and the quantum Yang-Baxter equation. J. Algebra **307**(1), 153–170 (2007)
100. W. Rump, The brace of a classical group. Note Mat. **34**(1), 115–144 (2014)
101. A. Smoktunowicz, Polynomial rings over nil rings need not be nil. J. Algebra **233**(2), 427–436 (2000)
102. A. Smoktunowicz, On some results related to Köthe's conjecture. Serdica Math. J. **27**(2), 159–170 (2001)
103. A. Smoktunowicz, *Some results in noncommutative ring theory, in International Congress of Mathematicians*, vol. II (European Mathematical Society, Zürich, 2006), pp.259–269
104. A. Smoktunowicz, A note on set-theoretic solutions of the Yang-Baxter equation. J. Algebra **500**, 3–18 (2018)
105. A. Smoktunowicz, On Engel groups, nilpotent groups, rings, braces and the Yang-Baxter equation. Trans. Amer. Math. Soc. **370**(9), 6535–6564 (2018)
106. A. Smoktunowicz, L. Vendramin, On skew braces (with an appendix by N. Byott, L. Vendramin). J. Comb. Algebra **2**(1), 47–86 (2018)
107. A. Soloviev, Non-unitary set-theoretical solutions to the quantum Yang-Baxter equation. Math. Res. Lett. **7**(5–6), 577–596 (2000)
108. L. Stefanello, S. Trappeniers, On bi-skew braces and brace blocks. J. Pure Appl. Algebra **227**(5), Paper No. 107295, 22 (2023)
109. A. Strojnowski, A note on u.p. groups. Comm. Algebra **8**(3), 231–234 (1980)
110. Y.P. Sysak, Products of almost abelian groups, in *Investigations of Groups with Restrictions for Subgroups (Russian)*, iii. Akad. Nauk Ukrain. (SSR Institute of Management, Kiev, 1988), pp. 81–85
111. C. Tsang, C. Qin, On the solvability of regular subgroups in the holomorph of a finite solvable group. Internat. J. Algebra Comput. **30**(2), 253–265 (2020)
112. P. Vojtěchovský, S.Y. Yang, Enumeration of racks and quandles up to isomorphism. Math. Comp. **88**(319), 2523–2540 (2019)
113. M. Wada, Group invariants of links. Topology **31**(2), 399–406 (1992)
114. A. Weinstein, P. Xu, Classical solutions of the quantum Yang-Baxter equation. Comm. Math. Phys. **148**(2), 309–343 (1992)
115. C.N. Yang, Some exact results for the many-body problem in one dimension with repulsive delta-function interaction. Phys. Rev. Lett. **19**, 1312–1315 (1967)
116. E. Zelmanov, *Nil Rings and Periodic Groups*. KMS Lecture Notes in Mathematics (Korean Mathematical Society, Seoul, 1992). With a preface by Jongsik Kim

Index

F. Cedó and L. Vendramin, *Groups, Radical Rings, and the Yang-Baxter Equation*, Progress in Mathematics 361, https://doi.org/10.1007/978-3-032-01524-2_1

MIX
Papier aus verantwortungsvollen Quellen
Paper from responsible sources
FSC® C105338

If you have any concerns about our products,
you can contact us on
ProductSafety@springernature.com

In case Publisher is established outside the EU,
the EU authorized representative is:
Springer Nature Customer Service Center GmbH
Europaplatz 3, 69115 Heidelberg, Germany

Printed by Libri Plureos GmbH
in Hamburg, Germany